M000011781

METHODS IN MOLECULAR BIOLOGY

Series Editor
John M. Walker
School of Life and Medical Sciences
University of Hertfordshire
Hatfield, Hertfordshire, AL10 9AB, UK

For further volumes:
http://www.springer.com/series/7651

Lung Innate Immunity and Inflammation

Methods and Protocols

Edited by

Scott Alper

*Department of Biomedical Research and Center for Genes, Environment and Health,
National Jewish Health, Denver, CO, USA; Department of Immunology and Microbiology,
University of Colorado School of Medicine, Aurora, CO, USA*

William J. Janssen

*Division of Pulmonary, Critical Care and Sleep Medicine, Department of Medicine,
National Jewish Health, Denver, CO, USA; Division of Pulmonary Sciences and
Critical Care Medicine, Department of Medicine, University of Colorado Denver,
Aurora, CO, USA*

 Humana Press

Editors
Scott Alper
Department of Biomedical Research
and Center for Genes
Environment and Health, National Jewish Health
Denver, CO, USA

Department of Immunology and Microbiology
University of Colorado School of Medicine
Aurora, CO, USA

William J. Janssen
Division of Pulmonary, Critical Care
and Sleep Medicine
Department of Medicine
National Jewish Health
Denver, CO, USA

Division of Pulmonary Sciences and
Critical Care Medicine
Department of Medicine
University of Colorado Denver
Aurora, CO, USA

ISSN 1064-3745   ISSN 1940-6029   (electronic)
Methods in Molecular Biology
ISBN 978-1-4939-8569-2   ISBN 978-1-4939-8570-8   (eBook)
https://doi.org/10.1007/978-1-4939-8570-8

Library of Congress Control Number: 2018946775

© Springer Science+Business Media, LLC, part of Springer Nature 2018
This work is subject to copyright. All rights are reserved by the Publisher, whether the whole or part of the material is concerned, specifically the rights of translation, reprinting, reuse of illustrations, recitation, broadcasting, reproduction on microfilms or in any other physical way, and transmission or information storage and retrieval, electronic adaptation, computer software, or by similar or dissimilar methodology now known or hereafter developed.
The use of general descriptive names, registered names, trademarks, service marks, etc. in this publication does not imply, even in the absence of a specific statement, that such names are exempt from the relevant protective laws and regulations and therefore free for general use.
The publisher, the authors and the editors are safe to assume that the advice and information in this book are believed to be true and accurate at the date of publication. Neither the publisher nor the authors or the editors give a warranty, express or implied, with respect to the material contained herein or for any errors or omissions that may have been made. The publisher remains neutral with regard to jurisdictional claims in published maps and institutional affiliations.

Printed on acid-free paper

This Humana Press imprint is published by the registered company Springer Science+Business Media, LLC part of Springer Nature.
The registered company address is: 233 Spring Street, New York, NY 10013, U.S.A.

Preface

The human lung ventilates approximately 10,000 L of air each day and as a result is exposed to billions of particles and pathogens. The majority of these inhaled agents are trapped in mucus lining the trachea and large airways and are eliminated via mucociliary and cough forces. The remainder deposit in the small airways and alveoli where innate immune cells comprise critical lines of host defense. Under most circumstances, these pathogens and particles are engulfed quickly and efficiently by resident lung macrophages and large-scale inflammatory responses are avoided. However, in cases where host defenses are compromised or overwhelmed, cells of the innate immune system are rapidly called in to further combat and contain infectious agents. Accordingly, the epithelial cells that line the airspaces in concert with leukocytes of the innate immune system comprise a critical mucosal barrier that is in constant contact with the ambient environment.

Chronic lower respiratory tract diseases are among the top causes of death in the United States and worldwide. For example, COPD is the third leading cause of death in the United States and affects more than 24 million people. Asthma is among the leading causes of missed work or school days and affects more than 15 million people in the United States. Acute lower respiratory tract infections, such as pneumonia, affect almost 300 million individuals worldwide each year and are responsible for approximately 3 million deaths. While these are complex diseases, they each share a significant inflammatory component, as do numerous other pulmonary disorders. Thus, the intersection of pulmonary medicine and innate immunity has become a key area of research. The complex interplay of genetics, environmental exposures, and multiple cell types involved in the lung innate immune response can make such studies challenging. Animal models have yielded tremendous insights into the pathogenesis of lung disease and will continue to provide critical knowledge for the foreseeable future. Similarly, studies with human lung tissue are becoming more common. The purpose of this book is to provide detailed methods to isolate, characterize, and investigate key lung innate immune cells and to provide methods for creating in vitro and in vivo model systems to study inflammatory lung diseases. While written for researchers focused on lung immunology, we expect that this guidebook generated by experts in the field also will be of value and interest to researchers investigating innate immunity and inflammation in other organs and tissues.

Denver, CO, USA   *William J. Janssen*
                                                                                             Scott Alper

Contents

Contributors

NEIL RAJ AGGARWAL • *Lung Biology and Disease Branch, Division of Lung Diseases, National Heart, Lung, and Blood Institute, Bethesda, MD, USA*

SCOTT ALPER • *Department of Biomedical Research and Center for Genes, Environment and Health, National Jewish Health, Denver, CO, USA; Department of Immunology and Microbiology, University of Colorado School of Medicine, Aurora, CO, USA*

JUAN R. ALVAREZ • *Division of Pulmonary, Critical Care and Sleep Medicine, Department of Medicine, Hastings Center for Pulmonary Research, Los Angeles, CA, USA*

KARIM BAHMED • *Department of Thoracic Medicine and Surgery, Temple University, Philadelphia, PA, USA; Center for Inflammation, Translational and Clinical Lung Research, Temple University, Philadelphia, PA, USA*

ZEA BOROK • *Division of Pulmonary, Critical Care and Sleep Medicine, Department of Medicine, Hastings Center for Pulmonary Research, Los Angeles, CA, USA; Department of Biochemistry and Molecular Medicine, Keck School of Medicine, University of Southern California, Los Angeles, CA, USA; Norris Comprehensive Cancer Center, Keck School of Medicine, University of Southern California, Los Angeles, CA, USA*

HONG WEI CHU • *Department of Medicine, National Jewish Health, Denver, CO, USA*

JOHN MATTHEW CRAIG • *Lung Biology and Disease Branch, Division of Lung Diseases, National Heart, Lung, and Blood Institute, Bethesda, MD, USA*

CHARMION CRUICKSHANK-QUINN • *Department of Pharmaceutical Sciences, University of Colorado Anschutz Medical Campus, Aurora, CO, USA*

FRANCO R. D'ALESSIO • *Division of Pulmonary and Critical Care Medicine, Johns Hopkins University School of Medicine, Baltimore, MD, USA*

BENJAMIN L. EDELMAN • *Program in Cell Biology, Department of Pediatrics, National Jewish Health, Denver, CO, USA*

KERRY M. EMPEY • *Department of Pharmacy and Therapeutics, University of Pittsburgh, Pittsburgh, PA, USA; Center for Clinical Pharmaceutical Sciences, University of Pittsburgh School of Pharmacy, Pittsburgh, PA, USA; Department of Immunology, University of Pittsburgh School of Medicine, Pittsburgh, PA, USA*

CHRISTOPHER M. EVANS • *Division of Pulmonary Sciences and Critical Care Medicine, University of Colorado Denver School of Medicine, Aurora, CO, USA*

XIAOHUI FANG • *Departments of Medicine and Anesthesia, Cardiovascular Research Institute, University of California, San Francisco, CA, USA*

KRISTOFER FRITZ • *Department of Pharmaceutical Sciences, University of Colorado Anschutz Medical Campus, Aurora, CO, USA*

ANETA GANDJEVA • *Program in Translational Lung Research, Division of Pulmonary Sciences and Critical Care Medicine, University of Colorado Denver, Anschutz Medical Campus, Aurora, CO, USA*

STAVROS GARANTZIOTIS • *Immunity, Inflammation and Disease Laboratory, National Institute of Environmental Health Sciences, Research Triangle Park, NC, USA*

SOPHIE L. GIBBINGS • *Department of Pediatrics, National Jewish Health, Denver, CO, USA*

MAGDALENA M. GORSKA • *Division of Allergy and Clinical Immunology, Department of Medicine, National Jewish Health, Denver, CO, USA; University of Colorado, Aurora, CO, USA*

JOSEPH A. HIPPENSTEEL • *Division of Pulmonary Sciences and Critical Care Medicine, Department of Medicine, University of Colorado Denver, Aurora, CO, USA*

NATHAN D. JACKSON • *Center for Genes, Environment and Health, National Jewish Health, Denver, CO, USA*

CLAUDIA V. JAKUBZICK • *Department of Pediatrics, National Jewish Health, Denver, CO, USA; Department of Microbiology and Immunology, University of Colorado, Denver, CO, USA*

NICOLE L. JANSING • *Division of Pulmonary, Critical Care, and Sleep Medicine, Department of Medicine, National Jewish Health, Denver, CO, USA*

WILLIAM J. JANSSEN • *Division of Pulmonary, Critical Care and Sleep Medicine, Department of Medicine, National Jewish Health, Denver, CO, USA; Division of Pulmonary Sciences and Critical Care Medicine, Department of Medicine, University of Colorado Denver, Aurora, CO, USA*

DI JIANG • *Department of Medicine, National Jewish Health, Denver, CO, USA*

HIDENORI KAGE • *Division of Pulmonary, Critical Care and Sleep Medicine, Department of Medicine, Hastings Center for Pulmonary Research, Los Angeles, CA, USA*

BEATA KOSMIDER • *Department of Thoracic Medicine and Surgery, Temple University, Philadelphia, PA, USA; Center for Inflammation, Translational and Clinical Lung Research, Temple University, Philadelphia, PA, USA; Department of Physiology, Temple University, Philadelphia, PA, USA*

FRANK FANG-YAO LEE • *Department of Biomedical Research and Center for Genes Environment and Health, National Jewish Health, Denver, CO, USA; Department of Immunology and Microbiology, University of Colorado School of Medicine, Aurora, CO, USA*

KENNETH LYN-KEW • *Division of Pulmonary, Critical Care, and Sleep Medicine, National Jewish Health, Denver, CO, USA*

KENNETH C. MALCOLM • *Department of Medicine, National Jewish Health, Denver, CO, USA*

THOMAS R. MARTIN • *Division of Pulmonary, Critical and Sleep Medicine, University of Washington School of Medicine, Seattle, WA, USA*

ROBERT J. MASON • *Department of Medicine, National Jewish Health, Denver, CO, USA*

MICHAEL A. MATTHAY • *Departments of Medicine and Anesthesia, Cardiovascular Research Institute, University of California, San Francisco, CA, USA*

JAZALLE MCCLENDON • *Division of Pulmonary, Critical Care, and Sleep Medicine, Department of Medicine, National Jewish Health, Denver, CO, USA*

ALEXANDRA L. MCCUBBREY • *Department of Medicine, National Jewish Health, Denver, CO, USA; Division of Critical Care Medicine and Pulmonary Sciences, Department of Medicine, University of Colorado, Denver, CO, USA*

WILLIAM MCKLEROY • *Division of Pulmonary, Critical Care, and Sleep Medicine, National Jewish Health, Denver, CO, USA; Department of Internal Medicine, University of Colorado School of Medicine, Aurora, CO, USA*

COLE MICHEL • *Department of Pharmaceutical Sciences, University of Colorado Anschutz Medical Campus, Aurora, CO, USA*

BETHANY B. MOORE • *Division of Pulmonary and Critical Care Medicine, Department of Internal Medicine, University of Michigan, Ann Arbor, MI, USA*

PAVAN MUTTIL • *Department of Pharmaceutical Sciences, College of Pharmacy, University of New Mexico, Albuquerque, NM, USA*

YASMEEN NKRUMAH-ELIE • *Department of Pharmaceutical Sciences, University of Colorado Anschutz Medical Campus, Aurora, CO, USA*

DAVID N. O'DWYER • *Division of Pulmonary and Critical Care Medicine, Department of Internal Medicine, University of Michigan, Ann Arbor, MI, USA*

IRINA PETRACHE • *Division of Pulmonary, Critical Care, and Sleep Medicine, Department of Medicine, National Jewish Health, University of Colorado Denver, Anschutz Medical Campus, Denver, CO, USA*

DOMINIQUE N. PRICE • *Department of Pharmaceutical Sciences, College of Pharmacy, University of New Mexico, Albuquerque, NM, USA*

DOROTA S. RACLAWSKA • *Division of Pulmonary Sciences and Critical Care Medicine, University of Colorado Denver School of Medicine, Aurora, CO, USA*

ELIZABETH F. REDENTE • *Program in Cell Biology, Department of Pediatrics, National Jewish Health, Denver, CO, USA; Department of Research, Veterans Affairs Eastern Colorado Health Care System, Denver, CO, USA; Division of Pulmonary Sciences and Critical Care Medicine, Department of Medicine, University of Colorado School of Medicine, Aurora, CO, USA*

NICHOLE A. REISDORPH • *Department of Pharmaceutical Sciences, University of Colorado Anschutz Medical Campus, Aurora, CO, USA*

RICHARD REISDORPH • *Department of Pharmaceutical Sciences, University of Colorado Anschutz Medical Campus, Aurora, CO, USA*

DAVID W. H. RICHES • *Program in Cell Biology, Department of Pediatrics, National Jewish Health, Denver, CO, USA*

LANDO RINGEL • *Center for Genes, Environment and Health, National Jewish Health, Denver, CO, USA*

NICCOLETTE SCHAEFER • *Department of Medicine, National Jewish Health, Denver, CO, USA*

ERIC P. SCHMIDT • *Division of Pulmonary Sciences and Critical Care Medicine, Department of Medicine, University of Colorado Denver, Aurora, CO, USA; Department of Medicine, Denver Health Medical Center, Denver, CO, USA*

MAX A. SEIBOLD • *Center for Genes, Environment and Health, National Jewish Health, Denver, CO, USA; Department of Pediatrics, National Jewish Health, Denver, CO, USA*

KARINA A. SERBAN • *Division of Pulmonary, Critical Care, and Sleep Medicine, Department of Medicine, National Jewish Health, University of Colorado Denver, Anschutz Medical Campus, Denver, CO, USA*

ADRIANNE L. STEFANSKI • *Division of Pulmonary Sciences and Critical Care Medicine, University of Colorado Denver School of Medicine, Aurora, CO, USA*

VANDY P. STOBER • *Immunity, Inflammation and Disease Laboratory, National Institute of Environmental Health Sciences, Research Triangle Park, NC, USA*

R. STOKES PEEBLES JR • *Division of Allergy, Pulmonary and Critical Care Medicine, Department of Medicine, Vanderbilt University, Nashville, TN, USA; Department of Pathology, Microbiology and Immunology, Vanderbilt University, Nashville, TN, USA*

MITSUHIRO SUNOHARA • *Division of Pulmonary, Critical Care and Sleep Medicine, Department of Medicine, Hastings Center for Pulmonary Research, Los Angeles, CA, USA*

BENJAMIN T. SURATT • *Department of Medicine, University of Vermont College of Medicine, Burlington, VT, USA*

ROBERT M. TIGHE • *Department of Medicine, Division of Pulmonary and Critical Care Medicine, Duke University School of Medicine, Durham, NC, USA*

RUBIN M. TUDER • *Program in Translational Lung Research, Division of Pulmonary Sciences and Critical Care Medicine, University of Colorado Denver, Anschutz Medical Campus, Aurora, CO, USA*

NIKI D. J. UBAGS • *Department of Medicine, University of Vermont College of Medicine, Burlington, VT, USA*

YIMU YANG • *Division of Pulmonary Sciences and Critical Care Medicine, Department of Medicine, University of Colorado Denver, Aurora, CO, USA*

YEN-REI A. YU • *Department of Medicine, Division of Pulmonary and Critical Care Medicine, Duke University School of Medicine, Durham, NC, USA*

RACHEL L. ZEMANS • *Division of Pulmonary, Critical Care, and Sleep Medicine, Department of Medicine, National Jewish Health, Denver, CO, USA; Division of Pulmonary Sciences and Critical Care Medicine, University of Colorado, Denver, CO, USA; Division of Pulmonary and Critical Care Medicine, Department of Medicine, University of Michigan, Ann Arbor, MI, USA*

Part I

Lung Innate, Immunity, and Inflammation

Chapter 1

500 Million Alveoli from 30,000 Feet: A Brief Primer on Lung Anatomy

William McKleroy and Kenneth Lyn-Kew

Abstract

The lungs are a complex organ that fulfill multiple life-sustaining roles including transfer of oxygen and carbon dioxide between the ambient environment and the bloodstream, host defense, and immune homeostasis. As in any biological system, an understanding of the underlying anatomy is prerequisite for successful experimental design and appropriate interpretation of data, regardless of the precise experimental model or procedure in use. This chapter provides an overview of human lung anatomy focused on the airways, the ultrastructure or parenchyma of the lung, the pulmonary vasculature, the innervation of the lungs, and the pulmonary lymphatic system. We will also discuss notable anatomic differences between mouse and human lungs.

Key words Lung, Anatomy, Airways, Alveoli, Parenchyma, Pulmonary vasculature, Pulmonary innervation, Pulmonary lymphatics

1 Introduction

1.1 Respiration, Vessels, and Immunity

The lungs are among the most complex organs in the human body. Their primary function of providing oxygen to the tissues for energy production is absolutely required for human life. They must actively draw in air and deliver enough oxygen to the circulation to meet the entire body's need for oxidative energy production and then quickly and efficiently eliminate the carbon dioxide produced by this reaction. This constant process maintains an acid-base equilibrium that is crucial for metabolic reactions throughout the body.

In mammals, gas exchange is facilitated by bidirectional, or tidal, ventilation. Tidal ventilation relies upon a highly coordinated system of muscles that surround a flexible, bony thoracic cavity. Contraction of the respiratory muscles causes the thorax to expand, creating a negative pressure in the thorax that draws ambient air into the lungs. Subsequent relaxation allows the system to return to a neutral, or slightly positive, pressure thereby resulting in exhalation. During the inspiratory phase of the cycle, the respiratory

Scott Alper and William J. Janssen (eds.), *Lung Innate Immunity and Inflammation: Methods and Protocols*,
Methods in Molecular Biology, vol. 1809, https://doi.org/10.1007/978-1-4939-8570-8_1,
© Springer Science+Business Media, LLC, part of Springer Nature 2018

circuit is open to any particulates or pathogens that are present in the inspired air as well as saliva, bacteria, and food particles from the mouth and oropharynx. This highlights the second major role of the lungs: maintenance of immunological integrity, protection from the threats of the outside world, and attenuation of the immune response to more minor threats.

This chapter will briefly discuss the gross and fine anatomy of the human lung as broken into five sections: (1) the airways, (2) the ultrastructure and parenchyma, (3) the vasculature of the respiratory and bronchial circulatory systems, (4) innervation, (5) and the lymph nodes and lymphatics. Given the ubiquity and utility of the mouse as a model organism for respiratory and immunological study, we have highlighted notable anatomical differences in the murine respiratory system where applicable.

2   The Airways

The airways refer to a system of tubes that allow the lungs to accomplish the task of exchanging fresh air for carbon dioxide produced as a by-product of metabolism. These tubes begin in the oropharyngeal and nasopharyngeal spaces before merging into the common pharynx, which then opens into the esophagus via the upper esophageal sphincter and the larynx via the glottis. The larynx consists of several cartilaginous and muscular components, including the vocal cords. Air continues distally into the trachea, which despite existing outside of the lung parenchyma is typically considered a part of the lung organ and the origin of the lower respiratory tract. The trachea is a firm tube bounded on the anterior and lateral sides by 16–20 cartilaginous rings to maintain patency and a strip of smooth muscle posteriorly. In mice, the contraction force of isolated ex vivo tracheal rings has been used to estimate airway contractility [1].

The trachea divides at the carina into the right and left mainstem bronchi, which continue dividing into lobar, segmental, and smaller bronchi for approximately five generations [2]. At this point, the cartilaginous superstructure of the bronchi ceases. The bronchioles comprise the next 10–12 generations of airways and are characterized by a circumferential lining of smooth muscle (Table 1). The airway smooth muscle (ASM) plays a major role in regulating airflow through the bronchioles and can contract or dilate the airway to allow less or more ventilation to distal airways and sites of gas exchange. Importantly, the ASM is exquisitely sensitive to the surrounding inflammatory signaling milieu and reacts by changing the airway caliber when exposed to pro- or anti-inflammatory cytokines [3, 4]; recent research also suggests that the ASM is a strong modulator of the airway inflammation itself [5]. ASM hypercontractility, which involves stronger or more eas-

Table 1
Airway generation, cross-sectional diameter, and the presence of cartilage, smooth muscle, and gas exchange in the human and mouse lung

Name	Generation (human)	Generation (mouse)	Cross-sectional diameter (human)	Cross-sectional diameter (mouse)	Cartilage	Smooth muscle	Gas exchange
Trachea	1	1	1.5–2 cm	1.2 mm	Y	Y	N
Mainstem bronchi	2	2	1.0–1.5 cm	1 mm	Y	Y	N
Bronchi	3–6	↓	↓	↓	Y	Y	N
Bronchioles	6–18	↓	↓	↓	N	Y	N
Terminal bronchiole	18	↓	600 μm	10 μm	N	Y	N
Respiratory bronchioles	19–21	↓	Present	Absent	N	N	Y
Alveolar ducts	22–25	↓	↓	↓	N	N	Y
Alveolar sacs and alveoli	25	13–17	150–200 μm	50–100 μm	N	N	Y

Downward arrows indicate a predictable progression to the next number in the column

ily triggered contraction and narrowing of airways, is a noted feature of asthma [6]. ASM dysfunction is also observed in chronic obstructive pulmonary disease (COPD), cystic fibrosis, and some fibrotic lung diseases. The role of the pulmonary nervous system in modulating airway smooth muscle contractility is discussed below.

The second to last generation of the bronchioles is the terminal bronchioles, which are the final airway generation that is lined by cuboidal epithelial cells. The terminal bronchioles open into respiratory bronchioles, which can be identified by the presence of alveoli that intermittently line their walls. After two to three more branches, the respiratory bronchiole gives way to an alveolar duct, which is similarly lined with alveoli but no longer contains any bronchiolar epithelial cells [7]. Alveolar ducts branch for two to three additional generations before ending in alveolar sacs, which directly open into anatomic alveoli. The anatomic unit defined by each terminal bronchiole and its subsequent branching generations is termed the acinus. Groups of three to five acini are typically referred to as lobules [8]. Notably, the emergence of respiratory bronchioles marks the beginning of the "respiratory zone" of the airways (given the presence of alveoli), whereas the airways that extend from the trachea to the terminal bronchioles comprise the "conducting zone" of the airways.

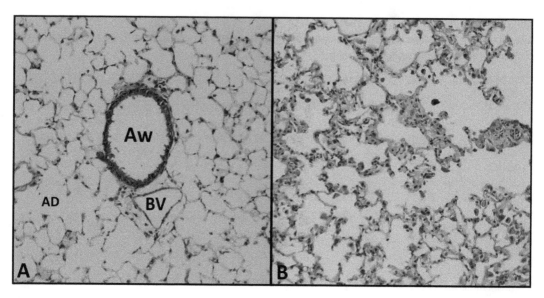

Fig. 1 Two samples of mouse lung tissue, inflated at a pressure of 20 cm H$_2$O, frozen and then stained with Masson's trichrome, viewed at 10× magnification. Masson's trichrome stains collagen green. (**a**) Normal mouse lung tissue. An airway (Aw), accompanying blood vessel (BV), and an alveolar duct (AD) are marked, and polygonal, thin-walled alveoli are observed throughout. (**b**) A sample of mouse lung tissue obtained 14 days after bleomycin instillation into the lungs. Bleomycin instillation induces patchy interstitial fibrosis causing distortion of the normal lung architecture and is a common model of fibrosis or acute lung injury in the mouse. Images courtesy of Alexandra McCubbrey

Taken as a whole, there are approximately 18–23 generations of branching airways in the human lung, with diameters that range from 1 to 1.5 cm in the main bronchi to 600 μm in the terminal bronchioles. In the mouse, the airways are diametrically smaller from main bronchi to the alveoli (from 1 mm to 50 μm, respectively), and there are fewer branching generations (approximately 13–17) due to the absence of respiratory bronchioles in the mouse [9, 10] (Table 1).

As noted above, the anatomic alveolus is the site of gas exchange with the blood. Alveoli are 100–200 μm-wide hexagonal structures lined by Type I (long, flat) and Type II (cuboidal) pneumocytes. Directly beneath the pneumocytes and a thin, shared basement membrane course the pulmonary capillaries, which are formed from a single layer of endothelial cells. This very delicate barrier permits easy diffusion of oxygen from the alveolus into the bloodstream and carbon dioxide in the opposite direction (Figs. 1, 2, and 3).

3   The Ultrastructure of the Lungs

The ultrastructure of the lungs is provided by the layers of connective tissues and cells that make up the very thin alveolar interstitial space (commonly referred to as the lung parenchyma),

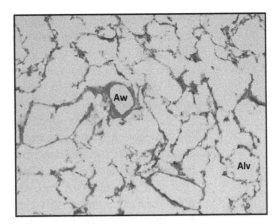

Fig. 2 Human lung tissue stained with hematoxylin and eosin, viewed at 10× magnification. A small airway (Aw) and one of the many alveoli (Alv) are highlighted. Human lung tissue is typically processed and stained in an uninflated state, leading to a degree of artifactual architectural distortion. Photo courtesy of the Patrick Hume

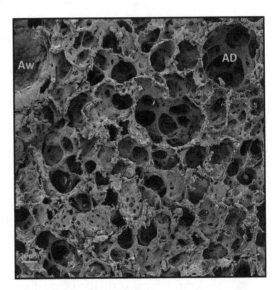

Fig. 3 Mouse lung tissue examined with electron microscopy. An airway (Aw), lined with columnar epithelial cells, is partially visualized in the upper left corner. An alveolar duct (AD) is also visible, along with many individual alveoli. Photo courtesy of Donald McCarthy, Frank Bos, Ann Zovein, and Kamran Atabai

the extraalveolar connective tissues, including larger bands of connective tissue that divide the lungs into multiple anatomic segments, as well as the visceral pleura, which forms an airtight border between the lungs and the pleural space. Bands of firm but flexible proteins, including collagen and elastin, interweave around groups of alveoli, forming basketlike networks throughout the entire lung.

This network enables the distal airways and alveoli to stretch open during inhalation and prevents their complete collapse during exhalation. Notably, when elastin is degraded by neutrophil elastase (e.g., in COPD), the airways lose the circumferential traction applied by elastin and are subject to collapse. This leads to diminished gas exchange, particularly during exhalation.

Collagen supplies approximately 60–70% of the total quantity of connective tissue in the lung [11] and is found primarily as Type I collagen, with additional contributions from Type III collagen. As in other organs, Type II collagen is found exclusively in the cartilage. The remainder of the extracellular matrix is made up of elastin fibers; collagen- and elastin-binding glycoproteins such as fibronectin, laminin, and others [12]; and "ground substance" that fills in the space between connective tissue and cellular elements of the lung. Proteoglycans, comprised of proteins attached to glycosaminoglycans such as hyaluronic acid, heparan sulfate, and dermatan sulfate, make up the majority of the ground substance in addition to water, electrolytes, cell and matrix degradation products, and some serum constituents [11]. It is now known that the extracellular matrix plays important roles in basement membrane and interstitium formation, cell anchoring, extracellular signaling, and other regulatory/homeostatic functions [12–14]. Disruption in the quantity and biochemical formation of these proteins are implicated in several respiratory disease states, such as pulmonary fibrosis, the acute respiratory distress syndrome, and infection.

On a gross level, the lungs are organized into anatomic lobes and segments that follow the major branches of the airways. In the human lung, there are three major lobes on the right and two on the left, not including the lingula, which is anatomically part of the left upper lobe. These lobes are covered by visceral pleura, which serves to separate the lobes. The areas between the lobes are termed fissures. The oblique fissures divide the upper and middle lobes from the lower lobe in the right lung and the upper from lower lobes in the left. In addition, the horizontal fissure divides the right upper and right middle lobe. In the mouse, the right lung is divided into five lobes, while the left lung is composed of one single lobe.

Within the lobes, the airways divide from lobar branches into segmental branches, entering segments of parenchyma bounded by thicker layers of connective tissue that anatomically divide the segments from one another. These individual segments may be surgically excised from one another with relatively little blood loss or trauma to the neighboring segments [15]. In total, the human lungs are organized into ten segments on the right (three upper, two middle, and five lower) and eight segments on the left (four upper, four lower). These segments are commonly identified by their classical anatomic nomenclature or by the abbreviated Boyden classification (Table 2, Fig. 4).

Table 2
The lobes and segments of the human lung

Right lung
• Right upper lobe
• Apical segment (B1)
• Posterior segment (B2)
• Anterior segment (B3)
• Right middle lobe
• Lateral segment (B4)
• Medial segment (B5)
• Right lower lobe
• Superior segment (B6)
• Medial segment (B7)
• Anterior segment (B8)
• Lateral segment (B9)
• Posterior segment (B10)
Left lung
• Left upper lobe (includes the lingula)
• Apicoposterior segment (B1, B2)
• Anterior segment (B3)
• Superior lingular segment (B4)
• Inferior lingular segment (B5)
• Left lower lobe
• Superior segment (B6)
• Anteromedial segment (B7, B8)
• Lateral segment (B9)
• Posterior segment (B10)

The segments are named by anatomical position. The Boyden classification, another system of nomenclature, of each segment is noted in parenthesis

The lungs are covered on their outer surfaces by the visceral pleura. This layer is made up of an innermost sheet largely continuous with the septal interstitial tissues dividing the lung segments and lobules; a thicker in-between layer that hosts collagen and elastic fibers as well as adipose tissue, blood vessels, lymphatics, and nerves; and a thin outer layer of epithelial cells referred to as mesothelial cells due to their mesenchymal origin [16, 17].

The visceral pleura is ultimately continuous with and opposes the inner lining of the thorax (called the parietal pleura) although the vascular, nervous, and lymphatic supply of these two layers is distinct. Between the visceral and parietal pleura is a potential space that is typically lubricated with approximately 1 mL of pleural fluid that allows smooth movement of the lung within the thoracic cavity. In disease states, this potential space can be filled with fluid (e.g., from inflammation, infection, bleeding, or malignancy) or air (primarily due to a tear in the visceral pleura or a penetrating thoracic wound that sucks air into the space). Given the delicate and easily compressible nature of the lungs, the pleural potential space

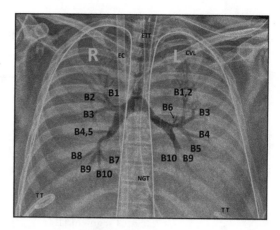

Fig. 4 Chest radiograph of a patient with severe acute respiratory distress syndrome, here causing filling of the smaller airways and alveoli with fluid (which is denser and thus appears white on a radiograph), while the main airways remain full of less-dense air (which appears black). In this image, the airways were further manipulated to appear darker than surrounding tissue using image processing software. The lungs are approximately outlined in green, and the airways are labeled according to Boyden's classification (Table 2). Bronchus B7 on the right (R) and Bronchus B7,8 on the left (L) are not well visualized. Note the presence of an extracorporeal membrane oxygenation catheter (EC), endotracheal tube (ETT), central venous line (CVL), nasogastric tube (NGT), and bilateral tube thoracostomies (TT) in this severely ill patient

can essentially accommodate the full volume of each hemithorax. Fluid (termed pleural effusion, or hemothorax if the fluid is blood) or air (termed pneumothorax) in this space can cause the air-filled lung to collapse under the weight of fluid and potentially limit gas exchange and pulmonary circulation.

4   Pulmonary Vasculature

The human lungs receive a dual supply of blood. Most blood comes from the pulmonary circulation, whereas the bronchial (i.e., systemic) circulation provides a minor contribution. In the mouse, only the pulmonary circulation exists. The main function of the pulmonary circulation (and lungs themselves) is to facilitate the acquisition of oxygen into the bloodstream and the offloading of carbon dioxide from the blood into the airspaces and out into the environment. The blood vessels in the lungs also fulfill the function of a vascular sieve that traps small particles and coagulation products generated in the venous systemic circulation. If these particles were to reach the systemic circulation, the risk of arterial blockage and tissue infarction would be high. The lung vasculature further serves as an expandable reservoir for blood returning to the left heart to prevent (up to a certain point) overdistention of the left heart. The endothelium of lung blood vessels is metabolically active and plays a role in elaboration and processing of circulating hormones, inflammatory markers, factors involved in thrombolysis, and other biologically active molecules [18–20].

The pulmonary arterial system begins with the pulmonary trunk, also known as the main pulmonary artery, which emerges from the right ventricle through the pulmonic valve. In healthy adults, the main pulmonary artery measures 2.5–3.0 cm in diameter [21]. The main pulmonary artery divides into two branches, one for each lung. These vessels emerge from the mediastinum through the hila along with the mainstem bronchi and follow the course and branching pattern of the bronchial tree. As the pulmonary arteries branch, they transition from elastin-containing elastic arteries to well-muscularized arteries and finally to partially muscularized pulmonary arterioles with a diameter of less than 100 microns. At this point, nonmuscular precapillaries and alveolar capillaries emerge to wrap around alveoli, covering up to 80–90% of each alveolus and creating an exceptionally large surface area for gas exchange, estimated at 143 square meters [22]. Oxygen-rich blood from the capillaries flows into the pulmonary veins which ultimately flow into the left atrium.

The second supply of blood reaching the lungs comes from the bronchial circulation. This system receives blood from the systemic circulation and thus supplies the lung tissue with an additional source of oxygenated blood. The sources of the bronchial circulation vary, but bronchial blood flow derives in most humans from the internal mammary arteries, from the intercostal arteries on the right, and from the aorta itself on the left. Although these arteries contain oxygenated blood, which suggests a major contribution of the bronchial circulation to tissue perfusion, lung tissue is able to receive adequate nutrients and oxygen from the pulmonary circulation alone. In human lung transplant patients, the bronchial circulation is completely lost, but the transplanted lungs remain viable. In normal conditions, anastomoses between the bronchial and pulmonary circulation exist [23], and the bronchial circulation can provide collateral blood flow to tissues in the event of embolism of the pulmonary circulation, including chronic thromboembolic disease [24]. Ultimately, the major role of the bronchial circulation may be its contribution to lung perfusion during fetal lung development [8].

5   Nerves

The respiratory cycle is powered by the coordinated contraction and relaxation of the diaphragm and accessory breathing muscles such as the intercostal muscles, pectoralis minor, and others. The rhythmic contraction and relaxation of these respiratory muscles are largely coordinated by oscillatory neurons of the pre-Botzinger complex in the ventrolateral medulla with contributions from other redundant oscillatory centers [25].

The lungs themselves contain both afferent and efferent parasympathetic and sympathetic fibers that have important roles in control of the airway and vascular smooth muscle, cough and stretch reflexes, and chemosensation. The parasympathetic nerve fibers of the lung originate in the vagal nucleus in the brainstem, while sympathetic fibers emerge from the thoracic sympathetic ganglia. These fibers form the pulmonary plexus that enter the lungs near the hila and branch along with the larger caliber airways and pulmonary vessels. Acetylcholine release from parasympathetic efferents causes airway smooth muscle-mediated bronchoconstriction as well as bronchial vasodilation and increased secretion of mucus from the airway epithelium. Sympathetic efferent fibers supply submucosal glands and pulmonary vessels leading to decreased mucus production and vasodilation. While adrenergic nerve fibers do not appear to actually synapse on airway smooth muscle in humans [26], airway smooth muscle nonetheless expresses β2 adrenergic receptors that cause bronchodilation in response to endogenous and exogenous β2 adrenergic agonists.

Several different types of afferent fibers are found in the mammalian lung. The majority of these fibers are thought to be parasympathetic in nature and are carried to the central nervous system via the vagal nerve. Parasympathetic afferents are broken into three groups: fibers containing rapidly adapting receptors (RARs), fibers containing slowly adapting receptors (SARs), and c-fibers. RAR- and SAR-containing fibers are found predominantly in the large airways of the lungs, including the trachea, and drive sensory reflexes in the lungs. They can be distinguished from each other by the speed of their adaptation to stimuli as well as their locations: RARs are likely found near the epithelium, while SARs are found close to airway and tracheal smooth muscle. RARs sense irritants, toxins, and inflammatory stimuli such as pathogens and mediate the cough reflex, mucus production, and bronchoconstriction [27]. SARs detect airway stretch or constriction by detecting changes in muscle and parenchymal tension and influence tidal volume, respiratory rate, airway smooth muscle contraction/relaxation, and even heart rate [28]. As might be expected, reflexes mediated by both RARs and SARs are lost in the denervated human lung after lung transplantation [29, 30]. Interestingly though, the cough reflex may slowly reappear in lung transplant patients [29].

C-fibers are unmyelinated (unlike RAR- and SAR-containing fibers) nerves that extend throughout the entire lung from airway to parenchyma and comprise up to 75% of the vagal nerve afferents coming from the lung [31, 32]. C-fibers have receptors that detect both stretch (weaker response) and chemical stimuli such as capsaicin, environmental irritants, and inflammatory chemokines (stronger response). Activation of C-fibers can result in tachypnea (rapid, shallow breathing), bronchoconstriction, cough, increased mucus secretion, and bradycardia.

It should be noted that another set of non-adrenergic, non-cholinergic (NANC) nerve fibers are found in the lung [33]. The NANC system is thought to cause bronchoconstriction via neuropeptides such as substance P, calcitonin gene-related peptide, neurokinin A, and bronchodilation via nitric oxide and vasoactive intestinal peptide.

6 Lymphatic System

The lungs are actively drained of lymphatic fluid by two networks of lymphatic vessels that direct lymph toward the hila and lymph nodes of the mediastinum. A subpleural network of lymphatic vessels runs in the connective tissues below the visceral pleura and drains the pleura and perhaps some more peripherally located parenchymal tissue. A deep peribronchovascular plexus of lymph vessels runs within the bronchovascular bundle and drains the remainder of the structures within the lungs. Fluid communicates directly with the lymphatic system of the parietal pleura via pulmonary stomata; regular drainage through these pores facilitates generation of new pleural fluid in normal states [34]. Pleural lymph vessels, like those found elsewhere in the body, are thin-walled and embryonically derived from the venous system. Lymphatic capillaries emerge blindly in the connective tissues around terminal bronchioles and, as such, do not extend directly to the alveolar ultrastructure. They merge to form vessels punctuated by frequent one-way valves and surrounded by a thin layer of smooth muscle. It is thought that the bellows action of the lung itself, in combination with bicuspid one-way valves, propels lymph toward the hila; there also may be a contribution of contraction of the lymphatic smooth muscle.

Lymph vessels move lymphatic fluid into and along networks of lymph nodes along the bronchi and trachea, eventually draining into the venous system via the right and left thoracic ducts at the level of the jugular-subclavian confluence [34]. Lymphatic drainage from one lung is usually into the ipsilateral hilar lymph nodes, although significant variability in lymphatic networks allows for connections directly to the mediastinum. In fact, in approximately 10–15% of patients with lung cancer and mediastinal lymph node metastasis, the cancer is found in contralateral, but not ipsilateral hilar nodes [35]. The extensive network of lymph nodes within the lung is a manifestation of the need for defense and clearance mechanisms in the face of broad exposure of the lung to pathogens and other environmental particles. These nodes will often detectably enlarge with recruited and active immune cells during infection or other inflammatory states: sampling these nodes is an oft-used technique by researchers studying the immunological responses to various lung insults in animal models [36].

Bronchus-associated lymphoid tissue, defined as patches of organized lymphocytes in the bronchial mucosa and periglandular areas, is found in most mammals in varying quantities. In adult humans and mice, this tissue is only found in some healthy control subjects but is seen more often and in greater quantity in response to inflammation or infection [37, 38]. These patches contain antigen-presenting cells, T-cells, B-cells, and some IgA- or IgM-positive cells. Other studies have found BALT present in all subject sample of human children without inflammatory or airway disease, further suggesting the role of this tissue in antigen processing and immune tolerance [39].

7 Conclusion

To fulfill the life-sustaining process of gas exchange, as well as provide an important immune barrier from the pathogens and antigens of the outside world, the lung has evolved five intricate and interrelated anatomical systems. The airways maintain a smooth and active conduit for oxygen and carbon dioxide to enter and exit the alveoli, the thin-walled sites of gas exchange. The parenchyma of the lung includes the network of approximately 500 million alveoli as well as the connective tissues that surround and support them. The vascular system of the lungs brings the entire cardiac output from the right side of the heart and spreads it throughout the alveolar capillaries, as well as provides blood flow to the tissues of the lung itself. A robust set of different types of neural fibers provide afferent and efferent innervation of multiple compartments in the lung, including the airway epithelium and smooth muscle, parenchyma, pleura, and vessels. Finally, the pulmonary lymphatics provide vessels for lymphatic return, while lymph nodes and extranodal lymphatic tissue host immunologically active cells involved in host defense and immune homeostasis. Taken as a whole, these structures provide a critical interface between the ambient environment and the organism and serve essential functions including host defense and gas exchange.

References

1. Chen H et al (2003) TNF-[alpha] modulates murine tracheal rings responsiveness to G-protein-coupled receptor agonists and KCl. J Appl Physiol (1985) 95(2):864–872. discussion 863

2. Weibel ER (1963) Morphometry of the human lung. Springer, Berlin, p 151

3. KleinJan A (2016) Airway inflammation in asthma: key players beyond the Th2 pathway. Curr Opin Pulm Med 22(1):46–52

4. Noble PB et al (2014) Airway smooth muscle in asthma: linking contraction and mechanotransduction to disease pathogenesis and remodelling. Pulm Pharmacol Ther 29(2):96–107

5. Koziol-White CJ, Panettieri RA Jr (2011) Airway smooth muscle and immunomodulation in acute exacerbations of airway disease. Immunol Rev 242(1):178–185

6. Bentley JK, Hershenson MB (2008) Airway smooth muscle growth in asthma: proliferation, hypertrophy, and migration. Proc Am Thorac Soc 5(1):89–96

7. Wang N-S (2002) Anatomy and ultrastructure of the lung. In: Bittar EE (ed) Pulmonary biology in health and disease. Springer-Verlag New York, Inc, New York, NY, pp 1–19

8. Murray JF (1986) The normal lung : the basis for diagnosis and treatment of pulmonary disease, 2nd edn. Saunders, Philadelphia. 377 p

9. West JB, Luks A (2016) West's respiratory physiology: the essentials, 10th edn. Wolters Kluwer, Philadelphia. 238 p

10. Albertine KH (2016) Anatomy of the lungs. In: Broaddus VC, Mason RJ (eds) Murray & Nadel's textbook of respiratory medicine. Elsevier Saunders, Philadelphia, PA

11. Hance AJ, Crystal RG (1975) The connective tissue of lung. Am Rev Respir Dis 112(5):657–711

12. McKleroy W, Lee TH, Atabai K (2013) Always cleave up your mess: targeting collagen degradation to treat tissue fibrosis. Am J Physiol Lung Cell Mol Physiol 304(11):L709–L721

13. Dunsmore SE, Rannels DE (1996) Extracellular matrix biology in the lung. Am J Physiol 270(1 Pt 1):L3–L27

14. White ES (2015) Lung extracellular matrix and fibroblast function. Ann Am Thorac Soc 12(Suppl 1):S30–S33

15. Leonard RJ (1995) Human gross anatomy: an outline text. Oxford University Press, New York

16. Sevin CM, Light RW (2011) Microscopic anatomy of the pleura. Thorac Surg Clin 21(2):173–175. vii

17. Hayek Hv (ed) (1960) The human lung. Rev. and augm. ed. Hafner Pub. Co., New York, NY. 372 p

18. Ryan US (1986) Pulmonary endothelium: a dynamic interface. Clin Invest Med 9(2):124–132

19. Ince C et al (2016) The endothelium in sepsis. Shock 45(3):259–270

20. Tuder RM et al (2001) The pathobiology of pulmonary hypertension. Endothelium. Clin Chest Med 22(3):405–418

21. Edwards PD, Bull RK, Coulden R (1998) CT measurement of main pulmonary artery diameter. Br J Radiol 71(850):1018–1020

22. Gehr P, Bachofen M, Weibel ER (1978) The normal human lung: ultrastructure and morphometric estimation of diffusion capacity. Respir Physiol 32(2):121–140

23. Charan NB, Turk GM, Dhand R (1984) Gross and subgross anatomy of bronchial circulation in sheep. J Appl Physiol Respir Environ Exerc Physiol 57(3):658–664

24. Ley S et al (2002) Bronchopulmonary shunts in patients with chronic thromboembolic pulmonary hypertension: evaluation with helical CT and MR imaging. AJR Am J Roentgenol 179(5):1209–1215

25. Feldman JL, Del Negro CA (2006) Looking for inspiration: new perspectives on respiratory rhythm. Nat Rev Neurosci 7(3):232–242

26. Richardson JB (1979) Nerve supply to the lungs. Am Rev Respir Dis 119(5):785–802

27. Sant'Ambrogio G, Widdicombe J (2001) Reflexes from airway rapidly adapting receptors. Respir Physiol 125(1-2):33–45

28. Schelegle ES, Green JF (2001) An overview of the anatomy and physiology of slowly adapting pulmonary stretch receptors. Respir Physiol 125(1-2):17–31

29. Duarte AG, Myers AC (2012) Cough reflex in lung transplant recipients. Lung 190(1):23–27

30. Iber C et al (1995) The Breuer-Hering reflex in humans. Effects of pulmonary denervation and hypocapnia. Am J Respir Crit Care Med 152(1):217–224

31. Agostoni E et al (1957) Functional and histological studies of the vagus nerve and its branches to the heart, lungs and abdominal viscera in the cat. J Physiol 135(1):182–205

32. Lee LY, Pisarri TE (2001) Afferent properties and reflex functions of bronchopulmonary C-fibers. Respir Physiol 125(1-2):47–65

33. Barnes PJ (1987) Neuropeptides in human airways: function and clinical implications. Am Rev Respir Dis 136(6 Pt 2):S77–S83

34. Brotons ML et al (2012) Anatomy and physiology of the thoracic lymphatic system. Thorac Surg Clin 22(2):139–153

35. Riquet M, Hidden G, Debesse B (1989) Direct lymphatic drainage of lung segments to the mediastinal nodes. An anatomic study on 260 adults. J Thorac Cardiovasc Surg 97(4):623–632

36. Kudo M et al (2013) Mfge8 suppresses airway hyperresponsiveness in asthma by regulating smooth muscle contraction. Proc Natl Acad Sci U S A 110(2):660–665

37. Richmond I et al (1993) Bronchus associated lymphoid tissue (BALT) in human lung: its distribution in smokers and non-smokers. Thorax 48(11):1130–1134

38. Moyron-Quiroz JE et al (2004) Role of inducible bronchus associated lymphoid tissue (iBALT) in respiratory immunity. Nat Med 10(9):927–934

39. Heier I et al (2011) Characterisation of bronchus-associated lymphoid tissue and antigen-presenting cells in central airway mucosa of children. Thorax 66(2):151–156

Chapter 2

Chapter 2

Overview of Innate Lung Immunity and Inflammation

David W. H. Riches and Thomas R. Martin

Abstract

The nasal passages, conducting airways and gas-exchange surfaces of the lung, are constantly exposed to substances contained in the air that we breathe. While many of these suspended substances are relatively harmless, some, for example, pathogenic microbes, noxious pollutants, and aspirated gastric contents can be harmful. The innate immune system, lungs and conducting airways have evolved specialized mechanisms to protect the respiratory system not only from these harmful inhaled substances but also from the overly exuberant innate immune activation that can arise during the host response to harmful inhaled substances. Herein, we discuss the cell types that contribute to lung innate immunity and inflammation and how their activities are coordinated to promote lung health.

Key words Innate immunity, Inflammation, Lung, Epithelium, Neutrophils, Macrophages

1 Overview of Innate Lung Immunity

The lung epithelium is the largest epithelial surface in the body with a surface area that approximates the size of a tennis court. Considering that humans have an average respiratory rate of 15 breaths/min and an average tidal volume of 500 mL, the lungs are exposed to over 10,000 L of ambient air per day. Inspired air consists of a complex mixture of gases with variable amounts of particulates that include inorganic and organic dusts, pollens, bacteria and viruses, as well as pollutants, oxidants and toxins. In addition, the conducting airways and gas-exchange surfaces of the lung can be exposed to aspirated acidic gastric contents, as well as infected mucus derived from the nasal sinuses and upper airways. Thus, the respiratory tract, from the nasal passages to the alveoli are constantly exposed to a spectrum of harmless, harmful and pathogenic agents, raising the question of how the lungs and airways discriminate between what is harmful and what is essentially harmless. To accomplish this range of protection, the respiratory system and the immune system coordinately utilize multiple mechanisms to survey and respond to inhaled particles based primarily

Scott Alper and William J. Janssen (eds.), *Lung Innate Immunity and Inflammation: Methods and Protocols*,
Methods in Molecular Biology, vol. 1809, https://doi.org/10.1007/978-1-4939-8570-8_2,
© Springer Science+Business Media, LLC, part of Springer Nature 2018

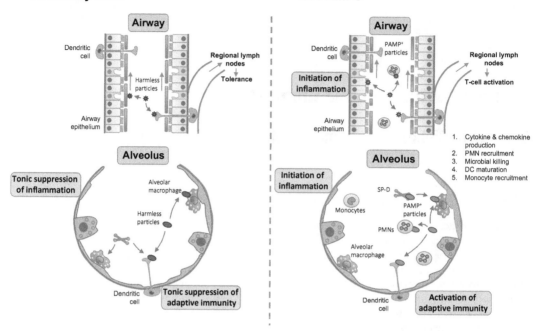

Fig. 1 Overview of innate immunity in the conducting airways and alveoli of the lung. (**a**) In the absence of pathogen-associated molecular patterns (PAMPs), the airways are protected by mucus which captures harmless particulates and transports them along the mucociliary escalator. Dendritic cells also capture particles, traffic to regional lymph nodes, and promote tolerance to commonly inhaled antigens. Also in the absence of PAMPs, the alveoli are maintained in an anti-inflammatory and immunosuppressed state to prevent unwanted inflammation and immune activation toward commonly inhaled particles and antigens. (**b**) Innate immunity is activated in the presence of PAMPs. PAMPs stimulate airway epithelial cells to express chemokines, cytokines and lipid mediators that attract neutrophils (PMNs) which in turn kill PAMP-expressing microbes. Airway dendritic cells respond to PAMPs by maturing, migrating to regional lymph nodes and stimulating T-cell proliferation. A similar program is activated in the alveoli upon PAMP detection by alveolar macrophages and alveolar epithelial cells resulting in the initiation of inflammation and activation of adaptive immunity

on their size and physicochemical properties. Thus, as conceptualized in Fig. 1, innate immune protection in the lung can be broadly divided into considerations of the conducting airways and the gas-exchange surfaces of the alveoli. At the cellular level, innate protection is mainly mediated by the coordinated functions of airway and alveolar epithelial cells, resident macrophages, recruited neutrophils (i.e., polymorphonuclear leukocytes [PMN]), and monocytes which respond immediately to inhaled material. In contrast, the adaptive immune system, which is primarily comprised by T lymphocytes, B lymphocytes and antigen-presenting cells, responds more slowly but with considerably greater specificity and with the development of immunologic memory.

An important component of airway host defense resides in the anatomic structure of the tracheobronchial tree. Air turbulence created by the nasal passages and the cartilaginous segmentation of the trachea and large airways ensures that particles in excess of 10 μm in diameter are deposited on the mucus-coated surfaces of the nose, pharynx, trachea, and conducting airways. Once captured, particulates and mucus are propelled toward the pharynx by the mucociliary system, which includes tracts of synchronously beating ciliated airway epithelial cells. Expulsion and clearance of the agglomerates of inhaled large particulates and mucus is also aided by coughing, sneezing and swallowing. In addition to its gel-like biophysical properties, the sol phase of mucus contains an array of epithelial cell-derived antimicrobial peptides, antioxidants, and antiproteases, as well as specific immunoglobulin A (IgA) antibodies. Principal among human airway antimicrobial peptides are salt-sensitive cysteine-rich cationic β-defensins and the cathelicidin, LL-37/hCAP-18 [1–5]. The β-defensin-1 is constitutively secreted by airway epithelial cells and accumulates in airway surface liquid at microgram per milliliter concentrations [6]. Other β-defensins and LL-37/hCAP-18 can be inducibly expressed following exposure to bacterial lipopolysaccharide (LPS) and other pro-inflammatory mediators [7–9]. Additional antimicrobial proteins, including lactoferrin and lysozyme, are also present in airway epithelial secretions and contribute to innate protection of the airways [10, 11]. Thus, the mucociliary system and its antimicrobial solutes have evolved to provide continuous surveillance and immediate protection of the upper respiratory tract and conducting airways from both harmful as well as relatively harmless inhaled particulates that deposit on their surfaces.

In contrast to larger particulates, particulates smaller than 5 μm in diameter are able to descend the entire tracheobronchial tree and lodge at bronchiolar-respiratory duct junctions or are deposited onto the surfactant-rich surfaces of the alveoli. While sharing many similarities with innate protection of the airways—for example, the presence of antioxidants, antiproteases, and anti-microbial enzymes—additional innate protection of the alveoli is afforded by the presence of resident alveolar macrophages and the lung-specific collectins, surfactant protein A (SP-A), and surfactant protein D (SP-D). In the absence of microbes, resident alveolar macrophages also play an important role in suppressing inflammation and adaptive immunity, thereby protecting the alveoli from unwanted responses to harmless inhaled particulates (Fig. 1). SP-A and SP-D also play a key role in suppressing inflammation through tonic signaling effects on resident alveolar macrophages. Thus, the innate immune system not only protects the lungs from harmful microbes, it also prevents inflammation, injury and activation of the adaptive immune system in the steady state, and in response to harmless inhaled particulates.

How then, do the airways and alveoli respond to the presence of potentially harmful microbes? As illustrated in Fig. 2, pathogens express various types of pathogen-associated molecular patterns

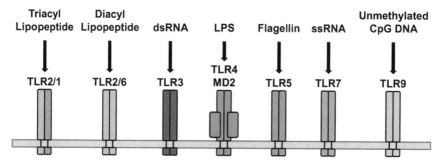

Fig. 2 Major classes of Toll-like receptors (TLRs) expressed by lung cells

(PAMPs) that include, but are not limited to complex lipids, carbohydrates derived from bacterial and yeast cell walls, unmethylated CpG DNA sequences, and double-stranded RNAs. In turn, PAMPs are recognized by an ancient, evolutionarily conserved system of secreted, cell surface, and intracellular pattern recognition receptors (PRRs) that are expressed by airway epithelial cells, resident alveolar macrophages, vascular endothelial cells and PMNs which, upon ligation by PAMPs, play a dominant role in alerting the host to the presence of microbes and triggering corresponding host responses. As illustrated in Fig. 1, once activated, PRR-induced signal transduction initiates an inflammatory response that promotes the expression of an array of innate response genes including chemokines (e.g., CXCL1, CXCL2, CXCL4) and pro-inflammatory cytokines (e.g., IL-1β, TNF-α), which collectively facilitate changes in endothelial and epithelial permeability to enhance inflammatory cell transmigration into the airspaces and induce the migration of PMNs and monocytes from the pulmonary circulation into the air spaces. In addition, pro-inflammatory cytokines and other mediators initiate the expression of specific genes involved in microbial phagocytosis, killing, and degradation. Finally, the innate inflammatory response promotes the homeostatic repair of damaged tissues and restoration of normal tissue structure and function. Based on this brief overview of lung innate immunity, we next consider the cell effector mechanisms that protect the mucosal surfaces of the lungs and undergo activation once they sense the presence of microbes.

2   Cellular Effector Mechanisms

2.1   Epithelium

The lung epithelium has an important role in host defense against microbes that pass through the glottis and reach the conducting airways and alveolar gas-exchange units. Briefly, the conducting airways of the descending respiratory tract are lined by ciliated columnar epithelial cells and goblet cells down to the level of the terminal airways, before becoming nonciliated in the respiratory bronchioles. By contrast, the alveolar epithelium consists of squamous type I alveolar epithelial cells that form most of the alveolar surface and less common cuboidal type II alveolar epithelial cells. Bacterial lipopolysaccharide (LPS), a PAMP contained in the cell walls of gram-negative bacteria, stimulates ciliated airway epithelial cells to produce CXC and CC chemokines, which recruit PMNs and monocytes, respectively, into the airway lumen [12]. Airway epithelial cells also produce IL-1β, IL-6, IL-8, RANTES (regulated on activation, normal T-cell expressed and secreted), granulocyte-macrophage colony-stimulating factor (GM-CSF), and transforming growth factor-β (TGF-β) [13]. As with other cells involved in innate immunity, cytokine production by airway epithelial cells occurs via the activation of transcription factors NF-κB, AP-1 and NF-IL-6, among others [14]. Interestingly, noninfectious environmental agents like ozone [15], asbestos [16], diesel exhaust particles [17] and air pollution particles [18] also lead to NF-κB activation in airway epithelial cells, which is typically followed by IL-8 production and release. Airway epithelial cells also recognize unmethylated bacterial DNA via Toll-like receptor (TLR) 9, leading to NF-κB activation and production of IL-6, IL-8 and β_2-defensin in the airways [19]. However, unlike airway epithelial cells, alveolar epithelial cells do not respond directly to microbial products, but produce chemokines in response to TNF-α and IL-1β, which are first produced by resident alveolar macrophages in response to LPS.

Studies with transgenic mice have provided further support for the critical role of the lung epithelium in innate immunity and microbial defense. Skerrett and coworkers [20] showed that mice expressing a dominant negative I-κB construct in distal airway epithelial cells, which prevents NF-κB activation, have impaired airway PMN recruitment in response to inhaled LPS. This finding supports the importance of bacterial recognition by distal airway epithelial cells in vivo and shows that epithelium-derived cytokines produced by the NF-κB pathway probably are just as important as macrophage-derived cytokines in driving innate inflammatory responses in the airways and alveolar spaces. Hajjar and colleagues [21] created bone marrow chimeras in which either myeloid (leukocytes) or nonmyeloid (epithelial) cells lacked MyD88, a key adapter molecule required for signaling via all TLRs except TLR3. The mice lacking MyD88 (and, therefore, deficient in TLR signaling) in nonmyeloid cells, including the airway and alveolar epithelium, had markedly impaired clearance of *P. aeruginosa*, whereas mice with or without MyD88 in myeloid cells had normal and equivalent bacterial clearance. This surprising result further supports the important role of the airway and alveolar epithelium in bacterial recognition and clearance from the lungs, a role that is perhaps equivalent to that of resident and recruited leukocytes. Thus, innate immune mechanisms in the airways stimulate endogenous defenses by enhancing production of airway defensins and antimicrobial products and by stimulating pro-inflammatory signaling and PMN and monocyte recruitment into the airways to augment antimicrobial defenses.

2.2 Neutrophils (PMNs)

PMNs serve as the immediate effector arm of the innate immune system [22, 23]. In the systemic circulation, PMN migration into tissues depends on the integrin CD11b/CD18 (integrin $\alpha M\beta 2$)on the PMN surface recognizing the counter ligand ICAM-1 on the endothelial surface. In the lungs, however, both CD18-dependent and -independent mechanisms exist and the signals that determine whether or not PMN migration depends on CD18 are not clear [24]. For example, PMN migration into the lungs in response to *E. coli* and *P. aeruginosa* is CD18-dependent, whereas neutrophils do not require CD18 to emigrate into the air spaces in response to *S. pneumonia*. TNF-α and IL-1β appear to direct CD18-dependent migration, whereas IFN-γ is more important for CD18-independent emigration.

Once in the air spaces, PMNs ingest bacteria and fungi that have been opsonized by complement and immunoglobulins that accumulate in the air spaces at sites of inflammation. PMNs contain a series of effector mechanisms to kill bacteria and fungi, including oxidant production, microbicidal proteins in primary azurophilic granules and extracellular traps. As microbes are recognized and the phagosome begins to form, the subunits of a reduced form of nicotinamide adenine dinucleotide phosphate (NADPH) oxidase are assembled in the cell membrane and the cell undergoes a respiratory burst, in which an electron is transferred to molecular oxygen to form superoxide anion, which is reduced again to form hydrogen peroxide (H_2O_2). This happens on the invaginating phagosomal membrane, focusing oxidants on the contents of the developing phagosome. As the phagosome forms, the primary granules fuse with the phagosomal membrane, adding myeloperoxidase and cationic antimicrobial peptides to the phagolysosome. Myeloperoxidase catalyzes the formation of hypochlorous acid from H_2O_2 and a halide, typically chloride (Cl^-) because of its high concentration in the cellular environment. Hypochlorous acid is a highly reactive oxidant that oxidizes methionines, tyrosines, and other amino acids on proteins, killing susceptible microbes. When phagocytosis is appropriately regulated, microbial killing occurs within the protected environment of the PMN phagolysosome. However, at sites of intense inflammation, PMNs release superoxide anion, H_2O_2, and granular contents directly into the extracellular environment, leading to oxidant formation in the alveolar spaces with oxidation of intracellular proteins in alveolar walls and leukocytes, and the accumulation of defensins and other granular contents in the alveolar spaces [25]. Release of these extracellular products by PMNs contributes to indirect tissue injury.

PMNs and their products are cleared largely by macrophages during the resolution of inflammation. PMNs die primarily by apoptosis, necrosis, or neutrophil extracellular trap (NET) formation, the latter of which are comprised of a mixture of neutrophil-derived DNA threads bound together with antimicrobial enzymes and other molecules. Apoptotic PMNs are rapidly ingested by macrophages, which express an array of cell surface receptors that recognize dead and dying cells. Consequently, large numbers of extracellular apoptotic PMNs are usually not seen. Macrophages that ingest apoptotic PMN produce TGF-β and IL-10, which have anti-inflammatory effects. This process results in the clearance of PMNs and their residual intracellular contents, ultimately dampening inflammation. PMNs that are not ingested by macrophages undergo secondary necrosis with loss of membrane integrity, cytoplasmic swelling and release of remaining intracellular contents into the inflammatory environment. These intracellular contents are recognized as danger signals by macrophages, some of which have been included as "alarmins," or damage-associated molecular patterns (DAMPs) [26]. DAMPs include the nuclear protein HMGB1, granular antimicrobial peptides and other intracellular products. Macrophages that recognize these danger signals produce IL-1β, TNF-α, IL-8 (CXCL8) and other pro-inflammatory cytokines, initiating or perpetuating inflammatory responses. Some PMNs recovered from the lungs of patients with acute respiratory distress syndrome (ARDS) have features of apoptosis, with small cytoplasmic features and nuclear pyknosis, whereas many have features of necrosis, including severe degranulation, membrane blebbing and cytoplasmic swelling. The signals that govern the balance between necrosis and apoptosis are incompletely understood. Furthermore, recent studies have provided new insights into an additional mechanism of PMN clearance from sites of injury. Using a model of sterile thermal liver injury, Wang and colleagues [27] confirmed an earlier report suggesting that reverse-migrated PMNs can accumulate in the bone marrow where they undergo apoptosis [28–30]. Using sophisticated fluorescence visualization approaches, Wang et al. [27] additionally show that a fraction of these reverse-migrated PMNs also accumulate in the lung where they adhere to the microvascular endothelium. Intriguingly, the authors speculate that lung accumulation of reverse-migrated activated PMNs in patients with severe trauma or hemorrhagic shock may contribute to the development of ARDS in these patients.

Thus, PMNs are important effector cells in innate immunity, ideally designed to circulate through the body and accumulate rapidly at tissue sites of acute inflammation. Under normal circumstances, they arrive in tissue ready to ingest and kill microbes, then go quietly away by programmed cell death. PMN-derived signals regulate local inflammation and stimulate adaptive immune responses, providing a broader role for PMNs in host defense.

2.3 Resident Alveolar Macrophages

Resident alveolar macrophages reside in the mixed environment of epithelial lining fluid and ambient inhaled air. The primary functions of resident alveolar macrophages are (1) to dispose of inhaled microbes and particulates, (2) to clear pulmonary surfactant, and (3) to tonically suppress the development of inappropriate inflammatory and immune responses. Resident alveolar macrophages are capable of phagocytosing a wide spectrum of harmless and harmful particulates and microbes. Under basal conditions, most ingested phagocytosed particulates are enclosed within phagosomes, which ultimately fuse with lysosomes, leading to their degradation by an array of acid-pH optimum hydrolytic enzymes. Some inhaled microbes (e.g., *M. tuberculosis*) and some environmental particulates (e.g., crystalline silica) are resistant to this process and become sequestered in secondary lysosomes where they remain for the lifespan of the macrophage. Most of these latter cells probably crawl into the airways and are cleared via the mucociliary escalator [31, 32]. Some particle-laden macrophages may remain in the lung for extended periods of time before either dying and releasing their particle burden, thereby rendering it available for phagocytosis by other macrophages, or undergoing apoptosis and being cleared by other phagocytes.

Resident alveolar macrophages also actively contribute to the normal homeostasis of pulmonary surfactant, as deduced from studies in patients with pulmonary alveolar proteinosis (PAP) in which a reduction in resident alveolar macrophage numbers and function due to the development of autoantibodies reactive against GM-CSF [33–35], leads to the accumulation of protein and lipid-rich surfactant in the alveoli, ultimately resulting in reduced gas exchange [34, 35].

Resident alveolar macrophages also play a vital role in basal suppression of alveolar inflammation and adaptive immunity, a function specific to resident alveolar macrophages [36, 37]. As noted earlier, the pulmonary collectins SP-A and SP-D inhibit PAMP-dependent production of pro-inflammatory cytokines through their interaction with the receptor, SIRPα [38]. In addition, TGF-β, produced by alveolar macrophages and activated at the surface of alveolar epithelial cells by the integrin αvβ6, has been shown to suppress alveolar inflammation [39, 40]. Other studies have emphasized the importance of IL-10, prostaglandin E_2, and nitric oxide in the tonic suppression of alveolar inflammation and adaptive immunity [41, 42]. The tonic suppressive activity of alveolar macrophages can be overcome in several different ways. One mechanism involves exceeding the threshold of phagocytic capacity of resident alveolar macrophages, thereby allowing microbes to interact with other lung cells, for example, pulmonary dendritic cells (DCs) or epithelial cells, which respond to PAMPs and other molecules by producing pro-inflammatory cytokines, of which GM-CSF and TNF-α have been shown to be responsible [43]. A second mechanism involves alveolar macrophages themselves. In this case, PAMP recognition by the globular head groups of SP-A and SP-D enables lung collectins to interact with macrophages via calreticulin and CD91 [38]. In contrast to the suppressive signal initiated when SP-A and SP-D interact with macrophages via SIRPα in the absence of PAMPs, CD91 signaling leads to the production of pro-inflammatory cytokines, which then augment inflammation and the recruitment of PMNs and monocytes. Thus, while resident alveolar macrophages are capable of suppressing alveolar inflammation and adaptive immune responses to commonly encountered environmental products that reach the alveolar spaces, mechanisms exist to overcome this suppression to allow the recruitment of inflammatory cells, especially PMNs and monocytes, and to promote adaptive immunity.

2.4 Monocytes and Inflammatory Macrophages

Circulating monocytes have the potential to migrate into inflamed tissues and differentiate into macrophages [44, 45]. Recruited monocytes have important pro-inflammatory and host defense activities, including microbial killing and amplification of inflammation [46]. Monocytes kill microbes with reactive oxygen species and reactive nitrogen species. Superoxide anion (O_2^-) is mainly produced by the NADPH oxidase on the phagocytic membrane [47]. The generation of reactive oxygen species is critical to host defense against commonly encountered and pathogenic bacteria. Patients with chronic granulomatous disease and mice bearing a targeted disruption of the p47[phox] component of the NADPH oxidase [48, 49] are deficient in their ability to control pulmonary and other infections [48–51]. However, inappropriate production of reactive oxygen species by monocytes (and other inflammatory cells) can result in epithelial injury that can lead to fibroproliferation as can be seen with survivors of ARDS and patients with idiopathic pulmonary fibrosis, asbestosis, or silicosis [52–55]. Thus, whereas reactive oxygen species play a vital role in the protection of the lung against microbes, inappropriate production in response to nonmicrobial particulates and pollutants including cigarette smoke [55] can be injurious to the airway and alveolar epithelium.

Recruited monocytes, which develop into macrophages, are capable of further differentiation or programming. Based on earlier work in which different PAMPs were found to induce distinct patterns of macrophage gene expression [56–59], the concept has evolved that uniquely different patterns of gene expression may be induced in response to the specific conditions or stimuli that prevail at the sites to which macrophages have been recruited [60–62]. For example, exposure of macrophages to Th1 cytokines and TLR ligands (e.g., LPS and poly[I:C]) results in increased expression of IL-12, IL-23, TNF-α, IL-1β, and NOS2, while IL-10 and arginase I expression are repressed. This response is generally referred to as "classical macrophage programming" [56–59, 61, 63]. In contrast, exposure to Th2 cytokines, including IL-4 and IL-13, leads to a different pattern of gene expression characterized by an increased expression of arginase I, FIZZ, YM1, and CCL19 together with reduced expression of NOS2 in an overall response that is often referred to as "alternative macrophage programming" [61–64]. The spectrum of cytokines identified as being capable of inducing alternative macrophage programming is growing and now includes IL-33, IL-21, IL-10, colony-stimulation factors, CCL2, CXCL4, glucocorticoids and TGF-β [65]. In addition, a so-called "restorative" macrophage programming state that is distinct from classically and alternatively programmed macrophages has recently been identified in self-resolving liver fibrosis [66]. Defining these responses as distinct "phenotypes" can be useful in thinking about how macrophages participate in innate immunity and influence effector cells in the adaptive immune response. However, macrophage programming is more complex than some of the simplistic schemes that have been proposed, because there are often overlapping patterns of gene expression in response to multiple stimuli [67]. Thus, it may be more appropriate to think about these adaptive responses as points on a continuum where the response generated is purely an adaptation to the microenvironment that macrophages encounter [60, 62, 68, 69]. In addition, ligation of distinct receptors and corresponding signal transduction pathways leads to differential activation of the transcription factors PU.1, STAT1, STAT6, interferon regulatory factors (IRF4, IRF5), PPARγ, cAMP-responsive element-binding proteins (CREB), NF-κB and AP-1, all of which contribute to diversity in macrophage programming [70]. Macrophage programming is also regulated and enhanced post-transcriptionally by microRNAs [71–73]. Thus, diversity in the macrophage programming response and the resulting unique patterns of gene expression contribute to their diverse roles in protection of the lungs [69].

In summary, macrophages play key roles in the maintenance of lung homeostasis through their abilities to suppress unwanted inflammation and immune responses to harmless, commonly encountered inhaled materials. However, they remain constantly poised to respond to harmful inhaled microbes and other substances. Part of this response involves calling in additional support through the recruitment of circulating blood PMNs and monocytes. In turn, recruited monocytes can differentiate into macrophages, which can then express appropriate responses to microenvironmental cues that they sense at the site to which they have been attracted to further influence the adaptive immune response.

3   Cell Cross Talk and System Integration

The innate immune system has evolved to protect the lungs from harm by environmentally acquired microbes and other inhaled substances as well as from host-derived harmful stressors such as unwanted inflammation. A core concept is that innate immune protection in the lungs is not a consequence of specific responses of individual cells but, represents coordinated responses and collective cooperation between resident and recruited lung cell types. These coordinated events result in homeostasis in the conducting airways and alveoli, tolerance to harmless inhaled substances, and the capacity to respond to harmful pathogenic microbes. In addition, innate immune mechanisms assist in the rapid resolution of injury and restoration of lung function.

From the nose to the alveolus, the respiratory epithelium, resident macrophages, and recruited PMNs and monocytes have evolved exquisite, diverse, and overlapping mechanisms to distinguish between harmful and harmless inhaled substances. In its simplest form, this distinction is achieved by sensing the presence of microbial PAMPs. In the absence of PAMPs, the epithelium remains largely ignorant of inhaled particulates. Similarly, resident macrophages phagocytose and dispose of particulates that reach the alveoli, while tonically maintaining an anti-inflammatory and immunosuppressive environment.

As illustrated in Fig. 1, integration of innate functions is dependent on cross talk and communication between the different cell types that contribute to lung innate immunity. In some settings, communication is fostered by proximity, e.g., resident alveolar macrophages use integrins to communicate with alveolar epithelial cells. In other settings, cell-cell communication is mediated through the secretion of a panoply of cytokines and lipid mediators, acting via autocrine or paracrine signaling mechanisms. Thus, different cell lineages can communicate with each other and among themselves. Communication between different cell types also plays an important role in the coordination of innate immunity and in the activation of adaptive immunity. For example, the finding that experimental or genetic depletion of resident alveolar macrophages results in the induction of alveolitis and lung injury while simultaneously diminishing tolerance to inhaled antigens [74, 75] emphasizes the importance of resident alveolar macrophages in suppressing inflammation and the activation of adaptive immunity. Integration of innate immunity in the lung is also associated with "coordinated burden sharing." For example, airway goblet cells and submucosal glands produce mucus, whereas ciliated epithelial cells propel mucus and entrapped particulates toward the pharynx. Similar coordinated burden sharing is exhibited by airway epithelial cells, which are poorly phagocytic but highly capable of sensing PAMPs and calling in PMNs and monocytes, which in turn are highly phagocytic and microbicidal and eliminate an array of microbes. Together, these integrated systems have evolved to provide the lungs with maximum protection against harmful microbes while minimizing harmful responses against harmless inhaled substances. The net result is an exquisite system for protecting the gas-exchange parenchyma of the lungs, which is critical for survival.

Conflicts of Interest

DWHR has no conflicts of interest to declare. TRM has no conflicts of interest to declare.

References

1. Bals R et al (1998) Human beta-defensin 2 is a salt-sensitive peptide antibiotic expressed in human lung. J Clin Invest 102(5):874–880
2. Bals R et al (1998) The peptide antibiotic LL-37/hCAP-18 is expressed in epithelia of the human lung where it has broad antimicrobial activity at the airway surface. Proc Natl Acad Sci U S A 95(16):9541–9546
3. Garcia JR et al (2001) Human beta-defensin 4: a novel inducible peptide with a specific salt-sensitive spectrum of antimicrobial activity. FASEB J 15(10):1819–1821
4. Goldman MJ et al (1997) Human beta-defensin-1 is a salt-sensitive antibiotic in lung that is inactivated in cystic fibrosis. Cell 88(4):553–560
5. Jia HP et al (2001) Discovery of new human beta-defensins using a genomics-based approach. Gene 263(1-2):211–218
6. Singh PK et al (1998) Production of beta-defensins by human airway epithelia. Proc Natl Acad Sci U S A 95(25):14,961–14,966
7. Alekseeva L et al (2009) Inducible expression of beta defensins by human respiratory epithelial cells exposed to Aspergillus fumigatus organisms. BMC microbiology 9:33
8. Beisswenger C, Bals R (2005) Functions of antimicrobial peptides in host defense and immunity. Curr Protein Pept Sci 6(3):255–264
9. Harder J et al (2000) Mucoid pseudomonas aeruginosa, TNF-alpha, and IL-1beta, but not IL-6, induce human beta-defensin-2 in respiratory epithelia. Am J Respir Cell Mol Biol 22(6):714–721
10. Thompson AB et al (1990) Lower respiratory tract lactoferrin and lysozyme arise primarily in the airways and are elevated in association with chronic bronchitis. J Lab Clin Med 115(2):148–158
11. Zhao YX et al (2000) Secretion of complement components of the alternative pathway (C3 and factor B) by the human alveolar type II epithelial cell line A549. Int J Mol Med 5(4):415–419
12. Becker MN et al (2000) CD14-dependent lipopolysaccharide-induced beta-defensin-2 expression in human tracheobronchial epithelium. J Biol Chem 275(38):29731–29736
13. Diamond G, Legarda D, Ryan LK (2000) The innate immune response of the respiratory epithelium. Immunol Rev 173:27–38
14. Jany B, Betz R, Schreck R (1995) Activation of the transcription factor NF-kappa B in human tracheobronchial epithelial cells by inflammatory stimuli. Eur Respir J 8(3):387–391
15. Jaspers I, Flescher E, Chen LC (1997) Ozone-induced IL-8 expression and transcription factor binding in respiratory epithelial cells. Am J Physiol 272(3 Pt 1):L504–L511
16. Janssen YM et al (1995) Asbestos induces nuclear factor kappa B (NF-kappa B) DNA-binding activity and NF-kappa B-dependent gene expression in tracheal epithelial cells. Proc Natl Acad Sci U S A 92(18):8458–8462
17. Takizawa H et al (1999) Diesel exhaust particles induce NF-kappa B activation in human bronchial epithelial cells in vitro: importance in cytokine transcription. J Immunol 162(8):4705–4711
18. Quay JL et al (1998) Air pollution particles induce IL-6 gene expression in human airway epithelial cells via NF-kappaB activation. Am J Respir Cell Mol Biol 19(1):98–106

19. Platz J et al (2004) Microbial DNA induces a host defense reaction of human respiratory epithelial cells. J Immunol 173(2):1219–1223

20. Skerrett SJ et al (2004) Respiratory epithelial cells regulate lung inflammation in response to inhaled endotoxin. Am J Physiol Lung Cell Mol Physiol 287(1):L143–L152

21. Hajjar AM et al (2005) An essential role for non-bone marrow-derived cells in control of Pseudomonas aeruginosa pneumonia. Am J Respir Cell Mol Biol 33(5):470–475

22. Cowburn AS et al (2008) Advances in neutrophil biology: clinical implications. Chest 134(3):606–612

23. Nathan C (2006) Neutrophils and immunity: challenges and opportunities. Nat Rev Immunol 6(3):173–182

24. Wang Q, Doerschuk CM, Mizgerd JP (2004) Neutrophils in innate immunity. Semin Respir Crit Care Med 25(1):33–41

25. Sittipunt C et al (2001) Nitric oxide and nitrotyrosine in the lungs of patients with acute respiratory distress syndrome. Am J Respir Crit Care Med 163(2):503–510

26. Oppenheim JJ et al (2007) Alarmins initiate host defense. Adv Exp Med Biol 601:185–194

27. Wang J et al (2017) Visualizing the function and fate of neutrophils in sterile injury and repair. Science 358(6359):111–116

28. Furze RC, Rankin SM (2008) The role of the bone marrow in neutrophil clearance under homeostatic conditions in the mouse. FASEB J 22(9):3111–3119

29. Suratt BT et al (2001) Neutrophil maturation and activation determine anatomic site of clearance from circulation. Am J Physiol Lung Cell Mol Physiol 281(4):L913–L921

30. Furze RC, Rankin SM (2008) Neutrophil mobilization and clearance in the bone marrow. Immunology 125(3):281–288

31. Doherty DE et al (1992) Prolonged monocyte accumulation in the lung during bleomycin-induced pulmonary fibrosis. A noninvasive assessment of monocyte kinetics by scintigraphy. Lab Invest 66(2):231–242

32. Harmsen AG et al (1985) The role of macrophages in particle translocation from lungs to lymph nodes. Science 230(4731):1277–1280

33. Kitamura T et al (1999) Idiopathic pulmonary alveolar proteinosis as an autoimmune disease with neutralizing antibody against granulocyte/macrophage colony-stimulating factor. J Exp Med 190(6):875–880

34. Martin RJ et al (1980) Pulmonary alveolar proteinosis: the diagnosis by segmental lavage. Am Rev Respir Dis 121(5):819–825

35. Nugent KM, Pesanti EL (1983) Macrophage function in pulmonary alveolar proteinosis. Am Rev Respir Dis 127(6):780–781

36. Holt PG (1978) Inhibitory activity of unstimulated alveolar macrophages on T-lymphocyte blastogenic response. The American review of respiratory disease 118(4):791–793

37. Toews GB et al (1984) The accessory cell function of human alveolar macrophages in specific T cell proliferation. Journal of immunology 132(1):181–186

38. Gardai SJ et al (2003) By binding SIRPalpha or calreticulin/CD91, lung collectins act as dual function surveillance molecules to suppress or enhance inflammation. Cell 115(1):13–23

39. Morris DG et al (2003) Loss of integrin alpha(v)beta6-mediated TGF-beta activation causes Mmp12-dependent emphysema. Nature 422(6928):169–173

40. Munger JS et al (1999) The integrin alpha v beta 6 binds and activates latent TGF beta 1: a mechanism for regulating pulmonary inflammation and fibrosis. Cell 96(3):319–328

41. Kawabe T et al (1992) Immunosuppressive activity induced by nitric oxide in culture supernatant of activated rat alveolar macrophages. Immunology 76(1):72–78

42. Roth MD, Golub SH (1993) Human pulmonary macrophages utilize prostaglandins and transforming growth factor beta 1 to suppress lymphocyte activation. Journal of leukocyte biology 53(4):366–371

43. Bilyk N, Holt PG (1993) Inhibition of the immunosuppressive activity of resident pulmonary alveolar macrophages by granulocyte/macrophage colony-stimulating factor. J Exp Med 177(6):1773–1777

44. Bromley SK, Mempel TR, Luster AD (2008) Orchestrating the orchestrators: chemokines in control of T cell traffic. Nat Immunol 9(9):970–980

45. Springer TA (1994) Traffic signals for lymphocyte recirculation and leukocyte emigration: the multistep paradigm. Cell 76(2):301–314

46. Mould KJ et al (2017) Cell origin dictates programming of resident versus recruited macrophages during acute lung injury. Am J Respir Cell Mol Biol 57(3):294–306

47. Vignais PV (2002) The superoxide-generating NADPH oxidase: structural aspects and activation mechanism. Cellular and molecular life sciences : CMLS 59(9):1428–1459

48. Chang YC et al (1998) Virulence of catalase-deficient aspergillus nidulans in p47(phox)−/− mice. Implications for fungal pathogenicity and host defense in chronic granulomatous disease. J Clin Invest 101(9):1843–1850

49. Jackson SH, Gallin JI, Holland SM (1995) The p47phox mouse knock-out model of chronic granulomatous disease. J Exp Med 182(3): 751–758

50. Kelly JK et al (1986) Fatal Aspergillus pneumonia in chronic granulomatous disease. Am J Clin Pathol 86(2):235–240

51. Tauber AI et al (1983) Chronic granulomatous disease: a syndrome of phagocyte oxidase deficiencies. Medicine 62(5):286–309

52. Castranova V (1994) Generation of oxygen radicals and mechanisms of injury prevention. Environ Health Perspect 102(Suppl 10):65–68

53. Fireman E et al (1989) Suppressive activity of alveolar macrophages and blood monocytes from interstitial lung diseases: role of released soluble factors. Int J Immunopharmacol 11(7):751–760

54. Gossart S et al (1996) Reactive oxygen intermediates as regulators of TNF-alpha production in rat lung inflammation induced by silica. J Immunol 156(4):1540–1548

55. Kondo T et al (1994) Current smoking of elderly men reduces antioxidants in alveolar macrophages. American J Respir Crit Care Med 149(1):178–182

56. Lake FR et al (1994) Functional switching of macrophage responses to tumor necrosis factor-alpha (TNF alpha) by interferons. Implications for the pleiotropic activities of TNF alpha. J Clin Invest 93(4):1661–1669

57. Laszlo DJ et al (1993) Development of functional diversity in mouse macrophages. Mutual exclusion of two phenotypic states. Am J Pathol 143(2):587–597

58. Noble PW et al (1993) Hyaluronate activation of CD44 induces insulin-like growth factor-1 expression by a tumor necrosis factor-alpha-dependent mechanism in murine macrophages. J Clin Invest 91(6):2368–2377

59. Riches DW et al (1988) Differential regulation of gene expression during macrophage activation with a polyribonucleotide. The role of endogenously derived IFN. J Immunol 141(1):180–188

60. Stout RD et al (2005) Macrophages sequentially change their functional phenotype in response to changes in microenvironmental influences. J Immunol 175(1):342–349

61. Gordon S (2003) Alternative activation of macrophages. Nat Rev Immunol 3(1):23–35

62. Riches DW (1995) Signalling heterogeneity as a contributing factor in macrophage functional diversity. Semin Cell Biol 6(6):377–384

63. Mosser DM, Edwards JP (2008) Exploring the full spectrum of macrophage activation. Nat Rev Immunol 8(12):958–969

64. Mantovani A et al (2002) Macrophage polarization: tumor-associated macrophages as a paradigm for polarized M2 mononuclear phagocytes. Trends Immunol 23(11):549–555

65. Mantovani A et al (2013) Macrophage plasticity and polarization in tissue repair and remodelling. J Pathol 229(2):176–185

66. Ramachandran P et al (2012) Differential Ly-6C expression identifies the recruited macrophage phenotype, which orchestrates the regression of murine liver fibrosis. Proc Natl Acad Sci U S A 109(46):E3186–E3195

67. Martinez FO et al (2006) Transcriptional profiling of the human monocyte-to-macrophage differentiation and polarization: new molecules and patterns of gene expression. J Immunol 177(10):7303–7311

68. Murray PJ, Wynn TA (2011) Obstacles and opportunities for understanding macrophage polarization. J Leukoc Biol 89(4):557–563

69. Sica A, Mantovani A (2012) Macrophage plasticity and polarization: in vivo veritas. J Clin Invest 122(3):787–795

70. Lawrence T, Natoli G (2011) Transcriptional regulation of macrophage polarization: enabling diversity with identity. Nat Rev Immunol 11(11):750–761

71. Banerjee S et al (2013) MicroRNA let-7c regulates macrophage polarization. J Immunol 190(12):6542–6549

72. Graff JW et al (2012) Identifying functional microRNAs in macrophages with polarized phenotypes. J Biol Chem 287(26):21816–21825

73. Liu G, Abraham E (2013) MicroRNAs in immune response and macrophage polarization. Arterioscler Thromb Vasc Biol 33(2):170–177

74. Shibata Y et al (2001) Alveolar macrophage deficiency in osteopetrotic mice deficient in macrophage colony-stimulating factor is spontaneously corrected with age and associated with matrix metalloproteinase expression and emphysema. Blood 98(9):2845–2852

75. Thepen T, Van Rooijen N, Kraal G (1989) Alveolar macrophage elimination in vivo is associated with an increase in pulmonary immune response in mice. J Exp Med 170(2):499–509

Part II

Isolation and Characterization of Lung Innate Immune Cells

Chapter 3

Isolation and Characterization of Mononuclear Phagocytes in the Mouse Lung and Lymph Nodes

Sophie L. Gibbings and Claudia V. Jakubzick

Abstract

There is a diverse population of mononuclear phagocytes (MPs) in the lungs, comprised of macrophages, dendritic cells (DCs), and monocytes. The existence of these various cell types suggests that there is a clear division of labor and delicate balance between the MPs under steady-state and inflammatory conditions. Here we describe how to identify pulmonary MPs using flow cytometry and how to isolate them via cell sorting. In steady-state conditions, murine lungs contain a uniform population of alveolar macrophages (AMs), three distinct interstitial macrophage (IM) populations, three DC subtypes, and a small number of tissue-trafficking monocytes. During an inflammatory response, the monocyte population is more abundant and complex since it acquires either macrophage-like or DC-like features. All in all, studying how these cell types interact with each other, structural cells, and other leukocytes within the environment will be important to understanding their role in maintaining homeostasis and during the development of disease.

Key words Macrophage, Monocyte, Dendritic cell, Mononuclear phagocyte, Lung, Flow cytometry, Pulmonary, Interstitial

Abbreviations

AM   Alveolar macrophage
BAL   Bronchoalveolar lavage
DC   Dendritic cell
IM   Interstitial macrophage
IN   Intranasal
IT   Intratracheal
LLN   Mediastinal lung-draining lymph node
MP   Mononuclear phagocyte

Scott Alper and William J. Janssen (eds.), *Lung Innate Immunity and Inflammation: Methods and Protocols*,
Methods in Molecular Biology, vol. 1809, https://doi.org/10.1007/978-1-4939-8570-8_3,
© Springer Science+Business Media, LLC, part of Springer Nature 2018

1   Introduction

The lung consists of three anatomic compartments. These include the airway lumen, the vasculature, and the space in between known as the interstitium. In the lower airways, alveolar macrophages serve as a first line of defense against invading pathogens [1]. If alveolar macrophages and incoming neutrophils are incapable of containing the pathogens, then an adaptive immune response is initiated to assist in pathogen clearance [2]. The airways and alveoli are bounded by epithelial cells, which form a protective barrier between the ambient environment and the interstitium. Underneath the epithelial barrier are dendritic cells (DCs), interstitial macrophages (IMs), and tissue monocytes [3–5].

DCs link innate and adaptive immunity by acquiring, processing, and trafficking foreign and self-antigens to the draining lymph nodes (LNs), where they present peptides on MHC molecules and activate cognate T cells [6–10]. A clear division of labor exists between the two overaching DC subtypes, known as classical DC1 (cDC1) and classical DC2 (cDC2). CDC2 can be divided into two additional subtypes: an IRF4, KLF4-dependent DC2 and an IRF4, Notch2-dependent DC2 [9, 11–22]. Interstitial macrophages appear to be most densely located in the bronchovascular bundles. Within the IM population, we have recently identified three unique subtypes [23]. Similar populations exist in the stroma of other organs; however their precise functions remain unclear. Lastly, during inflammation, monocytes can differentiate and assume either macrophage-like or DC-like properties. For example, during acute lung injury monocytes can replenish tissue-resident macrophages if a niche is open, or they can differentiate into inflammatory or "recruited" macrophages (different from tissue-resident macrophages), thus contributing to wound repair and clearance of cellular debris and microbes [24–27]. Alternatively, monocytes can differentiate into a DC-like cell, migrate to the draining LNs, and induce adaptive immunity [23, 28–30].

For decades, monocytes have been viewed as precursors to tissue-resident macrophages. However, we now know that monocytes continuously traffic through nonlymphoid and lymphoid tissue without becoming bona fide long-lived, self-renewing macrophages or classical dendritic cells [31]. Most interestingly, in both mice and humans, extravascular lung and LN monocytes are as abundant as DCs in the steady state and are even more abundant during inflammation [28, 31–33]. The role of monocytes play in adaptive immunity is underappreciated, and although it has been shown that they can induce CD4 and CD8 T cell proliferation, how they preferentially activate lymphocytes and under what conditions is unclear [28]. In this chapter, we describe methods to identify and isolate murine pulmonary MPs from whole lungs and lung-draining LNs.

2   Materials

2.1   Antibody Clones Used for Staining

The following antibody clones (Table 1) are the ones most commonly used in our laboratory. In our hands they give reliable results; however similar antibodies can be purchased from eBioscience, BD Pharmingen, BioLegend, or other sources. Alternative fluorescent conjugates can also be used.

2.2   Intravenous Injection of Anti-CD45 Antibody

1. Fluorochrome-conjugated anti-CD45 antibody. Fluorochromes such as FITC or PE provide the best results in our experience. In addition, the use of the correct congenic anti-CD45, specifically anti-CD45.1 or anti-CD45.2, also provides better results than general anti-CD45.

2. 1× PBS without calcium and magnesium.

3. 1 mL syringe and 27-gauge needle for tail vein or retro-orbital (requires experience for consistency) injections.

4. Heat lamp and mouse restrainer.

2.3   Single-Cell Suspension of BAL, Lung, and Lung-Draining Lymph Node

1. Scissors.

2. Blunt-curve and fine-point straight-tip forceps.

3. Needles: 18-gauge and 26-gauge.

4. Syringes: 1 mL, 3 mL, and 30 mL.

5. 35 mm × 10 mm round culture dishes or 12-well tissue culture plate.

6. FACS tubes.

7. Case of glass Pasteur pipettes and a rubber bulb.

8. Buffers: 1× PBS without calcium and magnesium, 1× Hank's buffered salt solution (HBSS) without calcium and magnesium.

9. Stock solution of 0.5 M EDTA pH 8.0.

10. EDTA 100 mM, pH 8.0 stored at room temperature.

11. Hank's buffered salt solution complete buffer (HBSS complete): In a 500 mL bottle of HBSS without calcium and magnesium, add 1 mL 30% BSA and 300 μL 0.5 M EDTA.

12. For tissue digestion, make 2 mg/mL stock solution of Liberase TM (Roche) and a 10× stock solution of collagenase D in RPMI, 37.5 mL of RPMI with 1 g of collagenase D (Roche). For lung digestion, one could either use 400 μg/mL of Liberase TM or 2× collagenase D in RPMI. For LN digestion, only use 1× collagenase D and not Liberase TM.

Table 1
Antibody clones used to stain pulmonary mononuclear phagocytes

Antigen	Clone	Company	Conjugate
CD11c	N418	eBioscience	PE-Cy7
CD11b	M1/70	eBioscience	eF450
MHCII	M5/114.15.2	eBioscience	APC-Cy7
CD64	X54-5/7.1	BD Biosciences	AF647
MerTK	Polyclonal	R&D Systems	Unconjugated/biotin
CD206	C06822	Biolegend	PE
Ly6C	HK1.4	Biolegend	BV510
F480	CI:A3-1	AbDSerotec	FITC
Lyve-1	ALY7	eBioscience	eF660
CD169	SER-4	eBioscience	PE
CD36	No.72-1	eBioscience	PE
CCR2	475301	R&D Systems	PE
FcER1	42430	eBioscience	FITC
CD14	Sa2-8	eBioscience	PerCP-Cy5.5
CD45	30-F11	eBioscience	FITC
CD45.1	A20	Biolegend	PE-Cy7
CD45.2	104	Biolegend	FITC
Siglec F	E50-2440	BD Biosciences	PE/BV421
CD 103	2E7	eBioscience	PE/eF450
BrdU	Bu20A	eBioscience	APC
CD43	S7	BD Biosciences	BV421
CD24	M1/69	eBioscience	eF450
CD115	AF598	Biolegend	PE
Ly6G	1A8	Biolegend	eF450
B220	RA3-6B2	eBioscience	eF450
CD3	17A2	eBioscience	eF450
NK1.1	PK136	eBioscience	eF450
Streptavidin		eBioscience	PerCP-Cy5.5

2.4 Fluorescence-Activated Cell Sorting (FACS) Staining for Single-Cell Suspension of BAL, Lung, and Lung-Draining Lymph Node

1. HBSS complete and FACS tubes.

2. Centrifuge and FACS machine.

3   Method

3.1 Intravenous Injection of Anti-CD45 Antibody to Eliminate Contaminating Intravascular Leukocytes from Extravascular Leukocytes in Lung Analysis

The lungs contain a vast network of pulmonary blood vessels through which leukocytes constantly traffic. Although vascular perfusion can help eliminate some leukocytes from the lung vasculature, a high number will always remain. In order to distinguish intravascular leukocytes from the leukocytes that reside in the lung tissue, anti-CD45 antibodies can be given intravenously.

1. Dilute 5 μL of anti-CD45 antibody in 200 μL of 1× PBS.

2. Place mouse under heat lamp to dilate tail veins.

3. Inject prepared antibody solution into the tail vein 4–5 min before sacrifice.

3.2 Single-Cell Suspension of Airway Cells from Bronchoalveolar Lavage (BAL)

1. Euthanize the mouse in a CO_2 chamber following standard protocols prescribed by your institution (*see* **Note 1**).

2. Expose the trachea by making a vertical midline incision and spread the skin.

3. Under the skin, there are two large masses of tissue. These are the submaxillary glands. Gently separate the glands from the midline with forceps (do not cut to avoid bleeding). The trachea will be easily seen and exposed.

4. Grab the outer fascial membrane covering the trachea with forceps. Carefully cut it away to expose the cartilage rings of the trachea.

5. Position the mouse upright and insert one side of a blunt forceps behind the trachea.

6. Through the largest uppermost cartilage ring, insert an 18-gauge needle, with the bevel facing outward. Attach a 3 mL syringe containing 1 mL of 1× HBSS (or PBS alone, no EDTA, if lung digestion follows). Do not insert the needle too deep into the trachea. Only insert it far enough to sufficiently cover the needle opening. After needle insertion, clamp down on the needle with the other side of the blunt forceps to hold the needle in place (*see* **Note 2**).

7. Lavage the lungs four times with 1 mL of 1× HBSS or PBS. Do not add more volume to the syringe (*see* **Note 3**).

8. Collect lavage fluid in a 5 mL FACS tube and centrifuge at $300 \times g$ for 5 min at 4 °C.

9. Remove supernatant. Rapidly tilt the FACS tubes to pour out supernatant or aspirate the supernatant using a syringe or pipette.

10. Place cells on ice, and add antibody cocktail for at least 45 min for optimal cell separation during FACS analysis (*see* Subheading 3.4).

3.3 Single-Cell Suspension of Lung and Lymph Nodes

1. To remove the lungs, place the euthanized mouse on its back, and cut open the thoracic cavity along the lower part of the rib cage. Next cut along the sides of the rib cage toward the axillary region of the mouse.

2. Pull back the sternum with blunt forceps, and locate the largest mediastinal lung-draining lymph node (LLN) under the right side of the heart below the thymus. A small blood vessel perpendicular to the trachea and superior vena cava indicates the location of the large LLN, which is slightly below this small blood vessel. There are also two very small lymph nodes slightly above the blood vessel, which are easily seen in an inflamed mouse but not in a steady-state mouse. Since the left side LLNs are very difficult to find, even during inflammation, only extract the one, large LLN on the right side for consistency and proper data analyses.

3. Place the LLN in a 35 mm culture dish containing 1 mL collagenase D for digestion.

4. Keep on ice until all samples have been collected.

5. Expose the heart. Nick the left atrium, and insert a 27-gauge needle with a 30 mL PBS-filled syringe into the right ventricle, and perfuse the lungs until they turn white. Note that even when the lung is white, there are still many intravascular leukocytes present. This highlights why intravenous injection of anti-CD45, which stains vascular leukocytes, is important (*see* Subheading 3.1), especially for extravascular lymphocyte, monocyte, and granulocyte analysis.

6. Remove lobes individually, and place them in a 35 mm culture dish containing 1 mL of Liberase TM or collagenase D for digestion. Keep on ice until all samples have been collected.

7. Lung digestion: Place lungs on a glass microscope slide, and cut them into very tiny pieces with scissors. Place the minced lungs back into the digestion buffer.

8. Lymph node digestion. Tease each sample apart with two 26-gauge needles attached to 1 mL syringes. To tease, hold down the lymph node with one needle while breaking open the lymph node with the other needle. When teasing is done correctly, concentrated cells bursting from the lymph node are easily observed in the media.

9. Place minced and teased cells in an incubator for 30 min at 37 °C.

10. Following incubation, add 100 μL of 100 mM EDTA to inhibit further tissue digestion.

11. Place culture dishes on ice, and homogenize the cell suspension by repeated pipetting with a glass Pasteur pipette and rubber bulb. Filter cells through 100 μm nylon filter (*see* **Note 4**), and collect cells into a 5 mL FACS tube. Wash the dish with HBSS complete to collect remaining cells. Filter the wash into the same FACS tube using the 100 μm nylon filter.

12. Centrifuge cells at $300 \times g$ for 5 min at 4 °C.

13. Decant or aspirate supernatant, leaving behind up to 200 μL volume with cells. Make sure not to double-tilt the FACS tube while pouring off the supernatant since this can lead to loss of cells.

3.4 Stain Single-Cell Suspension of Airway, Lung, and Lymph Nodes for FACS

1. Place cells on ice and make an antibody master mix for FACS staining. Use approximately 1 μL of antibody per sample in a final volume of 100 μL of HBSS complete for all antibodies except the MHC II (IA/IE) antibody, which is very strong. For the MHC II antibody use 0.3 μL for each lung and LLN sample and 1 μL for each blood sample. As an example, if there are five samples that require the same stain, add 5 μL of each antibody into 500 μL HBSS complete, and then add 100 μL of the antibody mix to each sample. Stain cells with antibodies for at least 45 min up to 1.5 h for optimal cell separation during FACS analysis.

2. The identification of DCs, macrophages, and monocytes in the lung is outlined in Figs. 1 and 2.

3. Identification of migratory DCs and monocytes in the LLN is outlined in Fig. 3.

   Migratory dendritic cells are gated using CD11c versus MHC II (Fig. 3): CD103+ DCs and CD11b+ DCs (alternatively, one could use XCR1 in place of CD103 and SIRPα in place of CD11b). Note: CD11b DCs can be divided further into two subpopulations using CD24 and Mgl2/CD301 (not shown, [22]).

4 Notes

1. We use CO_2 as our preferred method of euthanasia due to its low cost and quick application. However, alternative methods of euthanasia, such as lethal injection of pentobarbital, can also be used. Cervical dislocation should be avoided since it can cause bleeding into the lungs. Euthanasia methods should be approved by your institutional animal care and use committee (IACUC).

2. If the needle appears to be unable to suction delivered fluid, rotate the tip a bit, or slightly expose needle opening to the air to release lack of suction.

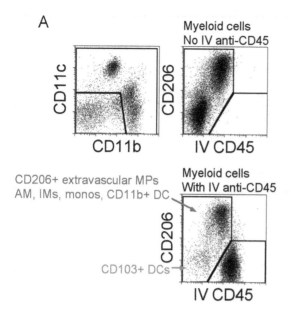

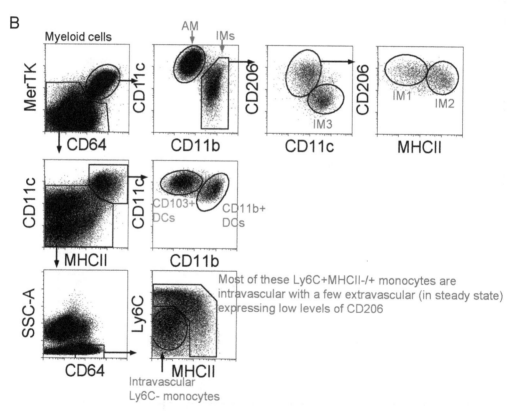

Fig. 1 Identification of mononuclear phagocytes (MPs) in steady-state mouse lung using flow cytometry. (**a**) The use of an IV injection of anti-CD45 antibody to differentiate intravascular leukocytes from pulmonary extravascular mononuclear phagocytes. Note that intravascular leukocytes stain positively with the anti-CD45 antibody. Mononuclear phagocytes from the lung tissue and airspaces can be highly autofluorescent, so a second stain, such as CD206, is helpful. Note that Fig. 2 illustrates in detail how to identify all the CD206+ MPs and CD206−CD103+ DC. (**b**) Gating strategy for pulmonary macrophages. Macrophages are double positive for MerTK and CD64 (Fig. 2). The macrophage-negative gate (MerTK−, CD64−) is used to identify DCs (MHCII^high, CD11c^high). Cells from the DC-negative gate can be further analyzed to identify extravascular monocytes, which are few and would require the use of IV anti-CD45 to identify them (not shown here) [31]

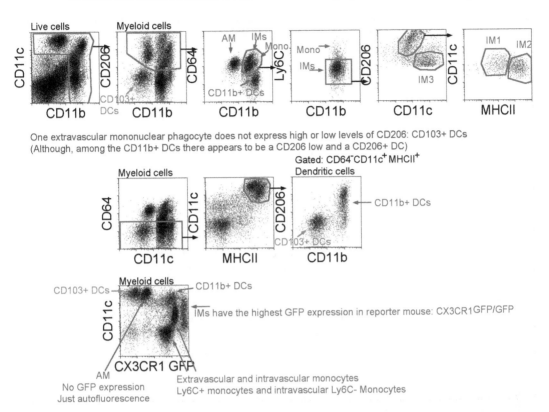

Fig. 2 Additional ways to identify mononuclear phagocytes (MPs) in steady-state mouse lung. Live cells are selected using FSC and SSC, followed by exclusion of doublet cell and dead cells (not shown). Myeloid cells are gated using CD11c versus CD11b, illustrated in the top left graph. Most extravascular MPs in the lung can be identified using CD206 (in both mice and humans [30]) with the exception of CD103$^+$ DCs, which are CD206 negative. Outlined above is the gating strategy to identify all the pulmonary MPs. In addition, in the lung, IMs have very high CX$_3$CR1 expression as reported in the CX$_3$CR1 GFP reporter mice [24]. Intravascular Ly6C$^-$ monocytes display higher GFP expression in CX$_3$CR1^{gfp} mice than intra- or extravascular Ly6C$^+$ monocytes

3. Never place more than 1 mL of HBSS or 1× PBS into the airways. Excess fluid forced into the airways will cause lung injury and bleeding will occur. Fluid extracted from airways of naïve mice should not contain red blood cells. If this happens, there are mainly two reasons: (1) The bronchoalveolar lavage was not extracted immediately after mouse euthanasia, or (2) lavage fluid is being injected into the mouse too aggressively. Also, do not add EDTA to BAL fluid since EDTA will inhibit lung digestion.

4. Do not use any filter smaller than 70 μm, because DCs and macrophages may not easily pass through a filter that is too fine. For example, a 40 μm nylon filter may result in reduced recovery of DCs and macrophages for FACs analysis. Lung cells and sometimes lymph nodes should be refiltered through a 70 μm or 100 μm nylon filter to remove clumped cells that could clog the FACS machine.

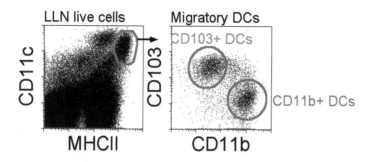

In steady state, there are a few monocyes in the LLN.
However during inflammation there are many more LN monocytes.
Use CD64 versus CD11b to identify Ly6C+ MHCII+ monocytes (make sure to exclude B cells using B220).

Fig. 3 Identification of mononuclear phagocytes (MPs) in steady-state mouse lung-draining LN (LLN). Live cells are gated using FSC, SSC, doublet cell, and DAPI exclusions. Migratory dendritic cells are gated using CD11c versus MHCII: CD103$^+$ DCs and CD11b$^+$ DCs (alternatively, one could use XCR1 in place of CD103 and SIRPα in place of CD11b). CD11b$^+$ DCs can be further divided into two subpopulations using CD24 and Mgl$_2$/CD301 (not shown [22]). Monocyte gating using CD64 and/or Ly6C versus CD11b was previously shown in reference (keep in mind that neutrophils express intermediate levels of Ly6C, express slightly more CD11b and have a higher SSC than LN monocytes) [31]. Lung-derived migratory DCs have higher MHC II expression levels than resident LN DCs and B cells

Acknowledgment

Grant support: C.V.J. NIH R01 HL115334 and R01 HL135001.

References

1. Janssen WJ, Bratton DL, Jakubzick CV, Henson PM (2016) Myeloid cell turnover and clearance. Microbiol Spectr 4(6). https://doi.org/10.1128/microbiolspec.MCHD-0005-2015

2. MacLean JA, Xia W, Pinto CE, Zhao L, Liu HW, Kradin RL (1996) Sequestration of inhaled particulate antigens by lung phagocytes. A mechanism for the effective inhibition of pulmonary cell-mediated immunity. Am J Pathol 148(2):657–666

3. Holt PG (2005) Pulmonary dendritic cells in local immunity to inert and pathogenic antigens in the respiratory tract. Proc Am Thorac Soc 2(2):116–120. https://doi.org/10.1513/pats.200502-017AW

4. Sung SS, Fu SM, Rose CE Jr, Gaskin F, Ju ST, Beaty SR (2006) A major lung CD103 (alphaE)-beta7 integrin-positive epithelial dendritic cell population expressing Langerin and tight junction proteins. J Immunol 176(4):2161–2172

5. Vermaelen K, Pauwels R (2005) Pulmonary dendritic cells. Am J Respir Crit Care Med 172(5):530–551. https://doi.org/10.1164/rccm.200410-1384SO

6. Vermaelen KY, Carro-Muino I, Lambrecht BN, Pauwels RA (2001) Specific migratory dendritic cells rapidly transport antigen from the airways to the thoracic lymph nodes. J Exp Med 193(1):51–60

7. Jakubzick C, Tacke F, Llodra J, van Rooijen N, Randolph GJ (2006) Modulation of dendritic cell trafficking to and from the airways. J Immunol 176 (6):3578–3584.

8. Jakubzick C, Helft J, Kaplan TJ, Randolph GJ (2008) Optimization of methods to study pulmonary dendritic cell migration reveals distinct capacities of DC subsets to acquire soluble versus particulate antigen. J Immunol Methods 337(2):121–131. https://doi.org/10.1016/j.jim.2008.07.005

9. Desch AN, Randolph GJ, Murphy K, Gautier EL, Kedl RM, Lahoud MH, Caminschi I,

Shortman K, Henson PM, Jakubzick CV (2011) CD103+ pulmonary dendritic cells preferentially acquire and present apoptotic cell-associated antigen. J Exp Med 208(9): 1789–1797. https://doi.org/10.1084/jem.20110538

10. Jakubzick C, Randolph GJ (2010) Methods to study pulmonary dendritic cell migration. Methods Mol Biol 595:371–382. https://doi.org/10.1007/978-1-60761-421-0_24

11. Guilliams M, Lambrecht BN, Hammad H (2013) Division of labor between lung dendritic cells and macrophages in the defense against pulmonary infections. Mucosal Immunol 6(3):464–473. https://doi.org/10.1038/mi.2013.14

12. Guilliams M, Ginhoux F, Jakubzick C, Naik SH, Onai N, Schraml BU, Segura E, Tussiwand R, Yona S (2014) Dendritic cells, monocytes and macrophages: a unified nomenclature based on ontogeny. Nat Rev Immunol 14(8):571–578. https://doi.org/10.1038/nri3712

13. Desch AN, Henson PM, Jakubzick CV (2013) Pulmonary dendritic cell development and antigen acquisition. Immunol Res 55(1–3):178–186. https://doi.org/10.1007/s12026-012-8359-6

14. Desch AN, Gibbings SL, Clambey ET, Janssen WJ, Slansky JE, Kedl RM, Henson PM, Jakubzick C (2014) Dendritic cell subsets require cis-activation for cytotoxic CD8 T-cell induction. Nat Commun 5:4674. https://doi.org/10.1038/ncomms5674

15. Atif SM, Nelsen MK, Gibbings SL, Desch AN, Kedl RM, Gill RG, Marrack P, Murphy KM, Grazia TJ, Henson PM, Jakubzick CV (2015) Cutting edge: roles for Batf3-dependent APCs in the rejection of minor histocompatibility antigen-mismatched grafts. J Immunol 195(1):46–50. https://doi.org/10.4049/jimmunol.1500669

16. Kim TS, Braciale TJ (2009) Respiratory dendritic cell subsets differ in their capacity to support the induction of virus-specific cytotoxic CD8+ T cell responses. PLoS One 4(1):e4204. https://doi.org/10.1371/journal.pone.0004204

17. Kim TS, Gorski SA, Hahn S, Murphy KM, Braciale TJ (2014) Distinct dendritic cell subsets dictate the fate decision between effector and memory CD8(+) T cell differentiation by a CD24-dependent mechanism. Immunity 40(3):400–413. https://doi.org/10.1016/j.immuni.2014.02.004

18. Kim TS, Hufford MM, Sun J, Fu YX, Braciale TJ (2010) Antigen persistence and the control of local T cell memory by migrant respiratory dendritic cells after acute virus infection. J Exp Med 207(6):1161–1172. https://doi.org/10.1084/jem.20092017

19. Wakim LM, Bevan MJ (2011) Cross-dressed dendritic cells drive memory CD8+ T-cell activation after viral infection. Nature 471(7340):629–632. https://doi.org/10.1038/nature09863

20. del Rio ML, Rodriguez-Barbosa JI, Kremmer E, Forster R (2007) CD103− and CD103+ bronchial lymph node dendritic cells are specialized in presenting and cross-presenting innocuous antigen to CD4+ and CD8+ T cells. J Immunol 178 (11):6861–6866

21. Hildner K, Edelson BT, Purtha WE, Diamond M, Matsushita H, Kohyama M, Calderon B, Schraml BU, Unanue ER, Diamond MS, Schreiber RD, Murphy TL, Murphy KM (2008) Batf3 deficiency reveals a critical role for CD8alpha+ dendritic cells in cytotoxic T cell immunity. Science 322(5904):1097–1100. https://doi.org/10.1126/science.1164206

22. Tussiwand R, Everts B, Grajales-Reyes GE, Kretzer NM, Iwata A, Bagaitkar J, Wu X, Wong R, Anderson DA, Murphy TL, Pearce EJ, Murphy KM (2015) Klf4 expression in conventional dendritic cells is required for T helper 2 cell responses. Immunity 42(5):916–928. https://doi.org/10.1016/j.immuni.2015.04.017

23. Gibbings SL, Thomas SM, Atif SM, McCubbrey AL, Desch AN, Danhorn T, Leach SM, Bratton DL, Henson PM, Janssen WJ, Jakubzick CV (2017) Three unique interstitial macrophages in the murine lung at steady state. Am J Respir Cell Mol Biol 57:66–76. https://doi.org/10.1165/rcmb.2016-0361OC

24. Janssen WJ, Barthel L, Muldrow A, Oberley-Deegan RE, Kearns MT, Jakubzick C, Henson PM (2011) Fas determines differential fates of resident and recruited macrophages during resolution of acute lung injury. Am J Respir Crit Care Med 184:547–560. https://doi.org/10.1164/rccm.201011-1891OC

25. Mould KJ, Barthel L, Mohning MP, Thomas SM, McCubbrey AL, Danhorn T, Leach SM, Fingerlin TE, O'Connor BP, Reisz JA, D'Alessandro A, Bratton DL, Jakubzick CV, Janssen WJ (2017) Cell origin dictates programming of resident versus recruited macrophages during acute lung injury. Am J Respir Cell Mol Biol 57:294–306. https://doi.org/10.1165/rcmb.2017-0061OC

26. McCubbrey AL, Barthel L, Mohning MP, Redente EF, Mould KJ, Thomas SM, Leach SM, Danhorn T, Gibbings SL, Jakubzick CV, Henson PM, Janssen WJ (2018) Deletion of c-FLIP from CD11bhi macrophages prevents

development of Bleomycin-induced lung fibrosis. Am J Respir Cell Mol Biol 58:66–78. https://doi.org/10.1165/rcmb.2017-0154OC

27. Misharin AV, Morales-Nebreda L, Reyfman PA, Cuda CM, Walter JM, McQuattie-Pimentel AC, Chen CI, Anekalla KR, Joshi N, Williams KJN, Abdala-Valencia H, Yacoub TJ, Chi M, Chiu S, Gonzalez-Gonzalez FJ, Gates K, Lam AP, Nicholson TT, Homan PJ, Soberanes S, Dominguez S, Morgan VK, Saber R, Shaffer A, Hinchcliff M, Marshall SA, Bharat A, Berdnikovs S, Bhorade SM, Bartom ET, Morimoto RI, Balch WE, Sznajder JI, Chandel NS, Mutlu GM, Jain M, Gottardi CJ, Singer BD, Ridge KM, Bagheri N, Shilatifard A, Budinger GRS, Perlman H (2017) Monocyte-derived alveolar macrophages drive lung fibrosis and persist in the lung over the life span. J Exp Med 214(8):2387–2404. https://doi.org/10.1084/jem.20162152

28. Jakubzick CV, Randolph GJ, Henson PM (2017) Monocyte differentiation and antigen-presenting functions. Nat Rev Immunol 17(6):349–362. https://doi.org/10.1038/nri.2017.28

29. Larson SR, Atif SM, Gibbings SL, Thomas SM, Prabagar MG, Danhorn T, Leach SM, Henson PM, Jakubzick CV (2016) Ly6C(+) monocyte efferocytosis and cross-presentation of cell-associated antigens. Cell Death Differ 23(6):997–1003. https://doi.org/10.1038/cdd.2016.24

30. Gibbings SL, Goyal R, Desch AN, Leach SM, Prabagar M, Atif SM, Bratton DL, Janssen W, Jakubzick CV (2015) Transcriptome analysis highlights the conserved difference between embryonic and postnatal-derived alveolar macrophages. Blood 126(11):1357–1366. https://doi.org/10.1182/blood-2015-01-624809

31. Jakubzick C, Gautier EL, Gibbings SL, Sojka DK, Schlitzer A, Johnson TE, Ivanov S, Duan Q, Bala S, Condon T, van Rooijen N, Grainger JR, Belkaid Y, Ma'ayan A, Riches DW, Yokoyama WM, Ginhoux F, Henson PM, Randolph GJ (2013) Minimal differentiation of classical monocytes as they survey steady-state tissues and transport antigen to lymph nodes. Immunity 39(3):599–610. https://doi.org/10.1016/j.immuni.2013.08.007

32. Jakubzick C, Bogunovic M, Bonito AJ, Kuan EL, Merad M, Randolph GJ (2008) Lymph-migrating, tissue-derived dendritic cells are minor constituents within steady-state lymph nodes. J Exp Med 205(12):2839–2850. https://doi.org/10.1084/jem.20081430

33. Yona S, Kim KW, Wolf Y, Mildner A, Varol D, Breker M, Strauss-Ayali D, Viukov S, Guilliams M, Misharin A, Hume DA, Perlman H, Malissen B, Zelzer E, Jung S (2013) Fate mapping reveals origins and dynamics of monocytes and tissue macrophages under homeostasis. Immunity 38(1):79–91. https://doi.org/10.1016/j.immuni.2012.12.001

Chapter 4

Isolation and Characterization of Mouse Neutrophils

Niki D. J. Ubags and Benjamin T. Suratt

Abstract

Isolation of murine neutrophils from several anatomical compartments allows for functional characterization and analysis of these cells. Here we describe the isolation of bone marrow, peripheral blood, and lung airspace and interstitial neutrophil populations, using density gradient separation, lavage, and flow cytometry techniques.

Key words Neutrophils, Isolation, Density gradient, Blood, Marrow, Lung

1 Introduction

The pulmonary innate immune response forms a first line of defense against invading pathogens through pathogen recognition and subsequent effector responses. Neutrophils, which are continuously generated in the bone marrow from myeloid precursors, are a key effector cell in this response [1–3]. In order to define the role of neutrophils in different diseases of the lung, including determining how molecular factors or altered homeostatic conditions affect neutrophil effector functions, it is often necessary to isolate and characterize this cell population. In this context, it is critically important to isolate neutrophils with as little manipulation as possible, so that isolation-associated effects on phenotype and survival are minimized.

Neutrophils can be isolated from several anatomical compartments of the mouse, each of which presents its own strengths and weaknesses in terms of both the physiological relevance and degree of isolation-associated alteration of the neutrophil populations. Morphologically mature bone marrow neutrophils represent postmitotic granulocytes that are retained in the marrow prior to normal homeostatic release or the accelerated "emergency" release that accompanies a wide range of physiological stresses [4]. Although these cells offer only a rough phenotypic approximation of circulating and tissue-migrated neutrophils, as they typically

Scott Alper and William J. Janssen (eds.), *Lung Innate Immunity and Inflammation: Methods and Protocols*,
Methods in Molecular Biology, vol. 1809, https://doi.org/10.1007/978-1-4939-8570-8_4,
© Springer Science+Business Media, LLC, part of Springer Nature 2018

show some functional immaturity and are less readily activated [5], they are easily isolated in large quantities ($10–20 \times 10^6$/mouse) and tend to be less susceptible to inadvertent activation and cell death related to cell isolation. In comparison, the arguably more physiologically relevant neutrophil populations (circulating and pulmonary neutrophils) are more susceptible to the effects of isolation techniques and hence are more activated and/or damaged following isolation. Furthermore, particularly in the case of airspace and tissue neutrophils, such cells have been "preselected" by their ability to transit into the compartment examined and may not reveal underlying functional defects (e.g., chemotaxis or integrin display) that may exist in the neutrophil pool as a whole.

In this protocol, we describe the isolation of bone marrow, peripheral blood, and lung airspace and interstitial neutrophils, using density gradient isolation (bone marrow-derived and peripheral blood), bronchoalveolar lavage (airspace), or tissue disaggregation (lung interstitial). Neutrophil function assays or phenotypic characterization by flow cytometry can be performed after isolation as described in Chap. 12 and has been shown previously [6–10].

2   Materials

It is critical to prepare all stock solutions in a sterile environment and store them at 4 °C unless indicated otherwise—do not freeze. Pre-warm all solutions to room temperature before use, unless indicated otherwise. In addition to manipulation, changes in temperature tend to activate neutrophils [11] which can lead to lower cell yields and less accurate approximation of neutrophil phenotype during subsequent characterization.

2.1 Bone Marrow Neutrophil Isolation

1. 3.8% sodium citrate buffer: Dissolve 38 g of sodium citrate in 1 L of sterile endotoxin-free water, and store in 50 mL aliquots at 4 °C.

2. 1× HBSS 0.38% sodium citrate buffer (250 mL): Add 25 mL of 10× HBSS (without calcium, magnesium, or phenol red) and 25 mL of 3.8% sodium citrate solution to 200 mL sterile endotoxin-free water.

3. Percoll working solution ("100% Percoll"): Add 1 mL 10× HBSS buffer (without calcium, magnesium, or phenol red) to 9 mL sterile Percoll stock solution (GE Healthcare). Invert several times but do not vortex.

4. 70% ethanol.

5. 5 mL syringe.

6. 25-gauge and 18-gauge needles.

7. 15 mL conical tube.

8. 50 mL conical tube.

9. Curved surgical scissors.

10. Sterile gauze.

11. 2 mL serological pipette.

12. Disposable transfer pipettes.

13. Centrifuge at room temperature with inserts for 15 and 50 mL tubes.

2.2 Peripheral Blood Neutrophil Isolation

1. 3.8% sodium citrate buffer: Dissolve 38 g of sodium citrate in 1 L of sterile endotoxin-free water and store in 50 mL aliquots at 4 °C.

2. 1× HBSS 0.38% sodium citrate buffer (250 mL): Add 25 mL of 10× HBSS (without calcium, magnesium, or phenol red) and 25 mL of 3.8% sodium citrate solution to 200 mL sterile endotoxin-free water.

3. PBS/0.1% BSA: Dissolve 0.05 g of sterile, high-quality BSA in 50 mL sterile 1× PBS.

4. Percoll working solution ("100% Percoll"): Add 1 mL 10× HBSS buffer (without calcium, magnesium, or phenol red) to 9 mL sterile Percoll stock solution (GE Healthcare). Invert several times but do not vortex.

5. Epinephrine solution (1:10,000 in sterile saline) (optional; *see* **Note 1**).

6. 18% sterile EDTA solution.

7. Sterile, pyrogen-free saline (0.9%).

8. 6% dextran solution: Prepare 6% dextran (Sigma-Aldrich) solution in 0.9% sterile saline.

9. 15 mL conical tube.

10. 1 mL syringes.

11. 25-gauge needles.

12. Curved surgical scissors.

13. Disposable transfer pipettes.

14. Centrifuge at room temperature with inserts for 15 and 50 mL tubes.

2.3 Alveolar and Interstitial Neutrophil Isolation and Characterization

1. PBS/0.1% BSA: Dissolve 0.05 g of sterile, high-quality BSA in 50 mL sterile 1× PBS.

2. 1× DNase/collagenase DMEM solution: 200 μg/mL DNase and 1 mg/mL collagenase in DMEM.

3. 5× DNase/collagenase DMEM solution: 1 mg/mL DNase and 5 mg/mL collagenase in DMEM.

4. FACS buffer: Add 1 mL of heat-inactivated fetal calf serum (FCS) to 49 mL sterile 1× PBS (*see* **Note 2**).

5. ACK lysis solution: Dissolve 150 mM NH_4Cl, 10 mM $KHCO_3$, and 0.1 mM Na_2EDTA in sterile ddH_2O, and adjust pH to 7.2–7.4. Sterile filter the solution.

6. 70% ethanol.

7. Live/dead flow cytometry stain such as PI or DAPI.

8. Fluorophore-labeled anti-CD45 antibody.

9. Fluorophore-labeled anti-CD11b antibody.

10. Fluorophore-labeled anti-F4/80 antibody.

11. Fluorophore-labeled anti-Ly6G antibody.

12. 50 mL conical tube.

13. 1.5 mL Eppendorf tubes.

14. 18-gauge needle.

15. 10 mL syringe.

16. 70 μm filter.

17. Petri dishes.

18. Curved surgical scissors.

19. Scalpel with #10 blade.

20. Incubator at 37 °C with shaker.

3   Methods

General cautions: (1) Never vortex cell solutions to resuspend them—this will activate and/or damage the neutrophils; (2) maintain cell solutions at room temperature until they are assayed or fixed—temperature change will activate neutrophils.

3.1   Bone Marrow Neutrophil Isolation: Gradient Preparation

1. Layer each dilution of Percoll (Table 1) onto the one beneath in a sterile 15 mL conical tube. This can be performed during the whole-marrow centrifugation steps in the bone marrow isolation protocol (Fig. 1).

2. Start with 2.5 mL of the 72% Percoll dilution, followed by 2 mL of the 64% dilution and then 2 mL of the 52% dilution, using a 2 mL serological pipette (*see* **Note 3**).

3. Control the speed of the "layering" with your thumb or index finger on the top of the pipette, ensuring a slow steady flow rate, which avoids mixing of the layers at the interfaces. Some people prefer "underlayering" of the Percoll solutions (reverse order of layering, placing the tip of the pipette at the very bottom of the conical tube)—both approaches will work if done with care.

Table 1
Percoll dilutions for density gradient (bone marrow neutrophil isolation)

	52% (mL)	64% (mL)	72% (mL)
100% Percoll	1.56	1.92	2.16
1× HBSS 0.38% sodium citrate	1.44	1.08	0.84
Total volume	3	3	3

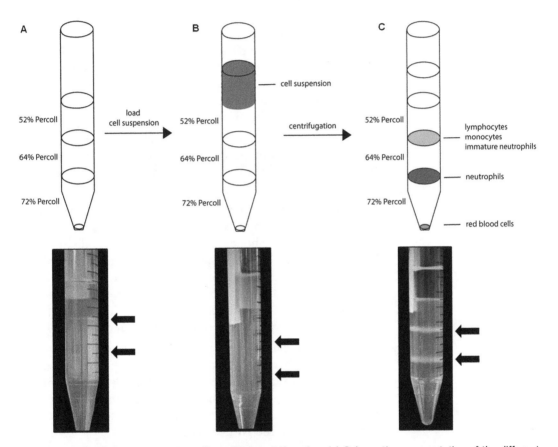

Fig. 1 Neutrophil isolation using Percoll gradient centrifugation. (**a**) Schematic representation of the different Percoll gradient layers (52%, 64%, and 72%) in a 15 mL conical tube (top) and photographic images of an actual gradient (bottom). Disaggregated whole bone marrow will be layered on top of the 52% Percoll layer as indicated (**b**). Gradient centrifugation will separate the different cell populations (**c**). The interface between the 52% and 64% layers contains lymphocytes, monocytes, and immature neutrophils. A pure, mature neutrophil population will be retained at the 64%–72% interface. For blood neutrophil isolation, the buffy coat of leukocytes from blood is layered similarly onto a gradient of 54%/66%/78% Percoll

3.2 Bone Marrow Neutrophil Isolation: Bone Marrow Extraction and Neutrophil Isolation

1. Euthanize mice, preferably using cervical dislocation (*see* **Note 4**), and then douse the animal with 70% ethanol.

2. Using curved scissors, make a small incision in the abdominal skin (try not to enter the peritoneum if possible). Pull the skin from the mouse in a downward motion to expose the abdomen and hind limbs, and then remove it entirely from both legs.

3. Carefully remove hamstring and quadriceps muscles from the femur using curved scissors. Avoid scoring or cutting the bone. Cut through the pelvis adjacent to the femur avoiding the joint, and separate the leg from the body. Repeat same procedure on the opposite leg.

4. Remove all remaining muscle tissue from the femur using blunt dissection with a piece of sterile gauze. Carefully separate the femur from the knee joint. Ensure that all muscle tissue is removed from the femur, and place the femur in 1× HBSS 0.38% sodium citrate buffer.

5. Remove all tissue from the tibia using gauze (be careful not to break the bone). Remove the knee joint from tibial plate, and remove the tibia from the ankle, and place the tibia in 1× HBSS 0.38% sodium citrate buffer. Repeat same procedure on the opposite leg.

6. Flush bone marrow from the bones (two femurs and two tibias per mouse): Use a 5 mL syringe and a 25-gauge needle to flush the marrow from the bones with 1× HBSS 0.38% sodium citrate buffer into a clean 50 mL conical tube (*see* **Note 5**). For tibias: Cut off the first 1–2 mm of the bone at the ankle end of the tibia, and insert needle through tibial plate to flush out the marrow plug. For femurs: Cut off the femoral head (the "ball"), and insert needle through the opposite end to flush out marrow.

7. Use an 18-gauge needle and a 5 mL syringe to gently disaggregate the bone marrow by repeatedly aspirating and expressing the solution to yield a single-cell solution—do not use much force or speed as this will shear the neutrophils and activate them. Transfer the disaggregated bone marrow to a clean 50 mL conical tube using the 18-gauge needle.

8. Centrifuge the bone marrow suspension at $230 \times g$ for 6 min at room temperature.

9. Discard the supernatant, and resuspend the pellet in 2 mL 1× HBSS 0.38% sodium citrate buffer. Add ~40 mL 1× HBSS 0.38% sodium citrate buffer to wash the cells, and centrifuge at $230 \times g$ for 6 min at room temperature.

10. Discard the supernatant, and resuspend the pellet in 2 mL 1× HBSS 0.38% sodium citrate buffer. Layer the cell suspension gently on top of the gradient using a 2 mL serological pipette (*see* **Note 6**).

11. Centrifuge the gradient at $1545 \times g$ for 30 min at room temperature (*see* **Note 7**).

12. Use a disposable transfer pipette, and carefully remove the first (top) and second cell layers (between top and 52% and between 52% and 64% layers). Do not mix the layers (you can retain these layers in order to determine your cell differentials for each layer using cytospin if the gradient needs troubleshooting—*see* **Note 3**). Use a clean transfer pipette to remove the third (neutrophil) layer, and transfer the cells to a clean 15 mL conical tube. Repeat this procedure with a clean transfer pipette until the third layer is entirely transferred or the layers begin to mix (stop if layers mix, as mixing will lead to lymphocyte contamination).

13. Add 1× HBSS 0.38% sodium citrate buffer to a total volume of 14 mL, and wash cells as described before.

14. Discard the supernatant, and resuspend the cells in 2 mL 1× HBSS 0.38% sodium citrate, and add additional 1× HBSS 0.38% sodium citrate buffer to a total volume of 10 mL. Cells are now ready to be counted, cytospin preparations can be made to determine purity of the cell population, and dye exclusion can be used to assess cell viability.

3.3 Peripheral Blood Neutrophil Isolation: Gradient Preparation

1. Layer each dilution of Percoll (Table 2) in a sterile 15 mL conical tube. This can be performed after resuspending the blood in the dextran solution (Fig. 1).

2. Start with 2.5 mL of the 78% Percoll dilution, followed by 2 mL of the 66% dilution and then 2 mL of the 54% dilution, using a 2 mL serological pipette (*see* **Note 3**).

3. Control the speed of the "layering" with your thumb or index finger on the top of the pipette, ensuring a slow steady flow rate, which avoids mixing of the layers at the interfaces. Some people prefer "underlayering" of the Percoll solutions (reverse order of layering, placing the tip of the pipette at the very bottom of the conical tube)—both approaches will work if done with care.

Table 2
Percoll dilutions for density gradient (peripheral blood neutrophil isolation)

	54% (mL)	66% (mL)	78% (mL)
100% Percoll	1.62	1.98	2.34
1× HBSS 0.38% sodium citrate	1.38	1.02	0.66
Total volume	3	3	3

3.4 Peripheral Blood Neutrophil Isolation: Peripheral Blood Neutrophil Isolation and Purification

1. Coat a 15 mL conical tube with PBS/0.1% BSA and then decant.

2. Fill the dead space of a 1 mL syringe with 18% EDTA solution (~70 μL). Collect blood via cardiac puncture from five to six mice using 25-gauge needles connected to the 1 mL syringe, and transfer the blood to the coated 15 mL tube (*see* **Notes 1** and **8**).

3. Centrifuge blood at $350 \times g$ for 20 min at room temperature. Discard the plasma (*see* **Note 9**), and resuspend the cell pellet in a 15 mL conical tube using a 1:5.25 dilution of 6% dextran: normal saline solution using the following equation—*total volume of 6% dextran/saline solution to add = (original blood volume * 1.75)—pellet volume* (*see* **Note 10**). Allow blood and dextran/saline mixture to settle for 30 min undisturbed at room temperature.

4. Using a transfer pipette, carefully aspirate the top layer of dextran/saline solution (containing the buffy coat of leukocytes) into a clean 15 mL conical tube, and avoid loosening the red blood cell pellet at the bottom half of the tube. Discard red cells. Bring the isolated leukocytes to a total volume of 14 mL with 1× HBSS 0.38% sodium citrate buffer.

5. Centrifuge cell suspension at $230 \times g$ for 6 min at room temperature.

6. Discard the supernatant, and resuspend the pellet in 2 mL 1× HBSS 0.38% sodium citrate buffer. Wash cells in 14 mL 1× HBSS sodium citrate buffer as described above (room temperature), and repeat this wash step.

7. Discard the supernatant, and resuspend the pellet in 2 mL 1× HBSS 0.38% sodium citrate buffer. Layer the cell suspension onto top of the prepared gradient (*see* **Note 6**).

8. Centrifuge the gradient at $1545 \times g$ for 30 min at room temperature (*see* **Note 7**).

9. Use a disposable transfer pipette to carefully remove the first and second cell layers (at the interfaces between the loading buffer and 54% Percoll and between 54% and 66% Percoll layers). Do not mix the layers. You can retain each of these layers and determine cell differentials using cytospins if the gradient needs troubleshooting (*see* **Note 3**).

10. Use a clean transfer pipette to remove the third (neutrophil) layer and transfer the cells to a clean 15 mL conical tube. Repeat this procedure with a clean transfer pipette until the third layer is entirely transferred or the layers begin to mix (stop if layers mix, as mixing will lead to lymphocyte contamination).

11. Wash cells in 14 mL 1× HBSS 0.38% sodium citrate buffer as described above (room temperature).

12. Discard the supernatant, and resuspend the cells in 2 mL 1× HBSS 0.38% sodium citrate, and bring to a total volume of 10 mL with 1× HBSS 0.38% sodium citrate. Cells are now ready to be counted. Centrifuge cell suspension at $230 \times g$ for 6 min at room temperature.

13. Discard supernatant and resuspend cells in 2 mL PBS/0.1% BSA. To determine cell purity, an aliquot can be taken for cytospin preparation at this point.

3.5 Alveolar and Interstitial Neutrophil Isolation and Characterization: Alveolar Neutrophil Isolation

1. Euthanize mouse and douse with 70% ethanol.

2. Carefully dissect the anterior neck of the mouse to expose the trachea. Cannulate the trachea with an 18-gauge angiocatheter, and secure with suture. Gently perform bronchoalveolar lavage (BAL) with four sequential 800 μL aliquots of cold PBS/0.1% BSA using a 1 mL syringe. Keep first aliquot in a separate 1.5 mL Eppendorf tube (supernatant can be used for cytokine analysis), and pool the remaining three aliquots in a 5 mL tube, and keep on ice.

3. Centrifuge aliquots at $350 \times g$ for 5 min. Save and snap-freeze supernatant from first aliquot if interested in analysis for cytokines, albumin, etc., and decant supernatant from the pooled aliquots.

4. Resuspend and pool cells from one mouse in 1 mL PBS/0.1% BSA.

5. Use cell counter or hemocytometer to determine cell count, and prepare cytospin to perform cell differential. BAL neutrophils may then be examined by flow cytometry (as detailed below) or separated from contaminating macrophages by plating the BAL cells for 1 h in a tissue culture well and then collecting the nonadherent cells (typically neutrophils) for further examination.

3.6 Alveolar and Interstitial Neutrophil Isolation and Characterization: Interstitial Neutrophil Isolation

1. Euthanize mouse as described in Subheading 3.5, **step 1**.

2. Optional: Perfuse the lungs with 10 mL PBS by cannulating the right ventricle and cutting either the inferior vena cava (IVC) or left atrium to allow drainage.

3. Cannulate the trachea as described in Subheading 3.5, **step 2**.

4. Instill 1 mL of 1× DNase/collagenase DMEM solution into lungs through tracheal cannula.

5. Incubate for 1 min and then remove the lungs one lobe at a time with curved scissors.

6. Place the lungs in a petri dish, and thoroughly mince with a scalpel (#10 blade), and place lung pieces in a 50 mL conical tube.

7. Wash petri dish with 4 mL DMEM to recover remaining lung pieces, and add to the rest of the minced lung.

8. Add 1 mL 5× DNase/collagenase DMEM solution to the minced lung.

9. Incubate in a shaking incubator at 37 °C, 200 rpm for 20 min.

10. Triturate the sample five times using a 10 mL syringe and an 18-gauge needle.

11. Incubate in a shaking incubator at 37 °C, 200 rpm for another 20 min.

12. Triturate the sample five times using a 10 mL syringe and an 18-gauge needle.

13. On the last trituration, pass sample through a 70 μm filter into a new 50 mL conical tube.

14. Rinse residual cells through filter with 10 mL PBS and collect in the same 50 mL tube.

15. Centrifuge samples for 5 min at 350 × g.

16. Resuspend samples well in 4 mL ACK lysis solution (pipette up and down 5×), and then add 8 mL PBS to inactivate lysis solution.

17. Centrifuge sample for 5 min at 350 × g.

18. Wash cells with 10 mL PBS (centrifuge samples for 5 min at 350 × g), and resuspend pellet in 4 mL FACS buffer.

19. Count cells and stain if performing neutrophil characterization by flow cytometry or sorting by FACS.

3.7 Alveolar and Interstitial Neutrophil Isolation and Characterization: Interstitial Neutrophil Characterization

1. Aliquot 0.5–1 × 10^6 cells per sample into a 96-well plate.

2. Stain samples with live/dead stain according to manufacturer's protocol.

3. Wash samples in FACS buffer (centrifuge at 1200 × g for 1 min).

4. Block samples with Fc Block for 10 min at room temperature.

5. Do not wash samples.

6. Add antibody cocktail consisting of anti-CD45, CD11b, F4/80, and Ly6G, in 100 μL FACS buffer, and incubate for 25 min at 4 °C. (Make sure to include single-stained controls and an unstained control to set up compensation. Compensation beads are recommended in case the amount of cells is insufficient to include such controls.)

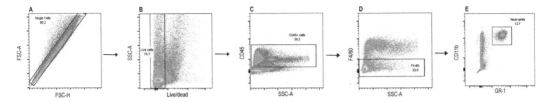

Fig. 2 Flow cytometry gating strategy for alveolar and interstitial pulmonary neutrophil characterization from the disaggregated whole lung. Neutrophil characterization from disaggregated lung tissue can be performed by flow cytometric analysis. The first step of analysis is to exclude remaining cellular aggregates by tightly gating the single-cell population on the FSC-H/FSC-A axis (**a**) and selecting the live cells (**b**). Next, the CD45+ cell population is selected from the live cell gate (**c**), followed by exclusion of F4/80+ cells (monocytes and macrophages) (**d**). Subsequent gating for CD11b+/Gr-1hi cells identifies the neutrophil population (**e**). Given the recent development of Ly6G-specific antibodies, it is recommended to use a CD11b+Ly6G+ population to characterize neutrophils, as indicated in Subheading 3.7. Additional gating using CD68 can be used to further characterize all myeloid populations (monocytes, macrophages, and dendritic cells) (*see* ref. [9])

7. Wash cells twice with FACS buffer (centrifuge at $1200 \times g$ for 1 min).

8. Resuspend cells in 200 μL FACS buffer and transfer to a FACS tube (*see* **Note 11**).

9. Use the following gating strategy to determine the neutrophil population (Fig. 2):

   • Select the single-cell population (by tightly gating the cell population lying on the FSC-H/FSC-A axis).

   • Exclude all dead cells using a live/dead stain.

   • Select CD45+ cells.

   • Exclude F4/80+ cells.

   • Gate on CD11b+ and Ly6G+ cells.

   The neutrophil population will be Ly6Ghi and CD11b+. Other cell populations (e.g., monocyte/macrophage, dendritic cells) can be similarly distinguished and quantitated/sorted using additional markers [9, 12, 13].

4 Notes

1. 150 μL of a 1:10,000 dilution of epinephrine can be injected IP 10–15 min before collecting blood to increase final neutrophil yield.

2. Heat inactivation of FCS is achieved by incubating FCS in a 56 °C water bath for 30 min. Aliquot FCS and store at −20 °C. Once thawed and opened, aliquots can be stored at 4 °C, maintaining sterility.

3. The Percoll gradient layer percentages for bone marrow neutrophil isolation may need to be adjusted depending on the mouse strain and background. The gradients listed were developed for C57Bl/6N mice. It is recommended to determine cell composition at each interface of the gradient for each new mouse strain/background and to adjust the layer percentages (typically the 72% layer) as needed to achieve optimal cell purity. For example, if isolating marrow neutrophils from C57Bl/6J mice, use 52%, 62%, and 72% layers.

4. Euthanasia with anesthetic overdose [14] or CO_2 exposure [15] can lead to neutrophil activation; therefore cervical dislocation is recommended if functional assays will be performed with the recovered cells.

5. Use as much buffer as needed when flushing the femurs and tibias. Typically, 20–25 mL is used per mouse. *Do not* use buffer already containing marrow to flush, as this will damage the cells.

6. Check whether the gradient interfaces are still intact before applying the cell suspension onto the gradient. You should see sharp interfaces between gradient layers. Gradient layers may diffuse (or be inadvertently mixed) into each other over time. If one or both of the interfaces have disappeared, prepare a new gradient to ensure proper separation of cells.

7. Make sure that the centrifugation inserts/buckets and gradient tubes are well balanced, even if using a "self-balancing" rotor. If the centrifuge is not balanced, gradient layers may be disrupted during centrifugation. Turn centrifuge break to "off." If brake is turned on, this may disrupt your gradient in the final steps. Make sure to check that gradient layers are still intact before placing gradient in centrifuge.

8. Remove needle from the syringe before transferring blood to tube. Pushing blood through the needle again may disrupt cells or cause cell activation.

9. Plasma may be saved for other purposes after centrifuging at $2000 \times g$ for 15 min to pellet and remove platelets.

10. Example calculation: If the blood volume was 5 mL, the pellet should be about 2.5 mL (5×1.75) − 2.5 = 6.25; this would mean that the pellet should be resuspended in 6.25 mL of 1:5.25 6% dextran: saline (1 mL 6% dextran and 5.25 mL normal saline).

11. If samples are acquired on the same day, no fixation is required. If cells are acquired on the next day, fix cells in 50 μL of BD FACS Fix solution (dilute in water) for 15 min at 4 °C. Wash samples twice with FACS buffer.

References

1. Amulic B, Cazalet C, Hayes GL, Metzler KD, Zychlinsky A (2012) Neutrophil function: from mechanisms to disease. Annu Rev Immunol 30:459–489. https://doi.org/10.1146/annurev-immunol-020711-074942

2. Ley K, Laudanna C, Cybulsky MI, Nourshargh S (2007) Getting to the site of inflammation: the leukocyte adhesion cascade updated. Nat Rev Immunol 7(9):678–689. https://doi.org/10.1038/nri2156

3. Mantovani A, Cassatella MA, Costantini C, Jaillon S (2011) Neutrophils in the activation and regulation of innate and adaptive immunity. Nat Rev Immunol 11(8):519–531. https://doi.org/10.1038/nri3024

4. Summers C, Rankin SM, Condliffe AM, Singh N, Peters AM, Chilvers ER (2010) Neutrophil kinetics in health and disease. Trends Immunol 31(8):318–324. https://doi.org/10.1016/j.it.2010.05.006

5. Suratt BT, Young SK, Lieber J, Nick JA, Henson PM, Worthen GS (2001) Neutrophil maturation and activation determine anatomic site of clearance from circulation. Am J Physiol Lung Cell Mol Physiol 281(4):L913–L921

6. Kordonowy LL, Burg E, Lenox CC, Gauthier LM, Petty JM, Antkowiak M, Palvinskaya T, Ubags N, Rincon M, Dixon AE, Vernooy JH, Fessler MB, Poynter ME, Suratt BT (2012) Obesity is associated with neutrophil dysfunction and attenuation of murine acute lung injury. Am J Respir Cell Mol Biol 47(1):120–127. https://doi.org/10.1165/rcmb.2011-0334OC

7. Palvinskaya T, Antkowiak M, Burg E, Lenox CC, Ubags N, Cramer A, Rincon M, Dixon AE, Fessler MB, Poynter ME, Suratt BT (2013) Effects of acute and chronic low density lipoprotein exposure on neutrophil function. Pulm Pharmacol Ther 26(4):405–411. https://doi.org/10.1016/j.pupt.2012.10.002

8. Ubags ND, Burg E, Antkowiak M, Wallace AM, Dilli E, Bement J, Wargo MJ, Poynter ME, Wouters EF, Suratt BT (2016) A comparative study of lung host Defense in murine obesity models. Insights into neutrophil function. Am J Respir Cell Mol Biol 55(2):188–200. https://doi.org/10.1165/rcmb.2016-0042OC

9. Zaynagetdinov R, Sherrill TP, Kendall PL, Segal BH, Weller KP, Tighe RM, Blackwell TS (2013) Identification of myeloid cell subsets in murine lungs using flow cytometry. Am J Respir Cell Mol Biol 49(2):180–189. https://doi.org/10.1165/rcmb.2012-0366MA

10. Abdulnour RE, Sham HP, Douda DN, Colas RA, Dalli J, Bai Y, Ai X, Serhan CN, Levy BD (2016) Aspirin-triggered resolvin D1 is produced during self-resolving gram-negative bacterial pneumonia and regulates host immune responses for the resolution of lung inflammation. Mucosal Immunol 9(5):1278–1287. https://doi.org/10.1038/mi.2015.129

11. Shalekoff S, Page-Shipp L, Tiemessen CT (1998) Effects of anticoagulants and temperature on expression of activation markers CD11b and HLA-DR on human leukocytes. Clin Diagn Lab Immunol 5(5):695–702

12. Plantinga M, Guilliams M, Vanheerswynghels M, Deswarte K, Branco-Madeira F, Toussaint W, Vanhoutte L, Neyt K, Killeen N, Malissen B, Hammad H, Lambrecht BN (2013) Conventional and monocyte-derived CD11b(+) dendritic cells initiate and maintain T helper 2 cell-mediated immunity to house dust mite allergen. Immunity 38(2):322–335. https://doi.org/10.1016/j.immuni.2012.10.016

13. Bedoret D, Wallemacq H, Marichal T, Desmet C, Quesada Calvo F, Henry E, Closset R, Dewals B, Thielen C, Gustin P, de Leval L, Van Rooijen N, Le Moine A, Vanderplasschen A, Cataldo D, Drion PV, Moser M, Lekeux P, Bureau F (2009) Lung interstitial macrophages alter dendritic cell functions to prevent airway allergy in mice. J Clin Invest 119(12):3723–3738. https://doi.org/10.1172/JCI39717

14. Yang KD, Liou WY, Lee CS, Chu ML, Shaio MF (1992) Effects of phenobarbital on leukocyte activation: membrane potential, actin polymerization, chemotaxis, respiratory burst, cytokine production, and lymphocyte proliferation. J Leukoc Biol 52(2):151–156

15. Pecaut MJ, Smith AL, Jones TA, Gridley DS (2000) Modification of immunologic and hematologic variables by method of CO_2 euthanasia. Comp Med 50(6):595–602

Chapter 5

Isolation and Characterization of Mouse Fibroblasts

Benjamin L. Edelman and Elizabeth F. Redente

Abstract

The lung parenchyma is comprised of many cells including the structurally important stromal fibroblasts. Fibroblasts function to produce extracellular matrix and are important in the maintenance of alveolar epithelial cells. To understand the role of fibroblasts both in homeostasis and disease, we isolate fibroblasts and grow them in culture. Two methods are presented here for the isolation and maintenance of mouse primary lung fibroblasts.

Key words Pulmonary fibroblasts, Isolation, Lung digestion, Cell culture

1 Introduction

Fibroblasts are the most abundant cell type in lung interstitium. Their principle function is production of type III collagen, elastin, and proteoglycans of the extracellular matrix, and accordingly, they play a critical role in maintaining lung structure and function [1–4]. The controlled accumulation of fibroblasts to sites of injury and inflammation is crucial to effective tissue repair [5]. However, either inadequate or excessive accumulation of fibroblasts results in abnormal tissue function and can lead to diseases including pulmonary fibrosis, chronic obstructive pulmonary disease, pulmonary hypertension, and asthma [5, 6]. For example, in pulmonary fibrosis, tissue injury followed by aberrant repair of the alveolar epithelium results in the persistent accumulation of fibroblasts, overabundant production of extracellular matrix, and ultimately the development of fibrosis [6]. In contrast, in normal wound repair, fibroblasts are present but undergo apoptosis at the completion of the repair process [4, 7, 8]. Much of our knowledge regarding the function of fibroblasts has derived from studies in which primary fibroblasts have been isolated from lung tissue and grown in culture. Fibroblasts are isolated from dissociated tissue (either mechanically or enzymatically) and allowed to proliferate from the tissue fragments [9–14]. Herein, we present two protocols for the

Scott Alper and William J. Janssen (eds.), *Lung Innate Immunity and Inflammation: Methods and Protocols*,
Methods in Molecular Biology, vol. 1809, https://doi.org/10.1007/978-1-4939-8570-8_5,
© Springer Science+Business Media, LLC, part of Springer Nature 2018

isolation of mouse lung fibroblasts. The first uses collagenase digestion and the second, a crawl-out method. These protocols can be applied to isolate fibroblasts from other tissues including ear and tail snips [9] as well as from human lung tissue [12].

2   Materials

2.1   Mouse Fibroblast Cell Isolation by Digestion

1. Mice.
2. Fatal-Plus (200 mg/mL) diluted in sterile saline.
3. Perfusion device: 10 mL disposable syringe attached to a 25-gauge needle.
4. Sterilized surgical scissors.
5. Wash solution: 1× Hank's balanced salt solution (HBSS).
6. Fibroblast culture media: 500 ml DMEM supplemented with 200 mM L-Glutamine (final concentration), 10 mL Pen-Strep (10,000 U/mL penicillin, 10,000 U/mL streptomycin), and 50 mL fetal bovine serum (filter to sterilize).
7. Phosphate-buffered saline (PBS), Ca^{2+} Mg^{2+}-free.
8. Collagenase A (Type XI-S—Sigma-Aldrich). Use at 1000–1500 U/mL diluted in HBSS.
9. Trypsins: 0.25% and 0.05% Trypsin-EDTA.
10. 1.5 mL microcentrifuge tubes.
11. 6-well tissue culture plate.
12. 100 mm tissue culture petri dish.
13. Tissue culture flasks (T75).
14. Razor blade.

2.2   Mouse Fibroblast Cell Isolation by Crawl-Out

1. Mice.
2. Fatal-Plus (200 mg/mL) diluted in saline.
3. Sterilized surgical scissors.
4. Wash solution: 1× Hank's balanced salt solution (HBSS).
5. Fibroblast culture media: 500 mL DMEM, 200 mM L-Glutamine and 10 mL Pen-Strep (10,000 U/mL penicillin, 10,000 U/mL streptomycin) containing 10% fetal bovine serum (filter to sterilize).
6. 0.05% Trypsin-EDTA.
7. 50 mL conical tube.
8. 100 mm cell culture petri dish.
9. Tissue culture flasks (T75).
10. Razor blade.
11. Sterilized forceps.

2.3 Characterization of Mouse Fibroblasts

1. Isolated fibroblasts (at least two passages).

2. Sterile glass coverslips to fit into the wells of 6- or 12-well tissue culture plates.

3. 6- or 12-well tissue culture plate.

4. 4% paraformaldehyde.

5. 0.1% Triton X-100.

6. Wash solution: 1× phosphate-buffered saline (PBS).

7. Microcentrifuge tubes.

8. Parafilm.

9. Phalloidin—conjugated to a fluorescent probe (Thermo Fisher Scientific).

10. DAPI mounting media (Vector Laboratories).

11. Glass microscope slides.

3 Methods

3.1 Mouse Fibroblast Cell Isolation by Lung Digestion

1. Euthanize mouse with 200 mg/kg Fatal-Plus in 500 μL saline by intraperitoneal injection (*see* **Note 1**).

2. Once the mouse is deceased, open the peritoneal cavity, and cut the descending aorta and the diaphragm. Open the thoracic cavity by cutting up next to the sternum, and pin back the walls of the thoracic cavity to expose the heart and lungs.

3. Perfuse the lungs with 10–20 mL of Ca^{2+} Mg^{2+}-free PBS by inserting a 10 mL syringe attached to a 25-gauge needle into the right ventricle until the lungs are flushed of blood and are white in appearance. Do not cut the left atrium.

4. Remove the lungs from the thoracic cavity, and place into a 100 mm tissue culture petri dish, and mince the tissue into very small pieces using a razor blade or surgical scissors (*see* Fig. 1 and **Note 2**).

5. Transfer the chopped lung into a 1.5 mL microcentrifuge tube, and add 0.5 mL collagenase (final concentration, 1000 U/mL).

6. Incubate at 37 °C for 30 min.

7. After incubation, centrifuge for 5 min at 24 °C and $1000 \times g$. Carefully decant and discard supernatant.

8. Wash by adding 1 ml of HBSS, and then centrifuge for 5 min at 24 °C and $1000 \times g$. Carefully decant and discard supernatant.

9. Add 0.5 mL of 0.25% Trypsin-EDTA, mix thoroughly with tissue, and incubate at 37 °C for 20 min.

10. After incubation, centrifuge for 5 min at 24 °C and $1000 \times g$. Carefully decant and discard supernatant.

Fig. 1 The lung after chopping with the razor blade or scissors. Minced perfused lung tissue that is ready for digestion has the consistency of a thick paste

11. Resuspend the pellet in 0.5 mL fibroblast culture media.

12. Gently pipette up and down to break up cell aggregates.

13. Plate the suspension into a well of a 6-well tissue culture plate, discarding any large tissue pieces.

14. Add 2 mL fibroblast culture media. Final culture volume will be about 2.5 mL.

15. Incubate cells at 37 °C and 5% CO_2.

3.2 Maintenance and Culture of Fibroblasts

1. Every 2–4 days, feed or split cultures of isolated fibroblasts.

2. For the initial passage from 1-well of a 6-well tissue culture plate (Fig. 2), use 250 μL of 0.05% Trypsin-EDTA to detach cells, and split from 1 well into 2–3 wells depending on confluence of initial culture.

3. Once 3 wells are confluent, cells can be split into and maintained in T75 flasks (*see* **Note 3**).

3.3 Mouse Fibroblast Cell Isolation by Crawl-Out

1. Euthanize mouse with 200 mg/kg Fatal-Plus in 500 μL saline by intraperitoneal injection (*see* **Note 1**).

2. Once the mouse is deceased, open the peritoneal cavity, and cut the descending aorta and the diaphragm. Open the thoracic cavity by cutting up next to the sternum, and pin back the walls of the thoracic cavity to expose the heart and lungs.

3. Remove the lungs without perfusing (*see* **Note 4**).

4. Place lungs in 10–15 mL of HBSS wash solution to rinse off excess blood.

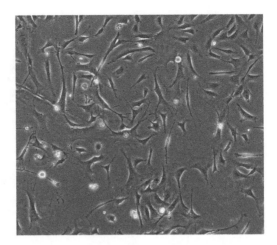

Fig. 2 Fibroblasts in a 6-well dish before the first passage. Mouse lung fibroblasts before the first passage. 20× magnification

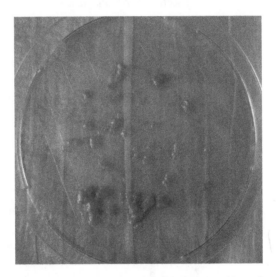

Fig. 3 Scored 100 mm tissue culture petri dish with the minced lung. Scored 100 mm tissue culture petri dish with the minced lung anchored before adding media

5. Under sterile conditions, score a 100 mm tissue culture petri dish in a grid pattern with a razor blade.

6. On the edge of the scored 100 mm tissue culture petri dish, mince the lung tissue into very small pieces using a razor blade or surgical scissors (Fig. 3).

7. Using sterilized forceps move the minced tissue over the grid so that it catches and anchors the tissue. Gently add 10 mL of fibroblast culture media so that tissue remains anchored to the grid on the 100 mm tissue culture dish.

8. Incubate at 37 °C and 5% CO_2.

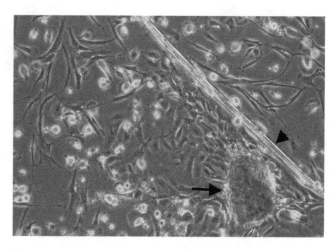

Fig. 4 Fibroblasts growing out of tissue pieces. Fibroblasts crawling out of an anchored piece of tissue (arrow) 1 week after harvest. The arrow head points to a score line. 20× magnification

9. After 24 h very gently remove media. Wash the dish 2–3 times with 10 mL HBSS wash solution to remove red blood cells, dead cells, and debris.

10. Replace with 10 mL fibroblast culture media.

11. Every 2–4 days change the medium to feed the cells. Be very careful not to disturb lung tissue pieces that have anchored to the tissue culture plate. This is where the fibroblasts will crawl out from.

12. After 1–1.5 weeks, clusters of fibroblasts should be visibly growing out of the remaining tissue pieces (Fig. 4). At this time during a media change, any remaining tissue can be removed. When the cells are confluent (in approximately 2–3 weeks), passage them as described in Subheading 3.2).

3.4 Characterization of Mouse Fibroblasts

1. Place sterile glass coverslips in the wells of a 6- or 12-well tissue culture plate.

2. Plate isolated fibroblasts (at least at passage 2) on the coverslips in the tissue culture plates. To obtain confluent cultures, seed fibroblasts at 250,000/well in a 6-well plate and 100,000/well in a 12-well plate.

3. After 24 h in culture (described in Subheading 3.2), fix cells in the cell culture plate using 1 mL of 4% paraformaldehyde for 15 min at room temperature.

4. Aspirate off paraformaldehyde and wash one time with 1 mL of 1× PBS for 5 min.

5. Aspirate off PBS and permeabilize cells with 0.1% Triton X-100 for 30 min at room temperature.

6. Aspirate off Triton X-100, and wash three times with 1 mL of 1× PBS for 5 min each wash.

7. Prepare the phalloidin. For each well that will be stained, add 5 μL phalloidin to 45 μL PBS in a microcentrifuge tube (e.g., if you are staining 10 wells, add 50 μL phalloidin and 450 μL PBS to the tube). Mix well.

8. On a piece of parafilm, pipette a 50 μL bubble of phalloidin solution. Place the glass coverslip on the parafilm with the cells facedown for staining. Stain in the dark for 20 min at room temperature.

9. At the end of the staining, return coverslip with cells to the tissue culture plate (cell side up), and wash three times with 1 mL of 1× PBS for 5 min each wash.

10. Place one drop of DAPI-containing mounting media on a glass slide, and place glass coverslip with cells (cell side down) onto the slide for mounting. The cells are now ready for fluorescence imaging (Fig. 5).

11. Fibroblasts will have a characteristic spindle shape (Fig. 2) and will contain actin stress fibers that stain positively for phalloidin (Fig. 5) (*see* **Note 5**).

4 Notes

1. Other methods of euthanasia may be used (including CO_2 followed by exsanguination) that follow current (American Veterinary Medical Association) AVMA guidelines.

2. This procedure will also work to isolate fibroblasts from an ear punch or tail snip.

3. Fibroblasts can be frozen in DMEM containing 10% DMSO and 20% FBS to maintain lines. Cells should be maintained and used between passages 2–10.

4. This procedure will also work for isolating fibroblasts from human lung tissue.

5. Isolated fibroblasts will also stain positively for alpha smooth muscle actin (anti-alpha smooth muscle actin—Sigma-Aldrich clone 1A4) as an additional method of characterization.

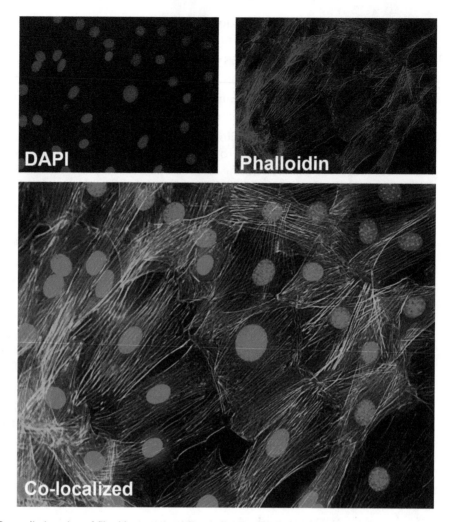

Fig. 5 Coverslip imaging of fibroblasts stained for phalloidin and DAPI. Isolated lung fibroblasts grown in culture have F-actin stress fibers stained with phalloidin (green). DAPI-stained nuclei (blue). 40× magnification

References

1. Pan T, Mason RJ, Westcott JY, Shannon JM (2001) Rat alveolar type II cells inhibit lung fibroblast proliferation in vitro. Am J Respir Cell Mol Biol 25:353–361

2. Sakai N, Tager AM (2013) Fibrosis of two: Epithelial cell-fibroblast interactions in pulmonary fibrosis. Biochim Biophys Acta 1832:911–921

3. Xu X, Dai H, Wang C (2016) Epithelium-dependent profibrotic milieu in the pathogenesis of idiopathic pulmonary fibrosis: current status and future directions. Clin Respir J 10:133–141

4. Selman M, Pardo A (2002) Idiopathic pulmonary fibrosis: an epithelial/fibroblastic cross-talk disorder. Respir Res 3:3

5. Darby IA, Laverdet B, Bonte F, Desmouliere A (2014) Fibroblasts and myofibroblasts in wound healing. Clin Cosmet Investig Dermatol 7:301–311

6. Katzenstein AL, Myers JL (1998) Idiopathic pulmonary fibrosis: clinical relevance of pathologic classification. Am J Respir Crit Care Med 157:1301–1315

7. Cha SI, Groshong SD, Frankel SK, Edelman BL, Cosgrove GP, Terry-Powers JL, Remigio

LK, Curran-Everett D, Brown KK, Cool CD, Riches DW (2010) Compartmentalized expression of c-FLIP in lung tissues of patients with idiopathic pulmonary fibrosis. Am J Respir Cell Mol Biol 42:140–148

8. Selman M, King TE, Pardo A, American Thoracic Society, European Respiratory Society, American College of Chest Physicians (2001) Idiopathic pulmonary fibrosis: prevailing and evolving hypotheses about its pathogenesis and implications for therapy. Ann Intern Med 134:136–151

9. Hansen S, Girirajan, S., Canfield, T., Eichler, E. Establishment and propagation of adult mouse fibroblast cultures. 2011.

10. Gauldie J, Cox G, Jordana M, Ohno I, Kirpalani H (1994) Growth and colony-stimulating factors mediate eosinophil fibroblast interactions in chronic airway inflammation. Ann N Y Acad Sci 725:83–90

11. Jordana M, Schulman J, McSharry C, Irving LB, Newhouse MT, Jordana G, Gauldie J (1988) Heterogeneous proliferative characteristics of human adult lung fibroblast lines and clonally derived fibroblasts from control and fibrotic tissue. Am Rev Respir Dis 137:579–584

12. Hogaboam CM, Carpenter KJ, Evanoff H, Kunkel SL (2005) Approaches to evaluation of fibrogenic pathways in surgical lung biopsy specimens. Methods Mol Med 117:209–221

13. Penney DP, Keng PC, Derdak S, Phipps RP (1992) Morphologic and functional characteristics of subpopulations of murine lung fibroblasts grown in vitro. Anat Rec 232:432–443

14. Seluanov A, Vaidya A, Gorbunova V (2010) Establishing primary adult fibroblast cultures from rodents. J Vis Exp. https://doi.org/10.3791/2033

Chapter 6

Isolation of Rat and Mouse Alveolar Type II Epithelial Cells

Nicole L. Jansing, Jazalle McClendon, Hidenori Kage,
Mitsuhiro Sunohara, Juan R. Alvarez, Zea Borok,
and Rachel L. Zemans

Abstract

The gas exchange surface of the lungs is lined by an epithelium consisting of alveolar type (AT) I and ATII cells. ATII cells function to produce surfactant, play a role in host defense and fluid and ion transport, and serve as progenitors. ATI cells are important for gas exchange and fluid and ion transport. Our understanding of the biology of these cells depends on the investigation of isolated cells. Here, we present methods for the isolation of mouse and rat ATII cells.

Key words Alveolar epithelium, Lung digestion, ATII

1 Introduction

The alveolar epithelium consists of ATI and ATII cells. ATI cells cover 95–98% of the alveolar surface and play a role in gas exchange and ion transport. ATII cells are critical for surfactant production and host defense and function as epithelial progenitors during homeostasis and repair. Abnormal ATII cell function has been implicated in a variety of diseases, including pulmonary fibrosis, emphysema, acute respiratory distress syndrome, and lung cancer. Much of our knowledge regarding the function of ATII cells derives from studies in which cells have been isolated and cultured [1]. Previously, ATII cells have been isolated using a variety of methods. Here, we present a protocol for the isolation and purification of mouse ATII cells using magnetic bead separation. Purification is accomplished in two stages: first, CD45+ cells are depleted using negative selection, and then EpCAM+ cells are purified using positive selection [2]. We also include an established protocol for the isolation of rat ATII cells using discontinuous density centrifugation [3–6].

Scott Alper and William J. Janssen (eds.), *Lung Innate Immunity and Inflammation: Methods and Protocols*,
Methods in Molecular Biology, vol. 1809, https://doi.org/10.1007/978-1-4939-8570-8_6,
© Springer Science+Business Media, LLC, part of Springer Nature 2018

2 Materials

<table>
<tr>
<td>

2.1 Mouse ATII Cell Isolation

</td>
<td>

1. C57BL/6 mice (*see* **Note 1**).

2. Pentobarbital sodium (Fatal-Plus 260 mg/mL stock diluted 1:20 in saline).

3. 20-gauge Luer stub adapter.

4. Perfusion device: 10 mL disposable syringe attached to a 27-gauge needle.

5. Suture: 2–0 silk suture.

6. Surgical scissors.

7. Media: 500 mL DMEM/F12 supplemented with 200 mM L-Glutamine (final concentration), 10 mL Pen-Strep (10,000 U/mL penicillin, 10,000 U/mL streptomycin), 10 mL MEM nonessential amino acids, and 10 mL 1 M HEPES. Filter to sterilize.

8. Ca^{2+} Mg^{2+}-free phosphate-buffered saline (PBS).

9. 1% low melting point agar in Ca^{2+} Mg^{2+}-free PBS. Warm to 45 °C.

10. Media with DNAse: 0.01% DNAse in media.

11. Buffer #1: Ca^{2+} Mg^{2+}-free PBS containing 2 mM EDTA and 0.5% BSA. Filter to sterilize. Degas and keep at 4 °C prior to use.

12. Buffer #2: Ca^{2+} Mg^{2+}-free PBS containing 0.5% BSA. Filter to sterilize. Degas and keep at 4 °C prior to use.

13. Dispase: Diluted 1:10 in Ca^{2+} Mg^{2+}-free Hank's balanced salt solution (HBSS). Equilibrate to room temperature prior to use.

14. Petri dish.

15. Razor blade.

16. Filters: 100 µM, 40 µM, and 20 µM.

17. Biotinylated anti-mouse CD326/EpCAM antibody (Miltenyi Biotec).

18. FcR Blocking Reagent (Miltenyi).

19. Anti-biotin MicroBeads (Miltenyi).

20. CD45 MicroBeads (Miltenyi).

21. LS Columns (Miltenyi).

22. Pre-Separation Filters (Miltenyi).

23. MACS multi-stand with QuadroMACS or MidiMACS separator (Miltenyi).

24. 14 mL round bottom tubes.

25. 15 mL conical tubes.

26. 50 mL conical tubes.

</td>
</tr>
</table>

2.2   Rat ATII Cell Isolation

1. Sprague-Dawley rat.

2. Fatal-Plus/heparin: Mix 0.15 mL Fatal-Plus (260 mg/mL) and 0.85 mL heparin (1000 U/mL).

3. Plastic beakers: three 100 mL beakers, two 250 mL beakers, and one 400 mL beaker (*see* **Note 2**).

4. Plastic Erlenmeyer flask: 250 mL (*see* **Note 2**).

5. Suture: 2–0 silk suture.

6. Surgical scissors.

7. Luer stub adapter: 15 gauge.

8. Perfusion syringe: 20 mL disposable syringe attached to a 14-gauge catheter.

9. BSS-A: 15.9 g NaCl, 0.8 g KCl, 0.1 g NaH_2PO_4 (monobasic), 0.64 g Na_2HPO_4 (dibasic), 5.06 g HEPES, and 2 g D-glucose (dextrose), adjust to pH 7.4, and adjust volume to 2 L with deionized water. Sterile filter. Warm to 37 °C.

10. BSS-A with EGTA: Mix 37.5 mL of BSS-A and 750 μL 0.1 M EGTA. Adjust to pH 7.4. Sterile filter.

11. Saline: Warm to 37 °C.

12. Board at angle to allow lavage to drain by gravity.

13. BSS-B: 15.52 g NaCl, 0.772 g KCl, 0.1 g NaH_2PO_4 (monobasic), 0.618 g Na_2HPO_4 (dibasic), 4.93 g HEPES, 2 g D-glucose (dextrose), 0.558 g calcium chloride dihydrate, and 0.638 g magnesium sulfate 7-hydrate, adjust to pH 7.4, and adjust volume to 2 L with deionized water. Sterile filter and place at 4 °C.

14. BSS-B with BSA: 15 mL BSS-B containing 1% BSA (final concentration) in a small plastic beaker. Adjust to pH 7.2–7.4 and warm to 37 °C.

15. BSS-B with elastase: Stir 172 U of elastase into 42 mL of BSS-B. Cover with saran wrap, and place in a shaking water bath at 37 °C. Make fresh.

16. Fluorocarbon-albumin emulsion: Mix and sonicate approximately 5 mL of BSS-B with BSA with 1.5 mL of 0.2 uM filtered fluorocarbon FC-40. Sonicate until completely mixed and emulsion turns white. Add sonicated FC-40 to remaining 10 mL of BSS-B with BSA. Swirl to mix.

17. 0.1 M sodium phosphate: Slowly add 0.1 M monobasic sodium phosphate to 0.1 M dibasic sodium phosphate until the pH is 7.4 at 25 °C.

18. Krebs/HEPES buffer preparation: Mix 1257.6 mL H_2O, 10.8 g NaCl (0.9% final concentration), 36 mL 0.1 M sodium phosphate, 48 mL 0.15 M KCl, and 14.4 mL 1 M HEPES.

19. Optiprep with HEPES: Add 2.5 mL of 1 M HEPES to a 250 mL bottle of Optiprep (Accurate Chemical).

20. Lightweight Optiprep: Add 8 mL of fetal bovine serum (FBS) to 45 mL of Optiprep with HEPES. Add a pinch of phenol red until solution is cherry red in color, and bring total volume of mixture to 400 mL with Krebs/HEPES buffer. Stir until completely mixed (*see* **Note 3**). The hydrometer reading should be between 1.038 and 1.045 (*see* **Note 4**).

21. Heavyweight Optiprep: Add 8 mL of FBS to 110 mL of Optiprep with HEPES. Bring mixture to 400 mL with Krebs/HEPES buffer. Stir until completely mixed (*see* **Note 3**). Hydrometer reading should be between 1.088 and 1.092 (*see* **Note 4**).

22. Optiprep gradient: Add 10 mL of lightweight Optiprep (red) to a 50 mL conical tube. Then slowly inject 10 mL of heavyweight Optiprep to the bottom of the conical tube (below the lightweight Optiprep); avoid mixing of layers. Store in the dark at 4 °C until use.

23. Mincing scissors and plastic mincing beakers (*see* **Note 2**).

24. Fetal bovine serum (FBS).

25. DNase I: 4 mg/mL in BSS-B.

26. Filtering device: 60 mL syringe with needle adaptor cutoff.

27. Gauze pads.

28. Rubber bands.

3   Methods

3.1   Mouse ATII Cell Isolation

3.1.1   Lung Digestion

1. Anesthetize mouse with intraperitoneal (i.p.) injection of 200 mg/kg Fatal-Plus in 400 μL saline (*see* **Note 1**). Once the mouse is fully anesthetized, perform the following steps at a rapid pace. Speed of the entire protocol is key for cell yield, purity, and viability.

2. Open the peritoneal cavity and cut the descending aorta/inferior vena cava. Open thoracic cavity, cut the diaphragm away from the walls, pin back the walls of the thoracic cavity, and remove the thymus.

3. Perfuse the pulmonary vasculature with 20 mL of Ca^{2+} Mg^{2+}-free PBS by inserting a 20 mL syringe attached to a 27-gauge needle into the right ventricle (*see* **Note 5**). Do not cut the left atrium.

4. After perfusing, remove the skin and muscle covering the trachea. Push under the trachea with forceps, and pull two separate sutures through. Cut a slit in the ventral surface of the

trachea making sure to avoid cutting all the way through. Insert a 20-gauge Luer stub adapter into the trachea, and securely tie it in place with one of the sutures. Leave the other suture untied.

5. Instill 1 mL of Ca^{2+} Mg^{2+}-free PBS through the Luer stub adapter, keeping the syringe attached. Next, withdraw the PBS back into the syringe to lavage the lung: approximately 800 μL will be returned.

6. Instill 3 mL of room temperature dispase diluted 1:10 in Ca^{2+} Mg^{2+}-free HBSS into lungs. Immediately following dispase, instill 0.3 mL of 45 °C 1% low melting agarose in Ca^{2+} Mg^{2+}-free PBS into the lungs leaving the syringe attached to the Luer stub adapter [7].

7. With the 1 mL syringe and Luer stub adapter still attached, securely tie the trachea off distal to the Luer stub adapter using the untied, remaining suture. This knot will prevent the dispase and agarose from leaking out once the syringe and Luer stub adapter are removed.

8. Remove the syringe and Luer stub adapter from the trachea. Remove the lungs from the thorax by holding the suture and cutting behind the trachea. Be careful not to cut the lungs. The heart can remain attached to the lungs at this point.

9. Place the lungs into 3 mL of dispase diluted 1:10 in Ca^{2+} Mg^{2+}-free HBSS in a 14 mL round bottom tube, and incubate for 30 min at room temperature.

10. After incubation, place lungs in a petri dish. Remove the heart. Dissect away the trachea and the bronchial tree.

11. Chop the lung with a razor blade back and forth from different directions until the tissue has the consistency of a thick soup/puree (Fig. 1). This requires about 3 min of continuous chopping.

12. Add 3 mL of media with 0.01% DNase to minced tissue and pipet up and down with a 5 mL pipet.

13. Filter the lung tissue sequentially through 100, 40, and 20 μm filters pre-wetted with media with 0.01% DNase. To aid with 100 μm filtering, stir solution around with a 5 mL pipet on top of the filter, pushing the tissue through very gently (Fig. 2). Let the solution drain by gravity through the 40 and 20 μm filters. Centrifuge the filtrate at $300 \times g$ for 5 min at 10 °C. Set acceleration and deceleration to medium.

14. Remove the supernatant and resuspend cells in 10 mL of media with 0.01% DNase. Count cells using a hemocytometer.

15. Transfer cells to a 15 mL conical tube and centrifuge at $300 \times g$ for 5 min at 10 °C.

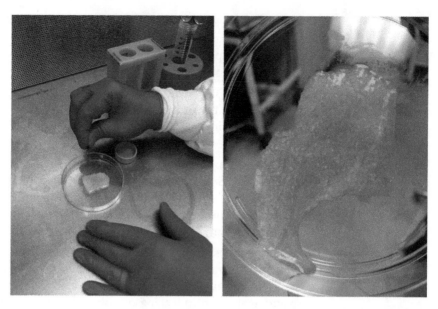

Fig. 1 The mouse lung after chopping with the razor blade. The lung should have the consistency of a thick soup/puree

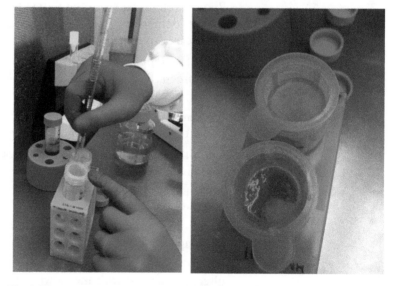

Fig. 2 Filtering mouse lung digest through the 100 μm filter. Notice the remains of the tissue that did not pass through the filter

3.1.2 Isolation of ATII Cells by Magnetic Sorting

1. Resuspend cells in 500 μL/mouse of Buffer #1.

2. Add 40 μL/mouse anti-CD45 MicroBeads to deplete leukocytes. Pipet up and down vigorously 5–10 times, and then incubate for 15 min at 4 °C.

3. Add 10 mL Buffer #1.

4. Centrifuge at $300 \times g$ for 5 min at 10 °C.

5. During the 5-min centrifugation, place LS column into a MidiMACS or QuadroMACS separator. Place pre-separation filter into LS column. Place a 15 mL conical tube under the column.

6. Rinse pre-separation filter/LS column with 3 mL of Buffer #1.

7. Place a new 15 mL conical tube under the column.

8. Resuspend centrifuged cells in 500 μL of Buffer #1.

9. Add cell suspension to pre-separation filter/LS column.

10. Wash three times with 3 mL Buffer #1.

11. Collect flow through, which contains CD45-negative cells. (Optional: count cells here.) Centrifuge flow through at $300 \times g$ for 5 min at 10 °C.

12. Resuspend cells in 100 μL/mouse of Buffer #1 in a 15 mL conical tube. Pipet up and down vigorously 10–20 times with a P1000 Pipetman.

13. Add 10 μL/mouse biotinylated anti-mouse CD326 (EpCAM). Mix and incubate for 10 min at 4 °C.

14. Add 2 mL/mouse of Buffer #1. Mix and centrifuge at $300 \times g$ for 5 min at 10 °C.

15. Remove the supernatant and resuspend cells in 100 μL/mouse of Buffer #1.

16. Add 20 μL/mouse anti-biotin MicroBeads. Pipet up and down vigorously 5–10 times and incubate for 15 min at 4 °C.

17. Add 2 mL/mouse of Buffer #1. Mix and centrifuge at $300 \times g$ for 5 min at 10 °C.

18. Rinse new pre-separation filter and LS column with 3 mL of Buffer #1 during centrifugation time. Leave the flow through in the 15 mL conical collection tube.

19. Remove the supernatant and resuspend cells in 500 μL of Buffer #1. Pipet up and down vigorously ten times.

20. Add cells to pre-separation filter on LS column. Do not apply pressure to the column.

21. Wash filter one time with 3 mL of Buffer #1. Then wash twice with 3 mL of Buffer #2.

22. Remove column from magnet, place into a new 15 mL conical, and add 5 mL of Buffer #2 to LS column. Using the plunger, strongly push out EpCAM+ cells with plunger. This should be fast, taking only 1 s for the whole solution to come through. The cells may be passed over another LS column to futher increase purity, if desired.

23. Dilute an aliquot of cells 1:10 in trypan blue and count live and dead cells.

24. Determine the purity of ATII cells (*see* **Note 6** and Fig. 3).

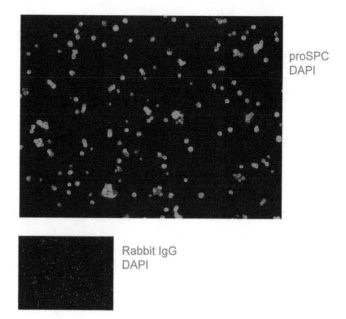

proSPC
DAPI

Rabbit IgG
DAPI

Fig. 3 Mouse cell isolation cytospins were fixed and stained for proSPC and DAPI

3.2 Rat ATII Cell Isolation[1]

1. Anesthetize a rat with a 1 mL intraperitoneal injection of Fatal-Plus/heparin solution prepared as described above. Once rat is fully anesthetized, perform the following steps at a rapid pace.[1]

2. Open the peritoneal cavity and cut the descending aorta/inferior vena cava. Remove the skin and muscle covering the trachea. Push under the trachea with forceps and pull a suture through. Cut a slit in the ventral surface of the trachea making sure to avoid cutting all the way. Insert a 15-gauge Luer stub adapter into the trachea, and securely tie it in place with a 2–0 silk suture in a double knot. It is important that this knot is tight, since the lungs will be held by this suture in later steps.

3. Open the thoracic cavity, cut the diaphragm carefully so as not to nick the lungs, pin back the walls of the thoracic cavity, and remove the thymus. Make an incision in the right ventricle. Cannulate the pulmonary artery with a 14-gauge catheter attached to a 20 mL syringe, and perfuse the lungs with 20 mL of BSS-A. Cut the left atrium to allow exit of the perfusate after the atrium distends and the lungs blanch (*see* **Note 5**).

4. Remove the lungs from the thorax by holding the sutures and cutting behind the trachea. Be careful not to cut the lung tissue. Do not remove the Luer stub adapter.

---

[1] Not used by Borok lab

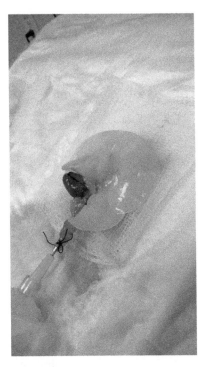

Fig. 4 Lungs on board angled to allow lavage to drain by gravity. The trachea is pointing down, and the heart is facing up

5. Place cannulated lung into a plastic beaker with 200 mL of 37 °C saline while performing surgery on subsequent animals. Remove the lung from saline, and lay it on an angled surface with the heart side facing up and the cannula pointing down (Fig. 4).

6. Lavage lung five times with BSS-A at 37 °C using a 20 mL syringe. Instill solution until lungs fill completely, using approximately 8–12 mL. Remove syringe and let lavage fluid drain by gravity between each lavage. The objective of this step is to remove some of the alveolar macrophages from the airspaces.

7. Pump sonicated fluorocarbon-albumin emulsion through syringe a few times to mix.

8. Hold lung upright by cannula, and slowly instill ~15 mL of sonicated fluorocarbon-albumin emulsion into the lung with a 20 mL syringe over a beaker of saline at 37 °C. Stop adding emulsion if the lung starts leaking.

9. Leaving the syringe attached and the lung distended, submerge the lungs into beaker of saline, and incubate the lung for 20 min in a 37 °C water bath.

10. After incubation, lavage the lung twice with 8–12 mL of BSS-A, and let it drain by gravity, as described above (Fig. 4).

11. Lavage the lung three more times with 8–12 mL of BSS-A with EGTA.

12. Load a 60 mL syringe with 40 mL of BSS-B with elastase. Lavage the lung one time with 8–10 mL. Let the lavage fluid drain by gravity.

13. Return the lung to the beaker of saline at 37 °C, and instill ~10 mL of BSS-B with elastase. Incubate at 37 °C for 15 min. Leave the syringe attached to the cannula. Every 5 min, instill an additional 10 mL of BSS-B with elastase.

14. Remove lung from the saline bath and place it in a plastic cup. Holding the cannula in one hand and the scissors vertically in the other hand, cut straight down to remove the heart, trachea, and mediastinal structures from the lung.

15. Transfer lung to mincing beaker and add 100 µL of DNase.

16. Mince for 200 strokes (30 s) with one pair of large, very sharp scissors until pieces are ~1 mm^3 in size (*see* **Note 7**). Do not over-mince.

17. Transfer the minced lung into a 250 mL plastic Erlenmeyer flask (*see* **Note 2**) containing 5 mL of FBS to quench the elastase. Rinse mincing beaker and scissors well with BSS-B, and add additional tissue to minced lung mixture.

18. Shake mixture vigorously in a water bath for 5 min at 37 °C.

19. Use a 60 mL syringe with the needle adaptor cutoff (Fig. 5). Then use a rubber band to attach gauze or Nitex filter to the

Fig. 5 Filtering device made from a 60 mL syringe with the needle adaptor cutoff. The filter, made of layered gauze, is attached using a rubber band

Fig. 6 To aid in filtering, apply pressure to the open end of the syringe with the palm of your hand

cutoff end of the syringe. Before filtering, wet each filter with BSS-B. Do not let the filter touch the filtrate in the cup. Pour half of the minced lung into one set of filters and the other half into the other. Filter sequentially through (1) two layers of gauze, (2) four layers of gauze, and (3) 20 μm Nitex nylon filtration fabric.

20. Make sure to rinse each cup with BSS-B after draining filtrate from syringe. Apply pressure to the open end of the syringe with the palm of your hand to aid in pushing the cells through the filter (Fig. 6).

21. With a 10 mL pipette, add filtrate to the top layer on the Optiprep gradient with the tip right above the lightweight (red) Optiprep layer. Slowly expel filtrate onto the gradient so that the layers do not mix (Fig. 7). Balance gradients with BSS-B and centrifuge for 20 min at 1000 rpm at 4 °C. Set acceleration and deceleration to low.

22. From this point on, handle cells in a sterile manner. Aspirate using a circular motion to ensure even aspiration down to the 15 mL level and discard. With a 10 mL pipette, remove the gradient contents between the 15 and 5 mL marks; this contains the ATII cells. Transfer to a sterile 50 mL conical tube.

23. Add sterile BSS-B at 4 °C to bring the volume up to 40 mL. Centrifuge 10 min at 1000 rpm at 4 °C.

24. Aspirate the supernatant and resuspend the cells in 10 mL of sterile BSS-B. Bring the volume up to 40 mL and centrifuge at 850–1000 rpm for 10 min at 4 °C.

25. Aspirate the supernatant, and immediately resuspend cells in 10 mL of 37 °C media. Any delay in resuspending the cells will cause them to clump.

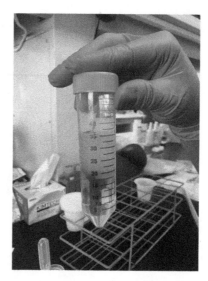

Fig. 7 Optiprep gradient showing clear defined layers. Slowly add filtrate to the top layer of the Optiprep gradient so that layers do not mix

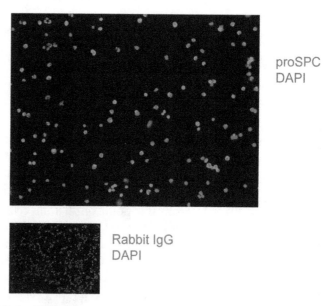

proSPC
DAPI

Rabbit IgG
DAPI

Fig. 8 Rat cell isolation cytospins were fixed and stained for proSPC and DAPI

26. Dilute an aliquot of cells 1:10 in trypan blue and count live and dead cells.

27. Determine purity of ATII cells (*see* **Note 6** and Fig. 8).

4   Notes

1. Other strains may be used. One or two mice will be needed. To increase the ATII cell yield, the lungs from two mice can be digested together, combining the digests before magnetic sorting.

2. Do not let cells come into contact with glassware. Cells can adhere to glass surfaces which can cause damage to the cells. Use plastic beakers for cells. If using a glass flask, you must siliconize the glass prior to adding cells.

3. When preparing the lightweight and heavyweight Optiprep gradients, mix with a stir bar until the Optiprep and Krebs buffers are completely mixed so that the two solutions cannot be seen separately at all.

4. To achieve the proper hydrometer reading, add Optiprep if the hydrometer reading is too low. Add Krebs/HEPES buffer if the hydrometer reading is too high.

5. A good perfusion is crucial to the ATII cell isolation. Make sure that lungs are white after perfusion.

6. Use an aliquot of cells to determine purity and viability after every cell isolation. To analyze viability, dilute cells 1:10 in trypan blue and count live and dead cells. To analyze purity, cytospin 100,000 cells in a cytofunnel at 600 rpm for 5 min, fix cells in 4% paraformaldehyde (PFA) for 10 min at room temperature, air-dry, and store at −20 °C. Immunostain slides with antibodies against proSPC.

7. To mince, hold the two pairs of scissors together, insert into mincing beaker, and pump scissors rapidly until pieces are ~1 mm in size. Move through this step quickly. This should take only 30 s.

Acknowledgments

This work was supported by NIH HL131608 (RLZ), the Boettcher Foundation (RLZ), and funds from the University of Colorado Denver Department of Medicine Outstanding Early Career Scholars Program (RLZ) and the Hastings Foundation (ZB).

Conflicts of Interest

None.

References

1. Dobbs LG (1990) Isolation and culture of alveolar type II cells. Am J Phys 258:L134–L147
2. Messier EM, Mason RJ, Kosmider B (2012) Efficient and rapid isolation and purification of mouse alveolar type II epithelial cells. Exp Lung Res 38:363–373
3. Gonzalez RF, Dobbs LG (2013) Isolation and culture of alveolar epithelial Type I and Type II cells from rat lungs. Methods Mol Biol 945:145–159
4. Dobbs LG, Gonzalez R, Williams MC (1986) An improved method for isolating type II cells in high yield and purity. Am Rev Respir Dis 134:141–145
5. Dobbs LG, Mason RJ (1979) Pulmonary alveolar type II cells isolated from rats. Release of phosphatidylcholine in response to beta-adrenergic stimulation. J Clin Invest 63:378–387
6. Dobbs LG, Geppert EF, Williams MC, Greenleaf RD, Mason RJ (1980) Metabolic properties and ultrastructure of alveolar type II cells isolated with elastase. Biochim Biophys Acta 618:510–523
7. Demaio L, Tseng W, Balverde Z, Alvarez JR, Kim KJ, Kelley DG, Senior RM, Crandall ED, Borok Z (2009) Characterization of mouse alveolar epithelial cell monolayers. Am J Physiol Lung Cell Mol Physiol 296:L1051–L1058

Chapter 7

Isolation and Characterization of Human Alveolar Type II Cells

Beata Kosmider, Robert J. Mason, and Karim Bahmed

Abstract

Alveolar type II (ATII) cells synthesize, store, and secrete pulmonary surfactant and restore the epithelium after damage to the alveolar epithelium. Isolation of human ATII cells provides a valuable tool to study their function under normal and pathophysiological conditions. Moreover, maintenance of their differentiated phenotype in vitro allows further study of their function. Here we describe a protocol for efficient ATII cell isolation, characterization, and culture.

Key words Lung, Alveolar type II cells, Purification

1 Introduction

The alveolar epithelium is composed of two main cell types, the alveolar type II (ATII) cell and the alveolar type I cell [1]. ATII cells cover about 5% of the alveolar surface area, comprise 15% of peripheral lung cells, and have an apical surface area of about 250 μm^2 per cell [2]. They have a cuboidal shape, lamellar bodies, and microvilli. ATII cells synthesize, store, and secrete pulmonary surfactant, which provides the low surface tension in the distal parts of the lung [1]. Moreover, ATII cells also transport sodium and fluid from the apical surface into the interstitium and play an important role in innate immunity. Furthermore, they have a stem cell potential and proliferate to restore the alveolar epithelium after damage to the very sensitive type I cells. ATII cells interact with other cells and have beneficial functions to the whole alveolus [3]. Consequently, the ATII cell expresses a number of receptors and signaling molecules involved in cell–cell and cell–matrix interactions. One of these is epithelial cell adhesion molecule (EpCAM), which can be used to isolate ATII cells, as described below.

Isolation and purification of ATII cells for culture and functional assays represent a powerful tool by which cellular and

Scott Alper and William J. Janssen (eds.), *Lung Innate Immunity and Inflammation: Methods and Protocols*,
Methods in Molecular Biology, vol. 1809, https://doi.org/10.1007/978-1-4939-8570-8_7,
© Springer Science+Business Media, LLC, part of Springer Nature 2018

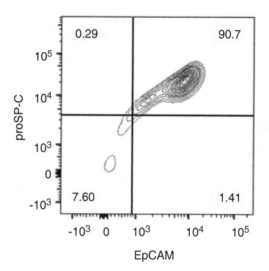

Fig. 1 Representative flow cytometry analysis of isolated ATII cells using co-staining for proSP-C and EpCAM

molecular mechanisms of respiratory diseases can be studied. This is highlighted by recent reports that investigated the function of human primary ATII cells in the pathogenesis of chronic obstructive pulmonary disease and idiopathic pulmonary fibrosis [4, 5]. Here we describe a protocol for ATII cell isolation from human lung tissue, in which EpCAM MicroBeads are used for positive selection of ATII cells. Our method yields ATII cells with up to 93% viability. Moreover, flow cytometry analysis shows 90% cell purity (Fig. 1). Importantly, the maintenance of the ATII cell phenotype in vitro depends not only on the method of isolation but also on culture conditions [6, 7]. Herein we describe these conditions and demonstrate the use of Western blotting to compare the levels of surfactant proteins: proSP-B, SP-B, proSP-C, and SP-C in both freshly isolated and cultured ATII cells (Fig. 2). Other investigators have used IgG adherence methods [8] or FACS to isolate ATII cells [9]. Notably, protocols for murine ATII cell isolation and culture are similar, although not identical [10]. The reader is referred to Chap. 6 for methods to isolate ATII cells from rodents.

2   Materials

2.1 Supplies and Equipment

1. Surgical scissors.

2. Surgical suture.

3. Forceps.

4. Perfusion tube.

5. 60 mL syringe.

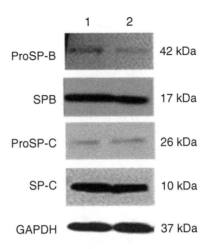

Fig. 2 Surfactant protein levels in ATII cells. ProSP-B, SP-B, proSP-C, and SP-C expression was detected by immunoblotting. *Lanes:* 1, freshly isolated cells; 2, cells cultured for 6 days

6. 20 µm and 100 µm nylon mesh.

7. 40 µm cell strainer.

8. Cheesecloth.

9. Graduated cylinder.

10. Containers.

11. Hydrometer.

12. LS columns (Miltenyi Biotec Inc.).

13. QuadroMACS Separator (Miltenyi Biotec Inc.).

14. Hemocytometer.

15. Shaking water bath.

16. Vacuum aspirator.

17. Neutator.

18. Refrigerated centrifuge.

2.2 Reagents

1. HEPES/saline buffer: 17.532 g NaCl; 2.383 g HEPES, 2 L final volume, pH 7.4.

2. HEPES/saline/EDTA buffer: 17.532 g NaCl; 2.383 g HEPES; 3.722 g EDTA, 2 L final volume, pH 7.4.

3. BSS-A buffer: 31.8 g NaCl; 1.6 g KCl; 0.2 g NaH_2PO_4; 1.28 g Na_2HPO_4; 10.12 g HEPES; 4 g dextrose, 10 µg/mL gentamycin, 4 L final volume, pH 7.4. Store at 4 °C.

4. BSS-B buffer: 31.04 g NaCl; 1.544 g KCl; 0.2 g NaH_2PO_4; 1.236 g Na_2HPO_4; 9.86 g HEPES; 4 g dextrose; 1.116 g $CaCl_2·2H_2O$; 1.276 g $MgSO_4·7H_2O$, 10 µg/mL gentamycin, pH 7.4, 4 L final volume. Store at 4 °C.

5. Column buffer in PBS: 10 g BSA; 2 mM EDTA, 2 L final volume, pH 7.4; degas and store at 4 °C.

6. 4 mg/mL DNase I in BSS-B buffer.

7. Krebs HEPES buffer: 0.25 M NaCl, 2.65 mM sodium phosphate buffer (pH 7.4), 5.31 mM KCl, 10 mM HEPES.

8. OptiPrep to make light and heavy gradients. Light gradient: 45 mL OptiPrep, 8 mL FBS, 1 mL phenol red; adjust with Krebs HEPES buffer to relative density 1.038–1.045, 400 mL final volume, and store at −20 °C. Heavy gradient: 110 mL OptiPrep, 8 mL FBS; adjust with Krebs HEPES buffer to relative density 1.088–1.092, 400 mL final volume, and store at −20 °C.

9. RTC mix: 10% 10× MEM and 90% of 2.5 mg/mL RTC, and adjust pH using 0.32 M NaOH to get pink color.

10. Pharm Lyse.

11. Elastase (Worthington Biochemical Corp.).

12. Epithelial cell adhesion molecule (EpCAM) MicroBeads (Miltenyi Biotec Inc.).

13. FcR Blocking Reagent (Miltenyi Biotec Inc.).

14. Matrigel Matrix.

15. 10× MEM.

16. Fetal bovine serum.

17. Fetal calf serum.

18. Dimethyl sulfoxide (DMSO).

19. Rat tail collagen (RTC).

20. DMEM.

21. Charcoal-stripped FBS (CS-FBS).

22. Keratinocyte growth factor (KGF).

23. Isobutylmethylxanthine (IBMX).

24. 8-Bromo-cyclic AMP (cAMP).

25. Dexamethasone.

26. Streptomycin.

27. Penicillin.

28. Amphotericin B.

29. Glutamine.

30. Gentamycin.

3   Methods

3.1   Lung Perfusion and Lavage

1. Select the lobe to be processed. The right middle lobe is the easiest to process but the lingula can also be used.

2. Identify the bronchus and corresponding pulmonary artery that supplies the lobe. Place a small tube in the pulmonary artery, and use a 60 mL syringe to inject HEPES/saline buffer to flush blood from the lobe's blood vessels. Repeat until blood no longer flushes out.

3. Put the cannula in the lobar bronchus. Tie off with surgical suture to secure the cannula in place.

4. Lavage the lobe by filling it completely with HEPES/saline/ EDTA four times to remove alveolar macrophages and debris from the airspaces. The volume depends on the size of the lobe and is usually between 100 and 250 mL.

5. Lavage the lobe with BSS-A buffer (37 °C) twice as described above. This will rinse EDTA from the lungs and remove additional alveolar macrophages.

6. Lavage the lobe with BSS-B buffer (37 °C) twice to neutralize any remaining EDTA. Drain this last lavage very well.

7. Instill BSS-B buffer with 12.9 U/mL elastase (37 °C) into the lobe. The volume depends on the size of the lobe and is usually between 200 and 250 mL.

8. Clamp the cannula that was used to lavage the lobe so that none of the elastase leaks out.

9. Place the lobe in a container and incubate in the water bath for 40 min at 37 °C (*see* **Note 1**).

10. Remove the lung from the water bath. Take out the cannula and trim away any surrounding bronchial tissue. Cut the lung into small pieces.

11. Homogenize lung pieces in 150 mL BSS-B buffer with 5 mL DNase I using a food processor (five bursts of 2–3 s).

12. Filter cell suspension through cheesecloth into a container with 90 mL FCS, and shake in a water bath for 5 min at 37 °C.

13. Sequentially filter the cell suspension through cheesecloth, 100 μm and 20 μm nylon mesh.

14. Centrifuge the suspension at $200 \times g$ for 10 min at 4 °C; aspirate and discard the supernatant (*see* **Note 2**).

15. Prepare six conical tubes (50 mL) with OptiPrep medium by pipetting a 10 mL light density gradient and underlying 10 mL heavy density gradient.

16. Resuspend the cell pellet in 150 mL BSS-B buffer, and load 25 mL onto OptiPrep medium (*see* **Note 3**).

17. Centrifuge at $200 \times g$ for 20 min at 4 °C.

18. Aspirate the liquid above cells located at the interface and discard.

19. Collect cells into separate tubes.

20. Wash cells by adding 45 mL BSS-B buffer and centrifuge at $200 \times g$ for 10 min.

21. Resuspend cell pellet with column buffer, and count total number of cells using a hemocytometer.

3.2 Optional Collection of Alveolar Macrophages from Lavage Fluid

1. Combine lavage fluid obtained from **Steps 4** and **5** in Subheading 3.1.

2. Filter lavage fluid through a 100 μm filter.

3. Centrifuge at $200 \times g$ for 10 min at 4 °C.

4. Resuspend cell pellet in 1× Pharm Lyse buffer to lyse red blood cells.

5. Incubate cells for 10 min in the dark at room temperature.

6. Centrifuge the cells at $200 \times g$ for 10 min at 4 °C.

7. Resuspend the cell pellets in DMEM with 10% FBS; centrifuge at $200 \times g$ for 10 min.

8. Freeze cells in a liquid nitrogen tank in freezing media (90% FBS and 10% DMSO).

3.3 ATII Cell Purification Using EpCAM MicroBeads

1. Prepare conical tubes based on the cell count obtained from **Step 21** in Subheading 3.1. Use 100×10^6–150×10^6 cells per tube.

2. Centrifuge cells at $200 \times g$ for 5 min at 4 °C.

3. Resuspend cell pellet with 900 μL column buffer per tube. Add 300 μL FcR Blocking Reagent and 300 μL EpCAM MicroBeads per conical tube.

4. Place conical tubes with cells on the neutator for 30 min at 4 °C.

5. Add 10 mL column buffer to each tube to wash cells.

6. Centrifuge the cells at $200 \times g$ for 10 min at 4 °C. Aspirate and discard supernatant.

7. Place LS columns in the QuadroMACS Separator, and rinse with 3 mL column buffer (*see* **Note 4**).

8. Resuspend cell pellet in 3 mL column buffer; filter cells through 40 μm cell strainer and add to the column. When this volume runs through the column completely, wash the column three times with 3 mL column buffer.

9. Elute ATII cells by adding 5 mL column buffer and then plunging with the provided plunger. Count EpCAM-positive ATII cells using a hemocytometer (*see* **Note 5**). The unlabeled fraction contains other cell types.

3.4 ATII Cell Culture

1. Use a mixture of 80% RTC mix and 20% Matrigel Matrix to coat inserts for 12-well plates. Incubate the plates at 37 °C for 10 min.

2. Plate 1×10^6 ATII cells on inserts in DMEM with 10% FBS, 2 mM glutamine, 100 U/mL penicillin, 100 μg/mL streptomycin, 2.5 μg/mL amphotericin B, and 10 μg/mL gentamycin (*see* **Note 6**).

3. After 2 days add DMEM with 1% CS-FBS, 10 ng/mL KGF, and antibiotics as described in **Step 2** to the basolateral side, and culture ATII cells under air/liquid interface conditions (*see* **Note 7**).

4. After 2 additional days, add DMEM with 1% CS-FBS, 10 ng/mL KGF, 0.1 mM IBMX, 0.1 mM cAMP, 10^{-8} M dexamethasone, and antibiotics to the basolateral side.

5. After 2 days start experiments.

4   Notes

1. We typically place the lobe in a plastic bag for the incubation step. This provides an easy way to collect any buffer that leaks out of the lobe. After 20 min you may re-instill the lobe with elastase that leaked out.

2. Use Pharm Lyse buffer as described in Subheading 3.2 if you see many red blood cells.

3. Add 1–2 mL DNase I if you see cell clumps.

4. Avoid air bubbles which may block the column.

5. Final ATII cell viability and yield depend on lung condition, how much time passed from lung recovery to start processing, and the speed of lung processing.

6. ATII cell plating efficiency may vary among lung donors.

7. ATII cells are plated on a filter insert coated with the mixture or RTC and Matrigel Matrix. After cell attachment for 2 days, the medium is placed outside the insert to culture cells under air/liquid interface conditions. The reader is referred to Chap. 8 for details.

Acknowledgments

This work was supported by NIH R01 HL118171 (BK) and FAMRI grant CIA130046 (BK). We thank Chih-Ru Lin for helping with experiments.

References

1. Mason RJ (2006) Biology of alveolar type ii cells. Respirology 11(Suppl):S12–S15

2. Stone KC, Mercer RR, Gehr P, Stockstill B, Crapo JD (1992) Allometric relationships of cell numbers and size in the mammalian lung. Am J Respir Cell Mol Biol 6(2):235–243

3. Fehrenbach H (2001) Alveolar epithelial type ii cell: defender of the alveolus revisited. Respir Res 2(1):33–46

4. Liang J, Zhang Y, Xie T, Liu N, Chen H, Geng Y, Kurkciyan A, Mena JM, Stripp BR, Jiang D et al (2016) Hyaluronan and TLR4 promote surfactant-protein-c-positive alveolar progenitor cell renewal and prevent severe pulmonary fibrosis in mice. Nat Med 22(11):1285–1293

5. Skronska-Wasek W, Mutze K, Baarsma HA, Bracke KR, Alsafadi HN, Lehmann M, Costa R, Stornaiuolo M, Novellino E, Brusselle GG et al (2017) Reduced frizzled receptor 4 expression prevents wnt/beta-catenin-driven alveolar lung repair in copd. Am J Respir Crit Care Med 196:172–185

6. Wang J, Edeen K, Manzer R, Chang Y, Wang S, Chen X, Funk CJ, Cosgrove GP, Fang X, Mason RJ (2007) Differentiated human alveolar epithelial cells and reversibility of their phenotype in vitro. Am J Respir Cell Mol Biol 36(6):661–668

7. Bahmed K, Messier EM, Zhou W, Tuder RM, Freed CR, Chu HW, Kelsen SG, Bowler RP, Mason RJ, Kosmider B (2016) Dj-1 modulates NRF2-mediated protection in human primary alveolar type ii cells in smokers. Am J Respir Cell Mol Biol 55:439–449

8. Bove PF, Grubb BR, Okada SF, Ribeiro CM, Rogers TD, Randell SH, O'Neal WK, Boucher RC (2010) Human alveolar type II cells secrete and absorb liquid in response to local nucleotide signaling. J Biol Chem 285(45):34939–34949

9. Fujino N, Kubo H, Ota C, Suzuki T, Suzuki S, Yamada M, Takahashi T, He M, Suzuki T, Kondo T et al (2012) A novel method for isolating individual cellular components from the adult human distal lung. Am J Respir Cell Mol Biol 46(4):422–430

10. Messier EM, Mason RJ, Kosmider B (2012) Efficient and rapid isolation and purification of mouse alveolar type II epithelial cells. Exp Lung Res 38(7):363–373

Chapter 8

Air–Liquid Interface Culture of Human and Mouse Airway Epithelial Cells

Di Jiang, Niccolette Schaefer, and Hong Wei Chu

Abstract

Air–liquid interface culture enables airway epithelial cells to differentiate into a pseudostratified cell layer, consisting of ciliated cells, goblet/secretory cells, and basal cells (Ghio et al., Part Fibre Toxicol 10:25, 2013). This technique is critically important for in vitro studies of lung diseases such as asthma, chronic obstructive pulmonary disease, and cystic fibrosis, since differentiated airway epithelial cells are more representative of the in vivo lung environment than non-differentiated cells (Derichs et al., FASEB J 25:2325–2332, 2011; Hackett et al., Am J Respir Cell Mol Biol 45:1090–1100, 2011;Schneider et al., Am J Respir Crit Care Med 182: 332–340, 2010). Here we describe the process of isolating and expanding human and mouse airway epithelial cells, as well as differentiation of airway epithelial cells by air–liquid interface culture.

Key words Air–liquid interface, Airway epithelial cells, Cell differentiation

1 Introduction

Air–liquid interface (ALI) is a type of culture system that allows epithelial cells to proliferate and differentiate much like they do in vivo [1]. ALI is the gold standard for human airway cell culture systems and can be used to study airway diseases such as asthma and COPD, diseases that involve cell proliferation and differentiation, and mucosal inflammatory and host defense responses [2–4]. In other culture systems, such as submerged culture, epithelial cells do not differentiate, making it difficult to study pathological conditions related to mucociliary cells.

ALI systems utilize an apical chamber and basolateral chamber that are separated by a thin polyester membrane on which the cells reside. In the most popular 12-well transwell ALI culture system, there is at most 50 µL of media on the apical side, and the basolateral

Di Jiang and Niccolette Schaefer contributed equally to this work.

Scott Alper and William J. Janssen (eds.), *Lung Innate Immunity and Inflammation: Methods and Protocols*,
Methods in Molecular Biology, vol. 1809, https://doi.org/10.1007/978-1-4939-8570-8_8,
© Springer Science+Business Media, LLC, part of Springer Nature 2018

chamber has approximately 1.2 mL of media [5, 6]. Human tracheobronchial epithelial cells, human nasal epithelial cells, and even mouse tracheal epithelial cells can be grown in ALI. In addition, ALI culture systems allow the researcher to manipulate the environment, for example, by treating cells with cytokines. Since it can be difficult to manipulate gene expression in human cells cultured at ALI, many investigators also utilize mouse airway epithelial cells to determine the functions of genes that are overexpressed or deleted in mice.

Epithelia in all species act as the first line of defense against environmental agents such as pathogens and allergens. Basal cells, ciliated cells, and secretory cells are present in all species, although there are differences in the frequency of each cell type present. Mucous goblet cells, a type of secretory cell, are very abundant in humans, but are uncommon in mice. Mice contain a relatively equal proportion of basal cells, ciliated cells, and secretory cells. The secretory cells that are present in mice are club cells. Therefore, the production of mucus is also minimal in mice [8].

This chapter will provide methods for isolating and expanding airway epithelial cells under submerged conditions and performing ALI culture to induce mucociliary differentiation. The protocols used for isolating airway epithelial cells from humans and from mice are similar. Airway tissues are first processed in a protease solution to liberate epithelial cells, which are then isolated and seeded on to collagen-coated culture dishes under submerged conditions for expansion. The expanded airway epithelial cells are then used fresh for ALI culture or frozen for future experiments (Fig. 1).

The protocol for mouse trachea epithelial cell isolation provided in this chapter has been refined from previously described protocols [7, 9–11]. It takes advantage of the irradiated fibroblast feeder system to generate large quantities of epithelial cells from only a few mice, thus enabling researchers to perform ALI cultures with few mice or to freeze the cells for banking for future experiments. Additionally, we provide an experimental example in which ALI cells are treated with IL-13 to mimic a diseased airway that is comparable to Th2 high asthma in humans.

2   Materials

All buffers must be prepared under sterile conditions unless specified otherwise.

2.1 Isolation of Airway Epithelial Cells from Human Donor Tracheal and Bronchial Tissue

1. Amphotericin B solution (Amph B). Store at −20 °C.

2. Bronchial Epithelial Cell Growth Medium (BEGM)—basal medium + BulletKit. Store base medium at 4 °C, BulletKit at −20 °C.

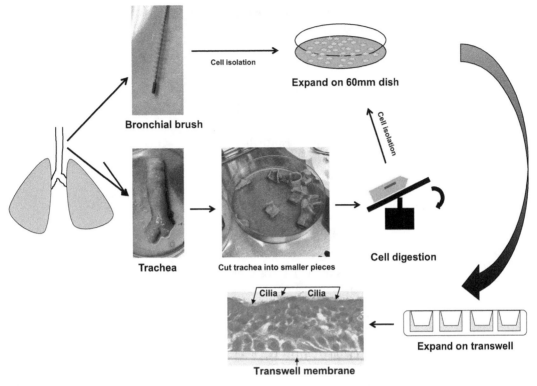

Fig. 1 Overview of human airway epithelial cell isolation, expansion, and air–liquid interface culture procedure

3. Fetal bovine serum (FBS). Store at −20 °C.

4. Dulbecco's Modified Eagles Medium (DMEM). Store at −20 °C.

5. Penicillin-Streptomycin-Amphotericin B, 100× (PSA). Store at −20 °C.

6. Protease from *Streptomyces griseus* (protease). Store at −20 °C.

7. Cell scraper.

8. Cell strainers.

9. Standard tissue culture dishes (100 mm dish).

10. Stock medium: Dissolve 5 mL of PSA into 500 mL (one bottle) of DMEM; mix well and store at 4 °C until time of use.

11. Protease: Completely dissolve 5 g of protease in 50 mL sterile PBS; mix well and store at −20 °C until time of use (*see* **Note 1**).

12. Digestion buffer: Dissolve 1 mL protease/PBS mix into 50 mL stock medium.

13. HEPES buffer: Dissolve 1 mL 0.5 M EDTA into 50 mL HEPES.

14. BEGM (culture medium): Dissolve the supplemented BulletKit into 500 mL basal medium (BEBM), and store at 4 °C until time of use (*see* **Note 2**).

15. Cell suspension medium: Dissolve 150 μL of PSA into 30 mL of BEGM.

2.2 Isolation of Airway Epithelial Cells from Human Bronchial Brushing

1. Bronchial Epithelial Cell Growth Medium—basal medium + BulletKit (BEGM). Store base medium at 4 °C, BulletKit at −20 °C.

2. Bronchial brushes.

3. Tissue culture dishes (100 mm dish).

4. Phosphate-buffered saline (PBS). Store at room temperature.

2.3 Expansion of Human Airway Epithelial Cells

1. Bronchial Epithelial Cell Growth Medium—basal medium + BulletKit (BEGM). Store base medium at 4 °C, BulletKit at −20 °C.

2. Collagen I, rat tail culture dish (collagen-coated dish).

2.4 Air–Liquid Interface of Human Airway Epithelial Cells

1. Bovine collagen (collagen). Store at 4 °C.

2. Transwell.

3. Trypsin-EDTA solution. Make 10 mL aliquots and store at −20 °C.

4. Trypsin neutralizing solution. Make 10 mL aliquots and store at −20 °C.

5. F6 medium (base medium): Mix 250 mL of BEBM and 250 mL of DMEM. Then add the following BulletKit supplements to 500 mL of BEBM/DMEM: insulin (0.4 mL, final concentration 0.4 μg/mL), transferrin (0.25 mL, final concentration 5 μg/mL), hydrocortisone (0.5 mL, final concentration 0.5 μg/mL), epinephrine (0.5 mL, final concentration 0.5 μg/mL), bovine hypothalamus extract (2.0 mL, final concentration 52 μg/mL), gentamicin/amphotericin (GA) (0.5 mL, final concentration 50 μg/mL). The F6 (base medium) is good for 2 months.

6. Full F6 medium (culture medium): Add the supplements below to 25 mL of the above BEBM/DMEM/F6 media on the day of use: albumin bovine (final concentration 0.5 μg/mL), ethanolamine (final concentration 50 mM), MgCl$_2$ (final concentration 3.0 mM), MgSO$_4$ (final concentration 4.0 mM), CaCl$_2$ (final concentration 1.0 mM), human EGF (final concentration 10 ng/mL). Incubate in a water bath at 37 °C for at least 10 min. Use within 2 h.

2.5 Mouse Epithelial Cell Media

1. Fibroblast growth media: Prepare 10 mL per 100 mm dish. To 500 mL DMEM with sodium pyruvate, add 50 mL of fetal bovine serum, 5 mL of L-glutamine, and 5 mL of 100× penicillin/streptomycin.

2. DMEM with 1× PSA: Prepare 1 mL per trachea. To 10 mL DMEM, add 100 μL of 100× penicillin/streptomycin.

3. Protease solution: Prepare 1 mL per trachea. To 10 mL DMEM, add 120 μL of 0.1% protease and 2 mL of 50 μg/mL amphotericin B.

4. DMEM with 2% FBS: To 490 mL DMEM, add 10 mL of fetal bovine serum.

5. DMEM with 10% FBS: To 450 mL DMEM, add 50 mL of fetal bovine serum.

6. Disaggregation solution: To 49 mL of 1× PBS with calcium and magnesium, add 500 μL of EGTA (0.5 M) and 500 μL of EDTA (0.5 M).

7. F-media: Make complete DMEM first, and then filter sterilize (500 mL DMEM without sodium pyruvate, 50 mL fetal bovine serum, 5.5 mL L-glutamine, and 5.5 mL 100× penicillin/streptomycin). Add 1 mL hydrocortisone/EGF mix, 8.6 μL cholera toxin, 16 mL adenine, 1 mL insulin, and 250 mL Ham's F12 to 730 mL of filtered complete DMEM. Then filter sterilize again. This recipe makes 1 L of F-media; store at 4 °C. Use 10 mL with supplements for each 100 mm dish.

8. Supplements for mL F-media: Add fresh every time before medium change. To 10 mL F-media, add 10 μL Y inhibitor (Y-27632) and 12.5 μL mouse epidermal growth factor (mEGF).

9. PneumaCult-ALI complete medium: 88.2 mL PneumaCult-ALI complete base medium, 10 mL 10× supplement, 1 mL 100× supplement, 200 μL of 0.2% heparin solution, 500 μL of 200× hydrocortisone solution, and 500 μL of 100× penicillin/streptomycin. For every 25 mL of PneumaCult-ALI complete medium, add 25 μL of Y inhibitor and 3.75 μL of mEGF prior to use.

10. Freezing medium: Prepare 1 mL aliquot per 5×10^5 cells. Mix 600 μL complete F-media, 300 μL fetal bovine serum, 100 μL DMSO, 1 μL Y inhibitor, and 1.25 μL mouse EGF.

3   Methods

3.1   Isolation and Expansion of Human Airway Epithelial Cells

This is a two-day procedure, including overnight protease digestion of tracheobronchial tissue. All procedures are performed at room temperature unless specified otherwise.

3.1.1   Isolation of Airway Epithelial Cells from Human Tracheobronchial Tissue

Day 1: Protease Digestion of Tracheobronchial Tissue

1. Put tracheobronchial tissue on a sterile 100 mm dish and keep it moist with stock medium.

2. Trim off all of the fat and connective tissue, and then cut the tissue into smaller pieces (1-inch diameter). Start cutting longitudinally to expose the lumen of the trachea/bronchus (Fig. 2).

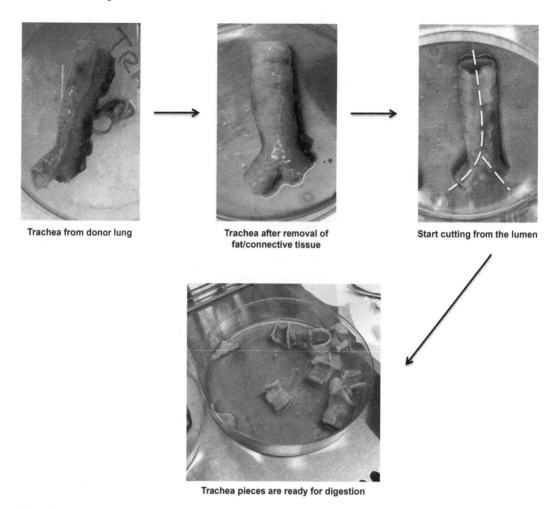

Trachea from donor lung

Trachea after removal of fat/connective tissue

Start cutting from the lumen

Trachea pieces are ready for digestion

Fig. 2 Overview of trachea preparation for overnight protease digestion

3. Rinse off the mucus on each piece of the tissue with stock medium.

4. Digest the tissue by putting up to five tissue pieces into each 50 mL tube filled with 25 mL of stock medium with 0.2% protease.

5. Incubate on a rotating rocker overnight at 4 °C.

Day 2: Cell Isolation

1. Neutralize proteases by adding 1 mL FBS into each tube containing the tracheobronchial tissue. Gently mix by inverting several times.

2. After the tissue sinks to the bottom of the tube, collect the tissue-free supernatant into cell collection tube 1 using a serological pipette, and store at room temperature.

3. Add 25 mL of HEPES solution with EDTA to each tracheobronchial tissue tube, and then incubate for 15 min at 37 °C (water bath).

4. After the tissue sinks to the bottom of the tube, collect the supernatant into cell collection tube 2 using a serological pipette. Store at room temperature.

5. Transfer the tracheobronchial tissue pieces to a sterile 100 mm dish with stock medium + 10% FBS, and then start scraping the luminal surface of tracheobronchial tissue using a cell scraper (up and down motion) (*see* **Note 3**).

6. Use a 10 mL serological pipet to collect all scraped cells/stock medium mixture into cell collection tube 3. Store at room temperature.

7. Centrifuge all three cell collection tubes at $230 \times g$ for 10 min at 4 °C.

8. Carefully collect the supernatant into a waste bottle and re-suspend the cell pellets in BEGM (10 mL/tube), and combine the cells from the three tubes in one tube.

9. Pour the cell suspension through a 70 μm cell strainer to filter out the mucus.

10. Count the total number of isolated live epithelial cells using a hemocytometer.

3.1.2 Expansion of Airway Epithelial Cells Isolated from Human Tracheobronchial Tissue

1. Seed 2×10^5 live airway epithelial cells in 3 mL BEGM + amphotericin B (5 μg/mL final concentration) per 60 mm collagen-coated dish.

2. Change the medium every other day, and expand the cells until they reach 80% confluence (Fig. 1). Remove amphotericin B after 48 h of culture (at first medium change).

3. An 80% confluent 60 mm dish will result in $1–1.5 \times 10^6$ cells. Do not let the cells overgrow.

3.1.3 Isolation of Airway Epithelial Cells from Bronchial Brushings

1. Bronchial brushings are collected from subjects during bronchoscopy. Brushes are submerged in PBS in a 15 mL tube until time of processing.

2. Pre-warm 10 mL BEGM, and store at 37 °C until use.

3. Transfer the brush, along with the PBS from the 15 mL tube onto a sterile 100 mm culture dish (*see* **Note 4**).

4. Using a sterile transfer pipet, gently rinse over the brush with PBS five to six times, and transfer the washed content into a cell collection tube.

5. Fill the dish with 10 mL of fresh PBS, and repeat an additional wash of the brush. Transfer the washed content into the cell collection tube.

6. Centrifuge the cell collection tube at $590 \times g$ for 10 min at 4 °C.

7. Discard the supernatant, and re-suspend the cell pellet in 2.5 mL BEGM (per brush) with a serological pipet.

8. Repeat the centrifugation of the cell collection tube.

9. Discard the supernatant and re-suspend the cell pellet in 1 mL BEGM for cell counting.

10. Determine total isolated live cell count using a hemocytometer.

3.1.4 Expansion of Airway Epithelial Cells Isolated from Human Bronchial Brushings

1. Seed 9×10^4 live airway epithelial cells in 3 mL BEGM per 60 mm collagen-coated dish to expand the cells.

2. Change the medium every other day, and expand the cells for 10 days or until 80% confluent, whichever comes first. Do not let the cells overgrow (Fig. 3).

3.2 Air–Liquid Interface Culture of Human Airway Epithelial Cells

3.2.1 Preparation of Collagen-Coated Transwell Membrane Inserts

PureCol bovine collagen must be kept on ice to prevent polymerization at room temperature. Preparation of transwells should be conducted under sterile conditions. The polymerized collagen should appear evenly distributed to the naked eye and can be inspected further under a microscope. Plates should be used the day of coating (*see* **Note 5**).

1. Add 0.5 mL of sterile 10× PBS and 30 μL of 1 N NaOH to 4 mL of bovine collagen (3 mg/mL).

2. Gently mix with a serological pipette (avoid foaming).

3. Add a drop of the collagen mixture onto pH paper.

4. The collagen mixture should be slightly acidic with pH around 6–7. If needed, adjust the pH by adding 1N HCL or 1N NaOH.

5. Add 160 μL of the collagen mixture to the center of each transwell insert.

6. In a small circular motion, shake the plate for ~10 s to spread the collagen evenly over the membrane.

7. Let the collagen polymerize for less than 10 s with the plate cover off.

8. After 10 s, tilt the plate and remove the excess collagen by pipetting.

9. With the plate lid off, let the collagen dry for 1 h at room temperature under sterile conditions.

10. Cover the plate after 1 h and keep under hood until time of use.

3.2.2 Trypsinization of Human Airway Epithelial Cells

Subculture epithelial cells when submerged cultures are 80% confluent, because cells may become trypsin resistant when more than 90% confluent.

1. Pre-warm the trypsin for 10 min at 37 °C.

2. Remove the culture medium from the 60 mm dish.

3. Wash cells with PBS once.

Fig. 3 Human airway epithelial cells grown under submerged culture on a 60 mm collagen-coated dish. Cell growth was monitored on days 2, 4, and 6. Pictures were taken on a phase-contrast microscope (magnification 40×)

4. Remove the PBS, and add 3 mL of trypsin to each 60 mm plate.

5. Incubate at room temperature for 2 min.

6. Closely observe the dish for detachment of cells under the light microscope (*see* **Note 6**).

7. When 50% of the cells have detached, neutralize the trypsin with 3 mL of trypsin neutralizing solution per dish.

8. Check the Petri dish under a microscope for remaining cells (*see* **Note 7**).

9. Collect the detached cells in a 15 or 50 mL sterile conical tube and spin at $230 \times g$ for 10 min at 4 °C.

10. After the centrifugation, discard supernatant and add 3 mL of cold BEGM to re-suspend the cell pellet.

11. Spin at $230 \times g$ for 10 min at 4 °C.

12. After the centrifugation, discard supernatant and add 2 mL of full F6 medium to re-suspend the cell pellet.

13. Determine total live cell count after trypsinization using a hemocytometer.

3.2.3 Seeding of Human Airway Epithelial Cells onto Collagen-Coated Transwell Membrane Inserts

1. Warm full F6 medium in a water bath at 37 °C for at least 10 min before use.

2. Dilute the 2 mL cell suspension with the appropriate volume of full F6 medium so that the final concentration of cells is 4×10^4 for 300 μL per transwell insert.

3. Add 300 μL of cells/media to the apical side of each transwell insert. Add 1.2 mL of full F6 medium to the basolateral side of each transwell and incubate at 37 °C/5% CO_2.

4. Change medium every other day until the cells reach 100% confluence (Fig. 3; *see* **Note 8**).

5. Start air–liquid interface culture by reducing the full F6 medium on the apical side of the transwell inserts to 50 μL while keeping the same volume (1.2 mL) on the basolateral side.

6. Change medium every other day, and culture the cells under air–liquid interface for at least 10 days to induce mucociliary differentiation (Fig. 4).

3.3 Isolation and Expansion of Mouse Airway Epithelial Cells (Fig. 5)

1. All surgical instruments should be sterile and wiped down with 70% ethanol prior to use.

2. Euthanize mice with intraperitoneal injection of pentobarbital sodium prepared in sodium chloride.

3.3.1 Trachea Harvest

3. Wet the neck and torso of the mice with 70% ethanol.

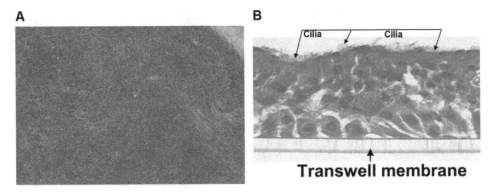

Fig. 4 Human airway epithelial cells were grown in the air–liquid interface culture for 14 days. Cell morphology (**a**) was observed with a phase-contrast microscope (magnification 40×). Mucociliary differentiation (**b**) is observed after H&E staining (magnification 400×). Black arrows = cilia; blue arrows = goblet cells

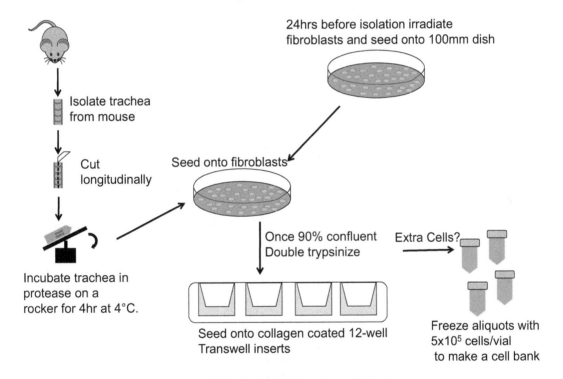

Fig. 5 Overview of mouse airway epithelial cell air–liquid interface culture steps

4. Cut inferiorly from the jaw to the sternum, exposing as much trachea as possible.

5. Make one cut at the proximal end of the trachea near the larynx and another cut at the distal end. Collect the isolated tracheas in a 15 mL conical tube prefilled with DMEM and 1× PSA (penicillin, streptomycin, amphotericin B).

<table>
<tr><td>

3.3.2 Trachea Protease Digestion

</td><td>

1. Transfer the trachea, along with the DMEM with 1× PSA, onto a 60 mm dish.

2. Using a surgical scalpel, cut the tracheas longitudinally to expose the lumen.

3. Transfer the tracheas into a 15 mL conical tube that contains the 0.1% protease solution in DMEM with 50 μg/mL of amphotericin B and put on a rotating rocker for 4 h at 4 °C.

</td></tr>
</table>

3.3.3 Airway Epithelial Cell Isolation

1. At the end of the digestion, transfer the digested trachea, along with the protease solution, onto a new 60 mm dish.

2. Immediately add 1 mL DMEM with 2% FBS to neutralize the protease reaction.

3. Using a P1000 pipette, forcefully wash the trachea using the 1 mL of DMEM with 2% FBS approximately 20 times (change orientation after 10 times), to detach the remaining airway epithelial cells. The tracheas can now be discarded.

4. Collect all of the remaining solution in the 60 mm dish and put into a new 15 mL conical tube.

5. Centrifuge the cells at $590 \times g$ for 10 min at 4 °C.

6. Discard the supernatant, and re-suspend the pellet in 1 mL of DMEM with 2% FBS using a transfer pipette. Then add fresh DMEM with 2% FBS to reach a total volume of 5 mL.

7. Repeat centrifugation as described in **step 5**.

8. Discard supernatant, and re-suspend the pellet in 1 mL of dis-aggregation solution, using a transfer pipette. Then, add more disaggregation solution to reach a total volume of 10 mL.

9. Centrifuge the tube at $162 \times g$ for 5 min at 4 °C.

10. Discard the supernatant and re-suspend the pellet in 1 mL of DPBS (with calcium and magnesium), using a transfer pipette. Once re-suspended, add more DPBS (with calcium and magnesium) to reach a total volume of 10 mL.

11. Repeat centrifugation as described in **step 9**.

12. Discard the supernatant. Re-suspend in 1 mL of pre-warmed F-media with mouse epidermal growth factor and Y inhibitor. Add more of this solution to reach a total volume of 10 mL per trachea.

3.3.4 Cell Expansion on Mouse 3T3 Irradiated Fibroblast Layer

1. Irradiate (5000 rads) NIH 3T3 fibroblasts and seed 1.7×10^6 cells/100 mm dish. Culture for 24 h in fibroblast growth medium prior to seeding airway epithelial cells.

2. Slowly remove the fibroblast growth media from the irradiated fibroblast dishes.

3. Slowly add the F-media containing mouse airway epithelial cells to the irradiated fibroblast dish (*see* **Note 9**).

4. Incubate at 37 °C with 5% CO_2.

5. Change media every other day, adding 10 mL of F-media with mEGF and Y inhibitor per dish slowly.

6. Colonies should be visible within 5–6 days of seeding.

7. When the cells are approximately 90% confluent, they must be removed from the irradiated fibroblasts using the double trypsinization method (*see* Subheading 3.4.2 and **Note 10**).

3.4 Air–Liquid Interface of Mouse Airway Epithelial Cells

3.4.1 Preparation of Double Collagen-Coated Transwell Membrane Inserts

PureCol bovine collagen must be kept on ice to prevent polymerization at room temperature. Preparation of transwells should be conducted under sterile conditions. The polymerized collagen should appear evenly distributed to the naked eye and can be inspected further under a microscope. Plates should be used the day of coating (*see* **Note 5**).

1. Add 0.5 mL of sterile 10 × PBS and 30 μL of 1 N NaOH to 4 mL of bovine collagen (3 mg/mL).

2. Gently mix with a serological pipette (avoid foaming).

3. Add a drop of the above collagen mixture onto pH paper.

4. The collagen mixture should be slightly acidic with pH around 6–7 (*see* **Note 11**).

5. Add 160 μL of the collagen mixture to the center of each transwell insert.

6. In a small circular motion, shake the plate for ~10 s to spread the collagen evenly over the membrane.

7. Let the collagen polymerize for less than 10 s with the plate cover off.

8. After 10 s, tilt the plate and remove the excess collagen by pipetting.

9. With the plate lid off, let the collagen dry for 50 min at room temperature under sterile condition.

10. At the end of the 50-min incubation, prepare a fresh collagen mixture and apply to each transwell following **steps 1–8**.

11. Air dry with the lid off for 1 h at room temperature.

12. At the end of the second 1-h incubation, cover the plate with the lid until ready to use (*see* **Note 12**).

3.4.2 Double Trypsinization of Mouse Airway Epithelial Cells

Initial Trypsinization for Removal of Fibroblasts

1. Aspirate the media off of the dish, and wash the cells with 10 mL of 1× HBSS (without calcium and magnesium) (*see* **Note 13**).

2. Add 5 mL of pre-warmed low-concentration trypsin (0.25%) to the cells and put in 37 °C incubator for approximately 1 min. Closely observe under light microscope, and make sure the irradiated fibroblasts are detaching.

3. Aspirate the trypsin/fibroblast mixture, and wash the dish with 10 mL of 1× HBSS twice to remove any loosely attached irradiated fibroblasts. The epithelial cells should remain attached.

Second Trypsinization for Removal of Epithelial Cells

1. Add 5 mL of pre-warmed low-concentration trypsin (0.25%) to the cells and incubate at 37 °C for approximately 3 min.

2. Remove the floating cells using a transfer pipette (*see* **Note 14**).

3. Transfer the trypsin/epithelial cell mixture to a 50 mL conical tube. Add heat-inactivated FBS to the conical tube (1.5 mL per dish).

4. Wash the plate with 10 mL of 1× HBSS to collect any remaining cells, and add to the 50 mL conical tube.

5. Centrifuge at $162 \times g$ for 5 min at 4 °C.

6. Re-suspend in 1 mL of F-media. Add more F-media if the solution is cloudy.

7. Determine total isolated cell count (*see* **Note 15**).

3.4.3 Cell Seeding onto Double Collagen-Coated Transwell Membrane Inserts

1. Pre-warm PneumaCult-ALI complete medium to 37 °C before seeding. It is easier to seed onto the transwell plate if the media is separated into apical and basolateral media. Basolateral compartments need 1.2 mL, and the apical compartment at the time of seeding needs 300 μL.

2. It is optimal to be able to seed 8×10^4 cells/transwell. To calculate how much of your cell mixture is needed to seed a full 12-well transwell plate, divide how many cells you need by how many cells you have. Then multiply that number by how much media you added to count (*see* **Note 16** for a sample calculation).

3. Once the cell suspension is made, slowly seed the cells onto the polyester membrane. Add 1.2 mL of medium to the basolateral chamber (*see* **Note 17**).

4. Check the cells every day to make sure that they are proliferating.

5. Every 48 h change the medium by removing the medium from the basolateral compartment with a vacuum aspirator or a P1000 pipette. Add 1.2 mL of fresh PneumaCult-ALI complete medium to the basolateral chamber. Slowly remove the medium from the apical compartment being careful not to touch the tip of the pipette to the polyester membrane, as this might scrape some of the cells off. Add 300 μL of fresh PneumaCult-ALI complete medium to the apical chamber slowly.

6. After the cells have become 100% confluent on the polyester membrane, it is time to transfer them to ALI where they can

begin to differentiate. When the cells go into ALI, PneumaCult-ALI complete medium with mEGF is still used, but the Y inhibitor is not added.

7. Remove the medium in the basolateral chamber and add 1.2 mL of fresh PneumaCult-ALI complete medium. Slowly remove the medium in the apical chamber, and add 50 μL of fresh PneumaCult-ALI complete medium. The cells are now in ALI.

3.4.4 Freezing Cells to Create a Mouse Airway Epithelial Cell Bank

If extra cells are left over after seeding onto the transwell membrane, the cells can be frozen in aliquots that can be expanded for future experiments. The number of cells per aliquot is based on the researcher's preference. Each aliquot has a total volume of 1 mL of freezing medium.

3.5 IL-13 Treatment of Well-Differentiated Mouse Airway Epithelial Cells

To illustrate the utility of airway epithelial cells at ALI, we describe IL-13 treatment and analysis of airway epithelial cells. IL-13 can be added to cells to create an environment that mimics Th2 high asthma. IL-13 up-regulates the production of mucus by increasing the amount of MUC5ac being produced [7, 12].

3.5.1 IL-13 Treatment

1. Begin IL-13 treatment when ALI is started so that the mouse tracheal epithelial cells can begin to mimic Th2 high asthma. Add IL-13 to the medium to reach a final concentration of 10 ng/mL. This cytokine must be added to both the apical and basolateral chambers of the wells. Medium containing IL-13 is added the same as if it was just a normal medium change.

2. At every medium change, IL-13 must be added to the medium and added to the appropriate wells.

3.5.2 Outcome Analysis

1. Analyze cells at three different time points: 7 days, 14 days, and 21 days post IL-13 treatment.

2. Wash the apical supernatant with 1× PBS three times to remove mucus.

3. Rinse the bottom of the chamber to remove any excess medium.

4. Move wells that will be used for histology to observe the progression of differentiation and cilia development into another new plate (Fig. 6).

5. Move wells that will be used for fluorescent staining to observe the production of mucus into a new plate (Fig. 7).

6. The basolateral medium can be collected for any additional analyses.

3.5.3 Evaluating Cell Morphology Using H&E Staining

Staining samples with hematoxylin and eosin (H&E) allows the progress of mucociliary differentiation and proliferation to be observed.

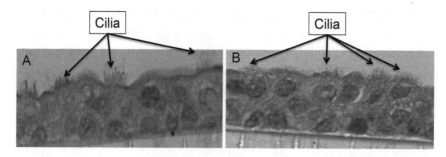

Fig. 6 Cultured mouse tracheal epithelial cells showing ciliary differentiation. Stained with hematoxylin and eosin, 400× magnification (**a**) 14 days post air–liquid interface. (**b**) 21 days post air–liquid interface

1. Fix by adding 1.2 mL of 10% formalin to the basolateral chamber and 300 µL to the apical chamber.

2. Seal the plate with parafilm.

3. Incubate at 4 °C overnight.

4. Process the samples for dehydration in alcohol, clearing in xylene, and infiltration in paraffin wax.

5. Embed in a paraffin block and cut into a 5 µM thick section.

6. Multiple slides can be made so that one can be stained with H&E and another can be left blank so immunohistochemistry (IHC) can be performed.

The H&E stained slides of each time point reveal the progression of ciliary differentiation. The untreated samples show cilia development as of day 14 and day 21 (Fig. 6a, b).

3.5.4 Analyzing Cells Using Fluorescent Staining

Fluorescent staining can be used to characterize things such as mucin expression, cell junctions, and other cell markers. Staining can be performed on fresh cells or paraffin-embedded samples, as above.

1. To stain fresh cells, wash them three times with 1× PBS to remove the mucus.

2. Fix with 10% formalin for 10 min at room temperature.

3. Wash three times with PBS.

4. Permeabilize with 0.1% Triton X-100 in PBS for 10 min at room temperature.

5. Wash three times with PBS.

6. Add the primary antibody; incubate for 1 h at room temperature.

7. Wash three times with PBS.

8. Add the secondary antibody; incubate for 1 h at room temperature.

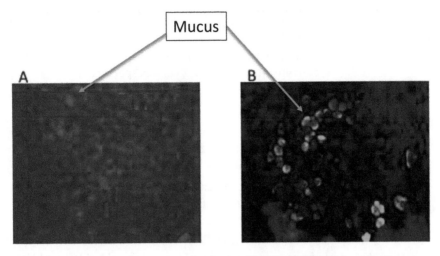

Fig. 7 IL-13 (10 ng/mL) induced mucus in mouse tracheal epithelial cells. MUC5AC marker is green, and nuclear DNA is stained with DAPI (blue). (**a**) Day 14 post air–liquid interface. (**b**) Day 21 post air–liquid interface

9. Wash three times with PBS.

10. Add DAPI for 1 min.

11. Wash three times with PBS.

12. Add fluorescent mounting medium to a charged slide.

13. Using a surgical scalpel, remove the polyester membrane from the insert.

14. Add the polyester membrane to the fluorescent mounting medium.

15. Add another drop of fluorescent mounting medium.

16. Add a cover slip, store at 4 °C, and protect from light.

17. The production of mucus can be seen in Fig. 7.

4   Notes

1. To avoid repeated freeze-thawing, make 1 mL aliquots of the protease/PBS mix. Do not sterile filter the protease/PBS mix.

2. Do not add hydrocortisone. This culture medium is good for up to 2 months after adding BulletKit.

3. Make sure the trachea pieces are immersed in the stock medium with 10% FBS while scraping.

4. The brush needs to be submerged in PBS.

5. Collagen coating on transwell membrane inserts takes approximately 1 h.

6. Occasionally agitating or turning the dish will help identify detaching cells.

7. If it appears that more than 10% of the cells are attached, re-trypsinize remaining cells as specified above.

8. For human airway epithelial cells growing on transwells, it usually takes 7 days for cells to reach 100% confluence on transwell inserts.

9. It is important to add medium slowly. If you add medium onto the fibroblasts too fast, you can dislodge the fibroblasts.

10. If the mouse tracheal epithelial cells are overgrown, then they will not adhere to the collagen-coated polyester membrane, and they will eventually die before they have a chance to reach ALI. Thus, 90% is the maximum confluence they should reach.

11. If needed, adjust the pH by adding 1N HCl or 1N NaOH.

12. If the plate will not be used within a few hours of coating, seal the plate with parafilm and store at 4 °C.

13. Do not be gentle on this step; if you are too gentle, then fibroblasts will remain adhered to the plate. Use a serological pipette and blow the fibroblasts off.

14. If you hold the dish a certain way, you can see the colonies that are adhered to the plate. When you are blowing the epithelial cells off, it will look like colonies are still adhered. The healthy cells will blow off easier, so if you are still blowing cells after 1–2 min, the ones that are still adhered are not healthy and should not be used.

15. After 25 μL has been removed for counting, store the rest of the cell suspension on ice.

16. Sample calculation: A total of 1.62×10^6 cells were counted, and they are re-suspended in 2 mL of media. A total of 9.6×10^5 cells are needed to seed a full 12-well plate. $(9.6 \times 10^5)/(1.62 \times 10^6) = 0.593$. 0.593×2 mL $= 1.185$ mL. This means that to seed a full 12-well transwell plate, 1.185 mL of your cell suspension needs to be added to 2.415 mL of PneumaCult-ALI complete medium so that a total of 300 μL can be added to the apical compartment of each transwell.

17. It is easiest to add 1.2 mL to the basolateral chamber by adding 600 μL of medium twice.

Acknowledgments

The authors thank Max A. Seibold, Reem Al Mubarak, Nicole Roberts, and Reena Berman for their technical assistance in cell culture methodology.

References

1. Ghio AJ, Dailey LA, Soukup JM, Stonehuerner J, Richards JH, Devlin RB (2013) Growth of human bronchial epithelial cells at an air–liquid interface alters the response to particle exposure. Part Fibre Toxicol 10:25

2. Derichs N, Jin BJ, Song Y, Finkbeiner WE, Verkman AS (2011) Hyperviscous airway periciliary and mucous liquid layers in cystic fibrosis measured by confocal fluorescence photobleaching. FASEB J 25:2325–2332

3. Hackett TL, Singhera GK, Shaheen F, Hayden P, Jackson GR, Hegele RG et al (2011) Intrinsic phenotypic differences of asthmatic epithelium and its inflammatory responses to respiratory syncytial virus and air pollution. Am J Respir Cell Mol Biol 45:1090–1100

4. Schneider D, Ganesan S, Comstock AT, Meldrum CA, Mahidhara R, Goldsmith AM et al (2010) Increased cytokine response of rhinovirus-infected airway epithelial cells in chronic obstructive pulmonary disease. Am J Respir Crit Care Med 182:332–340

5. Fulcher ML, Gabriel S, Burns KA, Yankaskas JR, Randell SH (2005) Well-differentiated human airway epithelial cell cultures. Methods Mol Med 107:183–206

6. Pezzulo AA, Starner TD, Scheetz TE, Traver GL, Tilley AE, Harvey BG (2011) The air–liquid interface and use of primary cell cultures are important to recapitulate the transcriptional profile of in vivo airway epithelia. Am J Physiol Lung Cell Mol Physiol 300:L25–L31

7. Robinson D, Humbert M, Buhl R, Cruz AA, Inoue H, Korom S (2017) Revisiting Type 2-high and Type 2-low airway inflammation in asthma: current knowledge and therapeutic implications. Clin Exp Allergy 47:161–175

8. Rock JR, Randell SH, Hogan BL (2010) Airway basal stem cells: a perspective on their roles in epithelial homeostasis and remodeling. Dis Model Mech 3:545–556

9. Davidson DJ, Gray MA, Kilanowski FM, Tarran R, Randell SH, Sheppard DN et al (2004) Murine epithelial cells: isolation and culture. J Cyst Fibros 3(Suppl 2):59–62

10. Horani A, Dickinson JD, Brody SL (2013) Applications of mouse airway epithelial cell culture for asthma research. Methods Mol Biol 1032:91–107

11. You Y, Richer EJ, Huang T, Brody SL (2002) Growth and differentiation of mouse tracheal epithelial cells: selection of a proliferative population. Am J Physiol Lung Cell Mol Physiol 283:L1315–L1321

12. Qin Y, Jiang Y, Sheikh AS, Shen S, Liu J, Jiang D (2016) Interleukin-13 stimulates MUC5AC expression via a STAT6-TMEM16A-ERK1/2 pathway in human airway epithelial cells. Int Immunopharmacol 40:106–114

Chapter 9

Isolation and Characterization of Human Lung Myeloid Cells

Yen-Rei A. Yu and Robert M. Tighe

Abstract

Multiparameter flow cytometry of human lungs allows for characterization, isolation, and examination of human pulmonary immune cell composition, phenotype, and function. Here we describe an approach to process lung tissues and then utilize a base antibody panel to define all of the major immune cell types in a single staining condition. This base antibody panel can also be used to identify major immune cell types in human blood and bronchoalveolar lavage (BAL) fluid.

Key words Bronchoalveolar lavage, Flow cytometry, Immune cells, Lymphocytes, Granulocytes, Mononuclear phagocytes, Macrophages, Monocytes, Dendritic cells, Immunophenotyping

1 Introduction

Over the past several decades, flow cytometry has become an essential immunologic tool that enables rapid identification of cell types and their functions [1–6]. To perform flow cytometry, single cells are tagged with fluorescent compounds/proteins or fluorescently labeled antibodies. Following labeling, single cells are excited by a sequence of lasers. Emitted signals collected by detectors are processed to yield detailed information regarding cell size, granularity, cell cycle, viability, and antigen expression density and patterns [7, 8]. This information can then be deciphered to identify specific immune cells, their subsets, and activation states [9]. Multiparameter flow cytometry represents a newer advance that enables simultaneous examination of multiple cell types and provides more comprehensive insight into immune cell repertoire and functions [10].

Here we present a method for tissue processing and describe a base antibody panel to simultaneously characterize all of the major immune cell types in human blood, bronchoalveolar lavage (BAL), and lung tissue [11, 12]. Cell suspensions can be analyzed in a

Scott Alper and William J. Janssen (eds.), *Lung Innate Immunity and Inflammation: Methods and Protocols*,
Methods in Molecular Biology, vol. 1809, https://doi.org/10.1007/978-1-4939-8570-8_9,
© Springer Science+Business Media, LLC, part of Springer Nature 2018

single tube to provide relative quantification of as many as 12 immune cell types including lymphocytes, granulocytes, and mononuclear phagocytes (i.e., monocytes, macrophages, and dendritic cells) [11]. Depending on the capacity of the flow cytometer, this base panel can be expanded to examine specific cell subsets of interest to the individual investigator.

2   Materials

2.1   Equipment

1. Dissection scissors.
2. Pair of very skinny sharp pointed tweezers.
3. Filter bottle, 0.2 μm.
4. 60 mm × 15 mm plastic dishes.
5. 50 mL conical tubes.
6. Cell strainer, 70 μm.
7. 3 mL syringe with rubber top plunger.
8. 96-well U-bottom plate.
9. Flow cytometer-compatible tubes.
10. Multichannel pipette with pipette tips.
11. Paraffin.
12. Aluminum foil.
13. Shaking incubator.
14. 4 °C centrifuge with adaptor for 96-well plates.

2.2   Stock Solutions

All solutions should be prepared in deionized water and, when indicated, sterile filtered. Use appropriate precautions for handling and disposing of hazardous and biologic materials. Stock solutions can be prepared and stored. However, working solutions should be freshly prepared on the day of the experiment.

1. Phosphate-buffered saline (PBS), 1X: 137 mM NaCl, 2.7 mM KCl, 4.3 mM Na_2HPO_4, 1.47 mM KH_2PO_4, pH 7.4, sterile filtered.

2. Hank's balanced salt solution (HBSS) without calcium or magnesium: 0.137 M NaCl, 5.4 mM KCl, 0.25 mM Na_2HPO_4, 1 g/L glucose, 0.44 mM KH_2PO_4, and 4.2 mM $NaHCO_3$, sterile filtered.

3. ACK red blood cell (RBC) lysis solution 1X: 8.29 g NH_4Cl, 1 g $KHCO_3$, 37.2 mg Na_2EDTA in 1 L deionized H_2O, pH 7.2–7.4, sterile filtered.

4. EDTA (Ethylenediaminetetraacetic acid) 120 mM, sterile filtered.

5. HEPES (4-(2-hydroxyethyl)-1-piperazineethanesulfonic acid) 1 M, pH 7.4.

6. Trypan blue.

7. Collagenase A (Roche) 10 mg/mL in PBS (*see* **Note 1**).

8. DNase I, grade II from bovine pancreas 40×10^3 Kunitz unit/mL (*see* **Note 2**).

9. Fetal calf serum (FCS), heat inactivated (*see* **Note 3**).

10. Bovine serum albumin (BSA), grade V.

11. Viability dye (*see* **Note 4**).

12. Base solution: HBSS 1X, 5% FCS, and 10 mM HEPES, sterile filter, store at 4 °C.

13. Re-suspension solution: 1% BSA, 5 mM EDTA, 10 mM HEPES in 1X PBS, sterile filter, store at 4 °C.

14. Wash solution: 3% FCS, 10 mM EDTA, 10 mM HEPES in 1× PBS, sterile filter, store at 4 °C.

15. Normal mouse serum (*see* **Note 2**).

16. Normal rat serum (*see* **Note 2**).

17. Normal human serum (*see* **Note 2**).

18. Purified rat antihuman CD32 (Clone FUN-2) 0.5 mg/mL.

19. Paraformaldehyde 16%.

2.3 Working Solutions

1. Digestion solution: 1.5 mg/mL collagenase A, 800 Kunitz unit/mL DNase I, in base solution.

2. Blocking solution: 5% normal mouse serum, 5% normal human serum, 5 µg/mL antihuman CD32, in wash solution.

3. Paraformaldehyde 3% in PBS.

4. Paraformaldehyde 0.4% in PBS.

5. Fluorochrome-conjugated antibodies (Table 1).

3 Methods

3.1 Cell Preparation

1. Divide human lung tissue into ~2 × 3 cm pieces.

2. Cut each 2 × 3 cm lung piece into 2–3 mm pieces and place the small pieces into 10 mL of digestion solution in 50 mL conical tube.

3. Incubate at 37 °C with continuous agitation (in shaking incubator) at ~250 rpm for 30 min.

4. Remove specimen from incubator every 10 min and vortex briefly at maximum speed.

Table 1
Base backbone antibody panel

Antibody	Clone	Isotype	Conjugate
CD11c	3.9	Mouse IgG1, κ	BV421
CD14	M5E2	Mouse IgG2a, κ	BV785
CD16	3G8	Mouse IgG1, κ	BV711
CD24	ML5	Mouse IgG2a, κ	PerCp-Cy5.5
CD45	HI30	Mouse IgG1, κ	AF700
CD123	6H6	Mouse IgG1, κ	BV650
CD169	7-239	Mouse IgG1, κ	PE
CD206	15-2	Mouse IgG1, κ	AF488
HLA-DR	LN3	Mouse IgG2b, κ	PECy7

(Reprinted with permission of the American Thoracic Society. Copyright© 2017 American Thoracic Society. Yu et al. Flow cytometric analysis of myeloid cells in human blood, bronchoalveolar lavage, and lung; Am J Respir Cell Mol Biol. 2016 Jan;54(1):13–24. The *American Journal of Respiratory Cell and Molecular Biology* is an official journal of the American Thoracic Society)

5. After 30 min incubation, pour tissues and digestion solution into a 60 mm × 15 mm dish. Gently tease tissues apart with a pair of very skinny sharp pointed tweezers.

6. Return the tissue and digestion solution back into the 50 mL conical tube. Rinse the dish with 3–5 mL of digestion solution and transfer the wash solution to the 50 mL conical tube.

7. Incubate tissue and digestion solution at 37 °C for an additional 20 min. Briefly vortex every 10 min at maximum speed. Once tissue is sufficiently digested (*see* **Note 5**), dilute to 30 mL with PBS. Vortex vigorously at maximum speed continuously for 30 s. Place the solution on ice.

8. Strain tissue digest through a 70 μm cell strainer into a new 50 mL conical tube. While straining, mix and grind tissues with a rubber tip plunger from a 3 mL syringe.

9. Wash the conical tube and strainer with 5 mL of PBS and add the wash to the new 50 mL conical tube.

10. Centrifuge strained cells at ~300 × g for 7 min at 4 °C. Discard the supernatant by decanting.

11. Re-suspend the cell pellet in 5 mL ACK RBC lysis buffer. Incubate on ice for 5 min. After 5 min, immediately dilute cells with 30 mL of cold PBS.

12. Centrifuge cell solution at 300 × g for 7 min at 4 °C. Decant supernatant.

13. Re-suspend cells in 5 mL of re-suspension solution by pipetting up and down.

14. Count cells and assess viability with trypan blue.

3.2 Cell Viability Staining

1. Transfer 1×10^6 cells/sample to 96-well U-bottom plate.

2. Wash cells two times with PBS by centrifugation at $300 \times g$ for 7 min at 4 °C. Discard the supernatant by decanting following each washing.

3. Re-suspend cells in 300 μL of PBS.

4. Add live/dead discrimination dye at appropriate concentration (*see* **Note 4**).

5. Incubate at room temperature in the dark for 20 min.

6. Centrifuge plate at ~$300 \times g$ at 4 °C. Decant supernatant (*see* **Note 6**).

3.3 Cell Fixation

1. Wash the cells once with PBS to remove live/dead discrimination dye. Centrifuge plate at ~$300 \times g$ at 4 °C. Decant supernatant.

2. Re-suspend cells in 125 μL of PBS, and then add 125 μL of 3% paraformaldehyde to each well. Incubate the plate for 10 min at 4 °C.

3. Wash cells twice with 200 μL of PBS. Following each wash, centrifuge plate at ~$300 \times g$ at 4 °C and then decant the supernatant.

4. Re-suspend cell pellet in 200 μL of staining solution. Wrap plate with paraffin and aluminum foil. Store cells at 4 °C until ready to proceed to surface marker staining. Cells can be stored up to 3 days. Longer storage time will need to be tested by the investigator.

5. Prior to surface staining, centrifuge cells at $300 \times g$ for 5 min at 4 °C. Decant supernatant.

3.4 Surface Marker Staining

1. Re-suspend cells in 100 μL of blocking solution in each well. Mix and incubate for 5–7 min at 4 °C.

2. Add base panel antibodies to each well (*see* Table 1 and **Note 4**).

3. Incubate cells in the dark at 25 °C for 30 min. Centrifuge at $300 \times g$ for 3–5 min. Decant the supernatant.

4. Wash cells twice with wash solution.

5. Re-suspend cells in 200 μL of 0.4% paraformaldehyde in PBS. Transfer cells to flow cytometer compatible tubes.

3.5 Flow Cytometry Acquisition Tips

1. Antibody panel design and data acquisition are highly dependent on the individual investigator's flow cytometer configuration. Of critical importance are the laser and filter configurations.

Table 2
Sample flow cytometry configuration

BD LSRII configuration details					
	Detector name	PMT	Long pass mirror	Band pass filter	Common fluor
488 nm Blue laser					
Sapphire, 20 mW	B-685LP_695_40	A	685	695_40	PerCP-Cy5.5
	B-495LP_525_50	B	495	525_50	FITC
	B-488_10	C	–	488_10	Side scatter (SSC)
532 nm Green laser					
Compass, 150 mW	G-735LP_780_60	A	735	780_60	PE-Cy7
	G-695LP_710_50	B	695	710_50	PE-Cy5.5
	G-655LP_660_20	C	655	660_20	PE-CY5
	G-600LP_610_20	D	600	610_20	PE-CF594
	G-550LP_575_26	E	550	575_26	PE
405 nm Violet laser					
Cube, 50 mW	V-750LP_780_60	A	750	780_60	BV786
	V-710LP_740_20	B	710	740_20	BV745
	V-695LP_710_50	C	695	710_50	BV711
	V-635LP_670_30	D	635	670_30	BV650
	V-595LP_610_20	E	595	610_20	BV605
	V-555LP_560_40	F	555	560_40	BV570
	V-505LP_515_20	G	505	515_20	BV510
	V-450_50	H		450_50	BV421
639 nm Red laser					
Uniphase, 70 mW	R-735LP_780_60	A	735	780_60	APC-Cy7
	R-690LP_730_45	B	690	730_45	AF700
	R-670 30	C	–	670 30	APC
Diode				Forward scatter (FSC)	Gain 175

Included is a sample configuration for BD LSRII flow cytometer with four lasers. This configuration is compatible with antibody and fluorochrome combination listed in Table 1

(Reprinted with permission of the American Thoracic Society. Copyright© 2017 American Thoracic Society. Yu et al. Flow cytometric analysis of myeloid cells in human blood, bronchoalveolar lavage, and lung; Am J Respir Cell Mol Biol. 2016 Jan;54(1):13–24. The *American Journal of Respiratory Cell and Molecular Biology* is an official journal of the American Thoracic Society)

As a reference, our cytometer and its configuration are included in Table 2. Investigators should become familiar with their individual flow cytometer before planning an experiment and perform appropriate testing of antibodies and configuration.

2. Care with cell preparation and digestion is essential for generating accurate and reproducible results. In addition, the use of a viability dye to eliminate dead cells is critical to minimize nonspecific signals. This is especially important for digested tissue specimens, because they contain significant amounts of dead cells and debris, which can increase nonspecific staining and background fluorescence.

3. Because human lung tissue contains cells with varied autofluorescent characteristics, photomultiplier tube (PMT) voltages and signal compensation should be performed using cells derived from the human lung. Compensation beads can be used to complement human lung cells in setting up the instrument but should not be a sole method of compensation.

4. When acquiring samples containing complex cell sizes and shapes, auto-compensation should to be avoided. Manual compensation is recommended. An experienced operator with knowledge of cell distribution pattern will ensure accurate and reproducible data.

5. To aid cell type identification, a representative gating strategy for human lung myeloid cells is shown in Fig. 1.

4   Notes

1. Collagenase A can be obtained from Roche, reconstituted in PBS, aliquoted and stored in −80 °C. Repeated freezing and thawing should be avoided, as the enzymatic activity will be reduced.

2. DNase I, normal mouse serum, normal rat serum, and normal human serum should be reconstituted in molecular grade water, aliquoted and stored at −20 °C. Repeated freezing and thawing should be avoided.

3. Fetal calf serum should be heat-inactivated by heating the serum to 56 °C for 30 min. Serum should be divided into aliquots and stored at −20 °C.

4. The final dilution of viability dye and antibodies are dependent on the source company. Dye and antibody dilutions should be tested by investigators during panel development. A list of tested reagents is included in the Supplemental Materials section in the following reference (11).

5. The teasing and incubation process may need to be repeated for fibrotic tissues.

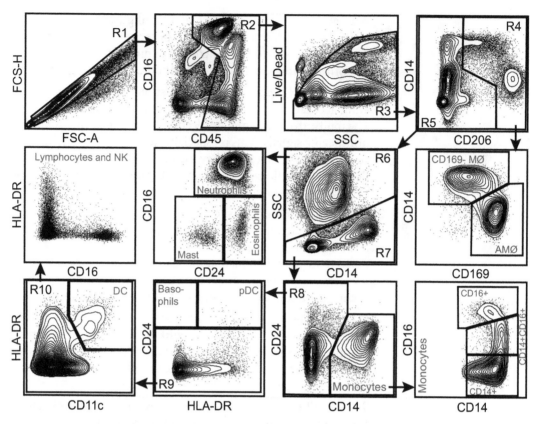

Fig. 1 Gating strategy for human lung myeloid cells. Single cells are identified by comparing forward-scatter area (FSC-A) to forward-scatter height (FSC-H). After identification of singlets (R1), proceed to delineate major human lung immune cells as CD45$^+$CD16$^+$ (R2). Dead cells are excluded using SSC-A vs. live/dead viability stain (R3). Macrophages are identified as CD206$^+$CD14$^+$ (R4); and subpopulations macrophages are defined as CD169$^+$CD14$^{-/lo}$ alveolar macrophages (AMØ) or CD169^{lo}CD14$^+$ interstitial macrophages (CD169− MØ). Other non-macrophage immune cells are defined as shown in R5. Granulocytes, including neutrophils, mast cells, and eosinophils are identified by their high SSC (R6), followed by their differential CD24 and CD16 expressions. Remainder of low SSC immune cells (R7) including monocyte subsets, lymphocytes, and dendritic cells are identified as demonstrated. (Reprinted with permission of the American Thoracic Society. Copyright$^©$ 2017 American Thoracic Society. Yu et al. Flow cytometric analysis of myeloid cells in human blood, bronchoalveolar lavage, and lung; Am J Respir Cell Mol Biol. 2016 Jan;54(1):13–24. The *American Journal of Respiratory Cell and Molecular Biology* is an official journal of the American Thoracic Society)

6. After live/dead discrimination staining, investigators can either proceed directly to cell surface staining (*see* Subheading 3.4) or the cells can be lightly fixed and stored for up to 3 days (*see* Subheading 3.3).

Acknowledgment

The work is supported by NIH grants K08 HL121185, K08 HL105537, Mandel Fellowship Award, and Proof of Concept Award from the Pulmonary Hypertension Association.

References

1. Desch AN, Gibbings SL, Goyal R, Kolde R, Bednarek J, Bruno T, Slansky JE, Jacobelli J, Mason R, Ito Y, Messier E, Randolph GJ, Prabagar M, Atif SM, Segura E, Xavier RJ, Bratton DL, Janssen WJ, Henson PM, Jakubzick CV (2016) Flow Cytometric analysis of mononuclear phagocytes in nondiseased human lung and lung-draining lymph nodes. Am J Respir Crit Care Med 193:614–626

2. Mandruzzato S, Brandau S, Britten CM, Bronte V, Damuzzo V, Gouttefangeas C, Maurer D, Ottensmeier C, van der Burg SH, Welters MJP, Walter S (2016) Toward harmonized phenotyping of human myeloid-derived suppressor cells by flow cytometry: results from an interim study. Cancer Immunol Immunother 65:161–169

3. Misharin AV, Morales-Nebreda L, Mutlu GM, Budinger GR, Perlman H (2013) Flow cytometric analysis of macrophages and dendritic cell subsets in the mouse lung. Am J Respir Cell Mol Biol 49:503–510

4. Bharat A, Bhorade SM, Morales-Nebreda L, McQuattie-Pimentel AC, Soberanes S, Ridge K, DeCamp MM, Mestan KK, Perlman H, Budinger GRS, Misharin AV (2016) Flow Cytometry reveals similarities between lung macrophages in humans and mice. Am J Respir Cell Mol Biol 54:147–149

5. De Grove KC, Provoost S, Verhamme FM, Bracke KR, Joos GF, Maes T, Brusselle GG (2016) Characterization and quantification of innate lymphoid cell subsets in human lung. PLoS One 11:e0145961

6. Freeman CM, Crudgington S, Stolberg VR, Brown JP, Sonstein J, Alexis NE, Doerschuk CM, Basta PV, Carretta EE, Couper DJ, Hastie AT, Kaner RJ, O'Neal WK, Paine R 3rd, Rennard SI, Shimbo D, Woodruff PG, Zeidler M, Curtis JL (2015) Design of a multi-center immunophenotyping analysis of peripheral blood, sputum and bronchoalveolar lavage fluid in the Subpopulations and Intermediate Outcome Measures in COPD Study (SPIROMICS). J Transl Med 13:19

7. Adan A, Alizada G, Kiraz Y, Baran Y, Nalbant A (2017) Flow cytometry: basic principles and applications. Crit Rev Biotechnol 37:163–176

8. Pozarowski P, Darzynkiewicz Z (2004) Analysis of cell cycle by flow cytometry. Methods Mol Biol 281:301–311

9. Baharom F, Thomas S, Rankin G, Lepzien R, Pourazar J, Behndig AF, Ahlm C, Blomberg A, Smed-Sörensen A (2016) Dendritic cells and monocytes with distinct inflammatory responses reside in lung mucosa of healthy humans. J Immunol 196:4498–4509

10. Ghanekar SA, Maecker HT (2003) Cytokine flow cytometry: multiparametric approach to immune function analysis. Cytotherapy 5:1–6

11. Yu YR, Hotten DF, Malakhau Y, Volker E, Ghio AJ, Noble PW, Kraft M, Hollingsworth JW, Gunn MD, Tighe RM (2016) Flow Cytometric analysis of myeloid cells in human blood, Bronchoalveolar lavage, and lung tissues. Am J Respir Cell Mol Biol 54:13–24

12. Yu Y-RA, O'Koren EG, Hotten DF, Kan MJ, Kopin D, Nelson ER, Que L, Gunn MD (2016) A protocol for the comprehensive flow Cytometric analysis of immune cells in normal and inflamed murine non-lymphoid tissues. PLoS One 11:e0150606

Measurement of Protein Permeability and Fluid Transport of Human Alveolar Epithelial Type II Cells Under Pathological Conditions

Xiaohui Fang and Michael A. Matthay

Abstract

Alveolar epithelial barrier dysfunction contributes to the influx of protein-rich edema fluid and the accumulation of inflammatory cells in the pathogenesis of acute lung injury (ALI) and acute respiratory distress syndrome (ARDS). To study the alveolar epithelial barrier function under pathological conditions, we developed an in vitro model of acute lung injury using cultured human alveolar epithelial type II (ATII) cells. Here we describe the methods that we use to measure protein permeability and fluid transport across human ATII cell monolayers under stimulated conditions. Both proinflammatory cytokines and edema fluid from ALI/ARDS patients can increase protein permeability and decrease fluid transport across the human ATII cells monolayer.

Key words Alveolar epithelia, Fluid transport, Protein permeability, Transwell, Inflammatory cytokines, Acute lung injury, ARDS

1 Introduction

Impaired alveolar fluid clearance and loss of epithelial integrity are common characteristics among patients with acute lung injury (ALI) and the acute respiratory distress syndrome (ARDS) [1]. The loss of epithelial integrity in ALI/ARDS is of critical importance, because the epithelium forms a tighter monolayer than the endothelium under normal circumstances [2]. Epithelial barrier dysfunction contributes to influx of protein-rich edema fluid into the injured alveoli and the accumulation of inflammatory cells. During alveolar injury, the capacity of alveolar epithelial type II (ATII) cells to remove pulmonary edema fluid (alveolar fluid clearance) is also impaired. This results in further extravascular lung water accumulation and is associated with higher mortality [3].

Human ATII cells are isolated from human lungs that are not used for transplantation as described in Chap. 7. After a series of

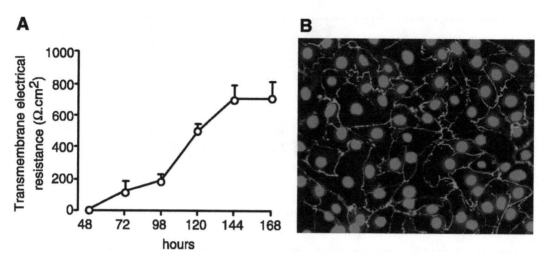

Fig. 1 Electrophysiological properties of cultured human ATII cells. (**a**) Time course of transmembrane electrical resistance of human ATII cells growing on collagen I-coated Transwell membranes. Data are mean ± SEM. (**b**) Immunostaining of zonula occludens (ZO)-1 protein in human ATII cells (red). Cell nuclei were counterstained with 4′,6-diamidino-2-phenylindole (blue). (This research was originally published in American Journal of Physiology Lung Cellular and Molecular Physiology. Fang X, Song Y, Hirsch J et al. Contribution of CFTR to apical-basolateral fluid transport in cultured human alveolar epithelial type II cells. Am J Physiol Lung Cell Mol Physiol 290:L242–249, 2005)

enzyme digestion and purification steps, human ATII cells are collected and seeded at a density of 1×10^6 cells/well on collagen I-coated Transwell plates with a pore size of 0.4 μm and a surface area of 0.33 cm². The cells are cultured in a 37 °C, 5% CO_2 incubator. Cells normally reach confluence after 48 h. At 72 h, the fluid in the top compartment is removed, and the cells are then left to grow in an air-liquid interface. Measurement of fluid transport and protein permeability from the apical to basolateral membrane of the type II cell monolayers is carried out 120–144 h following the initial isolation and 48–72 h after the air-liquid interface is achieved. We have found that transmembrane electrical resistance reached a plateau at 120–144 h with morphological evidence of tight junctions [4] (Fig. 1a, b). Generally, cell preparations with transmembrane electrical resistance of <700 ohms are excluded from the experiments.

We and others have reported pulmonary edema fluid isolated during the early phase of ARDS contains high levels of proinflammatory cytokines, including IL-1β, IL-8, and TNFα [5, 6]. To mimic the pathophysiological condition of ARDS, we measure the net fluid transport and protein permeability across human ATII cell monolayers exposed to cytomix (a mixture of IL-1β, TNFα, and IFN-γ) or edema fluid from ARDS patients [7, 8]. Both proinflammatory cytomix and pulmonary edema fluid from ARDS patients significantly increase the protein permeability across human ATII cell monolayers

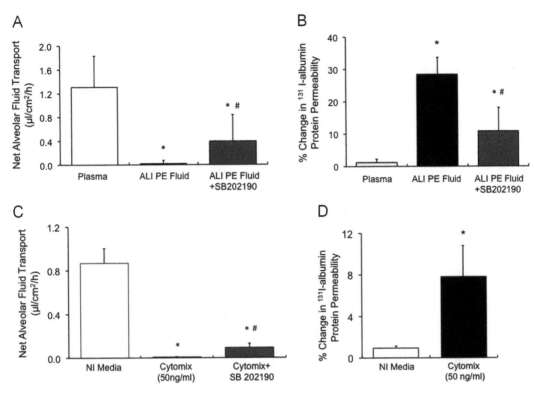

Fig. 2 ALI pulmonary edema fluid or cytomix diminishes net fluid transport and enhances paracellular protein permeability of human ATII cells. Human ATII cells cultured for 5 days (48 h after the air-liquid interface was achieved) were exposed to ALI pulmonary edema fluid, plasma from healthy donors, or cytomix for 24 h, and net fluid transport (**a** and **c**) and paracellular protein permeability (**b** and **d**) were monitored. Another set of human ATII cells were pretreated with 10.0 μM SB202190, a p38 MAPK inhibitor, for 30 min prior to exposure with ALI pulmonary edema fluid or cytomix. p38 MAPK inhibition prevented some of the damage induced by the ALI pulmonary edema fluid or cytomix. Each sample is run in triplicate. * = $p < 0.02$ compared with plasma controls; # = $p < 0.01$ compared with ALI pulmonary edema or cytomix-exposed cells. *PE* pulmonary edema, *NI* normal. (This research was originally published in Journal of Biochemistry. Lee JW, Fang X, Dolganov G et al. Acute lung injury edema fluid decreases net fluid transport across human alveolar epithelial type II cells. J Biol Chem 282 (33):24109–24,119, 2007)

and decrease the fluid transport by human ATII cells (Figs. 2 and 3). To demonstrate the utility of these assays, we show that inhibition of p38 MAPK activity (Fig. 2) or the addition of mesenchymal stem cells (Fig. 3) is able to restore epithelial cell integrity (i.e., decrease protein permeability and increase fluid transport).

2   Materials

1. Culture medium for ATII cells: DMEM high glucose 50%/F-12 50%. To this 50/50 mixture, add 2.438 g/L NaHCO$_3$, 3.151 g/L glucose, 0.055 g/L Na-pyruvate, 0.365 g/L L-

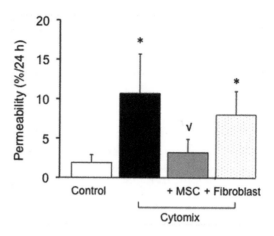

Fig. 3 Mesenchymal stem cells (MSC) diminish protein permeability across human ATII cells injured by cytomix. Epithelial permeability to protein was measured by the unidirectional flux of labeled ^{131}I-albumin from the upper to the lower compartment of the Transwell plate over 24 h. The addition of cytomix (50 ng/mL) increased epithelial protein permeability across the ATII cells. The simultaneous addition of MSC (but not normal human lung fibroblasts) to the bottom chamber restored alveolar epithelial barrier integrity. Data displayed are mean ± SD (% change/24 h, $n = 15$; * = $p < 0.02$ versus control; $\sqrt{}$ = $p < 0.03$ versus cytomix-injured). (This research was originally published in Journal of Biochemistry. Fang X, Neyrinck AP, Matthay MA et al. Allogeneic human mesenchymal stem cells restore epithelial protein permeability in cultured human alveolar type II cells by secretion of angiopoietin-1. J Biol Chem 285 (34):26,211–26,222, 2010)

glutamine, fetal bovine serum (FBS) (use refiltered, heat-inactivated FBS; add FBS to culture medium to make a final concentration of serum 10%), and antibiotics (Penicillin-streptomycin 1:100 dilution, gentamicin 1:1000 dilution, and amphotericin B 1:100 dilution). Insulin-transferrin-selenium (1:100 dilution in culture medium) can also be added to the culture medium.

2. Transwell plate: Prepare 24-well plate Transwell PTFE membrane, coated with collagen I. Pore size 0.4 μm, surface area 0.33 μm².

3. EVOMX: STX2 electrode set. Ag-AgCl electrode, sterilized with 70% alcohol for 10 min and then washed twice with PBS. The electrode can be dipped in the culture medium for 10 min for equilibration.

4. [125]-I albumin or [131]-I albumin stock: 100 mCi/10 mL. [125]-I albumin or [131]-I albumin is used as a volume tracer to be added to the top compartment. It is diluted to a final concentration of 0.3 μCi/mL after calculating the decay factor.

5. Undiluted pulmonary edema fluid is collected from ARDS and hydrostatic control patients and stored in cryovials.

6. Cytomix proinflammatory cytokine mixture: Add recombinant human IL-1β, TNFα, and IFN-γ to culture medium to achieve a final concentration of 50 ng/mL for each cytokine.

7. Hemocytometer for counting cell numbers.

8. Inverted microscope for observing cell growth.

3 Methods

Perform all procedures at room temperature inside a biosafety cabinet unless otherwise specified.

3.1 Culturing of Human ATII Cells

1. Resuspend freshly isolated human ATII cells in culture medium (*see* **Notes 1** and **2**). Count the total number of ATII cells with a hemocytometer, and calculate the cell concentration. Determine the volume of cell suspension that will be added to the top compartment of the Transwell so that each well has one million cells.

2. Bring the pre-hydrated collagen I-coated Transwell membrane into the biosafety cabinet (*see* **Note 3**). Add 600 μL of culture medium supplemented with 10% FBS in the bottom compartment of Transwell.

3. Add 1×10^6 human ATII cells (according to the calculation of Subheading 3.1, **step 1** above) into the top compartment of the Transwell. Let the cells sit for about 10 min, and then transfer the Transwell plate into a 37 °C incubator.

4. Check the cell growth under the microscope the next day. The culture medium will change into light orange color. Replace the culture medium in the bottom compartment with fresh medium only if necessary.

5. 48 h after cell seeding, gently wash the cells with fresh medium to remove unattached cells and debris. Replace the medium in the bottom compartment with fresh medium.

6. 72 h after cell seeding, wash the cells again with culture medium to remove debris or unattached cells. Aspirate all the fluid in the top compartment.

7. Monitor the cells daily. Remove the fluid in the top compartment of the Transwell every day if there is leakage of fluid from the bottom compartment. If the cell monolayers grow well, they will reach air-liquid interface 72–120 h after seeding (*see* **Note 4**).

3.2 Measurement of Transmembrane Electrical Resistance

1. Add 150 μL of culture medium back to the top compartment of the Transwell.

2. Put the long leg of EVOMX in the bottom compartment of the Transwell and the short leg in the top compartment. Be sure not to touch the cells.

3. Read the display of transmembrane electrical resistance (TER). Usually if the cells grow well, TER can be detected 48–72 h after seeding and reaches a plateau at 120–144 h.

3.3 Measurement of Fluid Transport and Protein Permeability

1. Gently remove fluid in the top compartment as in Subheading 3.1, **step 7** above until the cell monolayers reach air-liquid surface (*see* **Notes 5–7**).

2. The ATII cell monolayers are usually ready for experiments 120–144 h after seeding and 48–72 h after the air-liquid interface is achieved. Add 150 μL of culture medium (containing 0.3 μCi/mL ^{131}I-albumin or ^{125}I-albumin) to the top compartment of the Transwell, and add 600 μL of culture medium to the bottom compartment. Add cytomix to the culture medium as needed to arrive at a final concentration of 50 ng/mL for each cytokine. Alternatively, if using pulmonary edema fluid from an ARDS patient, add 150 μL of undiluted edema fluid (containing 0.3 μCi/mL ^{131}I-albumin or ^{125}I-albumin) to the top compartment, and add 600 μL of edema fluid to the bottom compartment (*see* **Note 8**).

3. Put the Transwell plate in a 37 °C, 5% CO_2 incubator with 100% humidity for 5 min. Then remove and save 20 μL of the medium from both the top and the bottom compartments as the initial samples.

4. Aspirate another 20 μL sample from the top compartment of the Transwell as the final sample after 24 h of incubation. Collect all the fluid in the bottom compartment. Weigh all samples, and monitor radioactivity in a gamma counter.

5. Calculate protein permeability and net fluid transport using the following formulae [7, 8] (*see* **Note 9**):

$$\text{Protein permeability} (\%) = 100 \times (\text{Rbf} - 29^* \text{Rbi}) / (6.5 \times \text{Rti})$$

where Rbf is the radioactivity of samples collected from bottom compartment at the end of the experiment, Rti is the radioactivity of the initial 20 μL sample from top compartment, and Rbi is the radioactivity of the initial 20 μL sample from bottom compartment.

$$\text{Net fluid transport} (\mu L/cm^2 / h) = [1 - (\text{Rti}/\text{Wti}) / (\text{Rtf}/\text{Wtf})] \times 130 / (24 \times 0.33)$$

where Rti is the radioactivity of the initial 20 μL sample from top compartment, Wti is the weight of the initial 20 μL sample from top compartment, Rtf is the radioactivity of the final 20 μL sample from top compartment, and Wtf is the weight of the final 20 μL sample from top compartment.

4   Notes

1. The right middle lobe or lingula is generally used to isolate human ATII cells. Ensure the lobe has no obvious injury, hemorrhage, or consolidation. Cells isolated within a short period of time after lung harvesting usually have better quality.

2. Adding insulin-transferrin-selenium to the culture medium in the first 2 days will help the cells grow better.

3. Pre-hydrate the Transwell membrane by adding culture medium 150 μL to the top and 600 μL to the bottom compartment before cell seeding. Pre-hydration will help the cells attach better.

4. Some wells are already dry after the fluid in the top compartment is removed at 72 h after seeding, so they may attain air-liquid interface earlier. However, some wells are still wet and might take a longer time to reach air-liquid interface.

5. Choose those wells that are dry in the top compartment for experiments. Check every well under the microscope to determine if there are any obvious holes in the monolayers before experiments. Exclude wells without intact monolayers.

6. Avoid washing the cells before experiments, because washing might cause a change in permeability and increase variation in the results.

7. If you need to measure TER before other experiments, pick a few wells for measurement, but try not to measure all of the wells before experiments for the same reason as in **Note 6**.

8. ^{125}-I albumin or ^{131}-I albumin is very sensitive and reliable tracer for measuring protein permeability. However, not every lab has a gamma counter to measure their activity. As an alternative, FITC-labeled albumin or dextran can be used.

9. Protein permeability is calculated according to the percentage of ^{131}I-albumin or ^{125}I-albumin across the epithelial monolayer (from top to bottom compartment). So the final amount of ^{131}I-albumin or ^{125}I-albumin recovered from the bottom compartment (less the initial amount in the bottom, usually close to zero) is divided by initial ^{131}I-albumin or ^{125}I-albumin in the top compartment. Fluid transport is calculated by comparing the fluid volume at the beginning and end of experiments in the top compartment over experimental time and surface area of the Transwell membrane.

Acknowledgment

This work was supported by NIH grant NHLBI R37HL51856.

References

1. Matthay MA, Ware LB, Zimmerman GA (2012) The acute respiratory distress syndrome. J Clin Invest 122(8):2731–2740. https://doi.org/10.1172/JCI60331

2. Bhattacharya J, Matthay MA (2013) Regulation and repair of the alveolar-capillary barrier in acute lung injury. Annu Rev Physiol 75:593–615. https://doi.org/10.1146/annurev-physiol-030212-183756

3. Ware LB, Matthay MA (2001) Alveolar fluid clearance is impaired in the majority of patients with acute lung injury and the acute respiratory distress syndrome. Am J Respir Crit Care Med 163(6):1376–1383. https://doi.org/10.1164/ajrccm.163.6.2004035

4. Fang X, Song Y, Hirsch J et al (2006) Contribution of CFTR to apical-basolateral fluid transport in cultured human alveolar epithelial type II cells. Am J Physiol Lung Cell Mol Physiol 290(2):L242–L249. https://doi.org/10.1152/ajplung.00178.2005

5. Pugin J, Verghese G, Widmer MC et al (1999) The alveolar space is the site of intense inflammatory and profibrotic reactions in the early phase of acute respiratory distress syndrome. Crit Care Med 27(2):304–312

6. Ware LB, Matthay MA (2000) The acute respiratory distress syndrome. N Engl J Med 342(18):1334–1349. https://doi.org/10.1056/NEJM200005043421806

7. Lee JW, Fang X, Dolganov G et al (2007) Acute lung injury edema fluid decreases net fluid transport across human alveolar epithelial type II cells. J Biol Chem 282(33):24109–24119. https://doi.org/10.1074/jbc.M700821200

8. Fang X, Neyrinck AP, Matthay MA et al (2010) Allogeneic human mesenchymal stem cells restore epithelial protein permeability in cultured human alveolar type II cells by secretion of angiopoietin-1. J Biol Chem 285(34):26211–26222. https://doi.org/10.1074/jbc.M110.119917

Chapter 11

Measuring Innate Immune Function in Mouse Mononuclear Phagocytes

John Matthew Craig and Neil Raj Aggarwal

Abstract

While serving as a conduit for gas exchange, the lung continually encounters potentially harmful airborne and bloodborne substances including particulate matter, allergens, toxins, and infectious agents. Resident alveolar and interstitial macrophages coordinate with neutrophils, dendritic cells, and recruited blood-derived monocytes to provide phagocytic host defense that aids in the removal and destruction of antigenic material following this myriad of exposures. Here we describe flow cytometric methods for specifically assessing phagocytic activity ex vivo in isolated mouse lung macrophages and monocytes utilizing fluorescently labeled *Streptococcus pneumoniae*.

Key words Macrophages, Monocytes, Bacteria, Innate immunity, Phagocytosis, *Streptococcus pneumonia*

1 Introduction

As key cellular components of the lung's innate immune system, mononuclear phagocytes such as monocytes, macrophages, and dendritic cells rapidly respond to and become activated by a series of conserved pathogen- and damage-associated molecular patterns (PAMPs and DAMPs) [1]. These molecular signatures are recognized by a variety of pattern recognition receptors (PRRs) including Toll-like receptors (TLRs), C-type lectins (CLRs), and scavenger receptors (SRs) that initiate signaling cascades, enhance phagocytosis, promote pathogen destruction, increase antigen presentation, and augment the transcription of inflammatory genes central to immune defense [2].

To promote phagocytosis, activated mononuclear phagocytes take advantage of both opsonic and non-opsonic pathways of antigen uptake that link this innate response to adaptive immunity through MHC presentation [3]. Opsonic phagocytosis utilizes Fc receptor- and complement receptor-mediated recognition of antigen complexed with antibody and/or complement components

Scott Alper and William J. Janssen (eds.), *Lung Innate Immunity and Inflammation: Methods and Protocols*,
Methods in Molecular Biology, vol. 1809, https://doi.org/10.1007/978-1-4939-8570-8_11,
© Springer Science+Business Media, LLC, part of Springer Nature 2018

including C3b. In contrast, non-opsonic mechanisms of antigen retrieval utilize integrins, lectins, and scavenger receptors to facilitate antigen engulfment. Once internalized, antigens encapsulated in phagosomes are delivered to lysosomes where fusion events trigger acidification, induction of the respiratory burst, and activation of hydrolytic enzymes that aid in pathogen degradation. Ultimately, phagocytosis and antigen processing enables the activation of CD4+ T cells through the canonical MHC II pathway and also elicits CD8+ T cell responses through cross-presentation. Given the critical role that mononuclear phagocytes play in coordinating both innate and adaptive immune responses to infection, the magnitude and quality of their phagocytic activity are often an important determinant of infection outcome and thus of interest experimentally. In particular, unlike dendritic cells that rapidly traffic to lung-draining lymph nodes following antigen exposure, recruited monocytes and resident macrophage populations are largely responsible for maintaining the in situ lung phagocyte system yet may differ in their phagocytic capacity. To this end, we outline standardized laboratory protocols to compare the phagocytic capacity of mouse lung macrophages and monocytes ex vivo using fluorescent *Streptococcus pneumoniae* challenge and flow cytometry.

2   Materials

2.1   Equipment

1. Dissecting scissors.
2. Forceps.
3. Surgical suture.
4. Hemocytometer.
5. Vacuum aspirator.
6. Micropipettors and tips.
7. Refrigerated centrifuge.
8. Spectrophotometer.
9. Flow cytometer/cell sorter.
10. 37 °C incubator infused with 5% CO_2.
11. Rodent anesthesia machine or other container for open-drop anesthetic exposure.

2.2   Disposable Supplies

1. 15 mL and 50 mL conical tubes.
2. 6-well tissue culture plates.
3. 18 gauge blunted needles with Luer lock.
4. 29 gauge needles.
5. 1 mL syringes.
6. 70 μm and 100 μm cell strainers.

2.3 Reagents

1. Lipopolysaccharides (LPS) purified from *Escherichia coli*.

2. *Streptococcus pneumoniae* stock.

3. Isoflurane.

4. Ketamine.

5. Xylazine.

6. Propylene glycol.

7. 1× phosphate buffered saline (PBS).

8. Dulbecco's Modified Eagle Medium (DMEM).

9. Complete DMEM solution: DMEM, 10% fetal bovine serum, 2 mM L-glutamine, 100 units/mL penicillin G, 100 μg/mL streptomycin.

10. Collagenase type II.

11. DNase I.

12. ACK (ammonium-chloride-potassium) red cell lysis buffer.

13. CD11b and CD11c magnetic bead isolation kits.

14. Anti-Fcγ II/III receptor block.

15. Fluorochrome-conjugated antibodies to CD45, CD3, CD19, Ly6C, CD11b, CD11c, Siglec-F, and CD64.

16. Viability dye for flow cytometry.

17. 5 mL polystyrene tubes for flow cytometry.

18. Blood agar plates (tryptic soy agar with 5% sheep blood).

19. Todd-Hewitt broth.

20. Hexidium iodide.

21. Trypsin-EDTA (0.25%).

22. PBS-EDTA (0.6 mM).

---

3 Methods

3.1 Optional: Pre-activation of Lung Macrophage Subpopulations by Oropharyngeal Aspiration of LPS [4]

Pre-activation with inflammatory stimuli delivered to the lung such as lipopolysaccharide (LPS) or recombinant IFN-γ can be used to increase the number and diversity of the lung macrophage population to include more inflammatory macrophages and blood-derived monocytes (*see* **Note 1**).

1. Anesthetize mouse by inhaled isoflurane using a 1–3% solution delivered via a rodent anesthesia machine or a 20% v/v solution in propylene glycol delivered by open-drop exposure until the breathing rate is reduced to ~40 breaths/min and the mouse becomes unresponsive to toe pinch.

2. While restraining the mouse supine with the head up at a 45° angle, retract the tongue using forceps and pipet LPS

(0.1–3 µg LPS/g body weight suspended in 50 µL of H_2O) into the oropharynx.

3. Obstruct the nares and allow the mouse to take several deep breaths through the mouth to adequately aspirate the 50 µL LPS solution into the lungs. Return the mouse to its cage, and monitor for arousal from anesthesia.

4. Between 1 and 4 days following LPS instillation, continue with the protocols below.

3.2 Optional: Bronchoalveolar Lavage (BAL) Collection [5]

See **Note 2** about BAL collection.

1. Euthanize mice by intraperitoneal injection of ketamine (100 mg/kg) and xylazine (15 mg/kg).
2. Expose the trachea by making a midline incision in the skin of the neck and separating the connective tissue. Make a small hole in the ventral surface of the trachea and introduce an 18-gauge blunt cannula containing a Luer lock hub into the hole. Slide suture material posterior to the trachea, and tie it around the trachea to secure the cannula tip in place.
3. Lavage the lungs via the cannula three separate times with a syringe containing 1 mL of PBS containing 0.6 mM EDTA.
4. Pool the three BAL volumes for each sample and centrifuge at 4 °C and $500 \times g$ for 5 min to pellet the cells.
5. After the supernatant is removed, resuspend the cells in 1 mL of PBS, count on a hemocytometer, and adjust to a final concentration of 1×10^6 cells/mL PBS on ice.

3.3 Isolation of Lung Mononuclear Cells

1. Euthanize mice by intraperitoneal injection of ketamine (100 mg/kg) and xylazine (15 mg/kg).

2. With dissecting scissors, remove the skin and serous membrane of the thoracic cavity to expose the rib cage.

3. Cut away the diaphragm to expose the lungs.

4. Remove and place the lungs in a 5 mL solution containing collagenase type II (1 mg/mL), DNase I (30 µg/mL), and DMEM medium in a single well of a 6-well plate.

5. Mince the lungs into fine pieces with the scissors.

6. Incubate the minced lungs in the collagenase/DNase/DMEM solution for 35 min at 37 °C with 5% CO2.

7. Flush a 100 µm cell strainer with the 5 mL of collagenase/DNase/DMEM media containing minced lung. Using the plunger of a syringe, grind the tissue through the strainer into a clean 50 mL conical tube.

8. Centrifuge the cells at $500 \times g$ for 5 min at 4 °C.

9. Aspirate the supernatant and discard.

10. Resuspend the cell pellet in 2 mL of ACK red blood cell lysing buffer for 5 min at room temperature.

11. Add 3 mL of PBS to the cell solution, and pass the suspension through a 70 μm cell strainer into a clean 15 mL conical tube.

12. Centrifuge the cells at $500 \times g$ for 5 min at 4 °C.

13. Aspirate the supernatant and discard.

14. Resuspend lung cells (1×10^6/mL) in PBS and keep on ice.

3.4 Optional: Magnetic Bead Enrichment and Fluorescence-Activated Cell Sorting of Lung Monocyte and Macrophage Subpopulations

See **Note 3** about enrichment and sorting of monocyte and macrophage subpopulations and Chapter 3 for additional details.

1. Myeloid cells from the lungs can be enriched using commercially available magnetic bead-conjugated CD11c and CD11b reagent kits. Follow the manufacturer's recommendations for isolating CD11c-positive cells, which will include alveolar and interstitial macrophages, and CD11b-positive cells, which will contain blood-derived monocytes, neutrophils, and some dendritic cells. Alveolar macrophages can also express CD11b if animals have been treated with LPS.

2. Once CD11c⁺ and CD11b⁺ cells have been isolated, incubate the cells in 100 μL of PBS with an antibody cocktail to block Fcγ II/III receptors for 10 min on ice.

3. Add fluorochrome-conjugated antibodies to CD45, CD3, CD19, Ly6G, Ly6C, CD11b, CD11c, Siglec-F, and CD64 according to the manufacturer's recommended concentrations for 20 min on ice in the dark.

4. Sort macrophages and monocytes on a flow cytometer using the following schema [6, 7].

5. Eliminate lymphocytes and neutrophils using a dump channel with CD3, CD19, and Ly6G.

6. Alveolar macrophages will be side-scatter (SSC) high/CD45⁺/CD11c⁺/Siglec-F⁺.

7. Interstitial macrophages will be SSC^mid/CD45⁺/CD64⁺/CD11b⁺/CD11c⁺

8. Monocytes will be SSC^low/CD45⁺/CD11b⁺/Ly6C⁺.

3.5 Preparation of Streptococcus pneumoniae [8, 9]

See **Note 4** about the choice of *Streptococcus pneumonia*.

1. Seed the surface of a sheep's blood agar plate with *Streptococcus pneumoniae* and incubate overnight with 5% CO_2 at 37 °C.

2. Harvest the bacteria from the plate and inoculate into 5 mL of buffered Todd-Hewitt broth (THB).

3. Gently shake the culture at for 2–3 h at 37 °C with 5% CO_2 until the OD_{600} reaches ~0.5.

4. Mix the tube by inversion and centrifuge at $5000 \times g$ for 10 min at 4 °C.

5. Aspirate and discard the supernatant.

6. Resuspend the cell pellet in 1 mL of PBS to achieve approximately 1×10^9 CFU/mL.

7. To confirm the exact CFU/mL, perform a serial dilution series ranging from 10^{-1} to 10^{-9} in PBS, and plate the dilutions for 10^{-4} to 10^{-9} on the sheep's blood agar plate, incubating the plate upside down overnight at 37 °C.

3.6 Fluorescent Staining of Streptococcus [10]

1. Add the 1 mL of 1×10^9 CFU/mL *Streptococcus pneumoniae* into a 50 mL conical tube.

2. Add 40 µL of 5 mg/mL hexidium iodide and mix well.

3. Incubate the bacteria for 15 min at room temperature in the dark.

4. Wash the bacteria twice with 50 mL of cold PBS by centrifugation at 4 °C and $5000 \times g$ for 10 min.

5. Resuspend the bacteria in complete DMEM medium at a concentration of 1×10^7 CFU/mL.

3.7 Streptococcus Phagocytosis by Adherent Lung Mononuclear Phagocytes [10, 11]

1. Resuspend 2×10^6 lung cells in 2 mL of complete DMEM media in a single well of a 6-well plate (*see* **Note 5**).

2. Let the cells adhere at 37 °C with 5% CO2 for 2 h and then remove the non-adherent cells by gentle washing with 2 mL of PBS three times (*see* **Note 6**).

3. Add 2 mL of the 1×10^7 CFU/mL solution of fluorescently labeled *Streptococcus pneumoniae* to each well of adherent mononuclear cells to achieve a multiplicity of infection (MOI) of 10.

4. Centrifuge the 6-well plate at room temperature and $150 \times g$ for 5 min.

5. Incubate the cells for 30 min with 5% CO_2 at 37 °C to allow for bacterial phagocytosis.

6. Add 200 µL of cold trypsin-EDTA to the cells without agitation, and incubate for no more than 10 min at room temperature to remove residual bacteria (*see* **Note 7**).

7. Gently wash the plate three times with PBS, using 2 mL of PBS for each wash.

8. Add 0.5 mL of 5 mM EDTA to each well for 5 min at room temperature to detach the cells.

9. Transfer the cells to a clean 15 mL conical tube and wash the cells twice in 2 mL of PBS by centrifugation at 4 °C and $500 \times g$ for 5 min.

10. Resuspend the cells in 1 mL of PBS and count using a hemocytometer.

3.8 Flow Cytometry-Based Assessment of Phagocytosis in Lung Monocytes and Macrophages [6, 7]

1. Stain 2×10^6 cells with a fluorescent viability dye in 1 mL of PBS in a 5 mL polystyrene tube suitable for flow cytometry for 30 min on ice.

2. Centrifuge the cells at $500 \times g$ for 5 min at 4 °C.

3. Aspirate the supernatant and discard.

4. Resuspend and incubate the cells in 100 μL of PBS with an antibody cocktail to block Fcγ II/III receptors for 10 min on ice.

5. If cells were not presorted, add fluorochrome-conjugated antibodies to CD45, CD3, CD19, Ly6C, Ly6G, CD11b, CD11c, Siglec-F, and CD64 according to the manufacturer's recommended concentrations for 20 min on ice in the dark.

6. Assess for *Streptococcus* fluorescence (excitation at 488 nm, emission at 575 nm) in the macrophage and monocyte subpopulations using the steps below after gating out dead cells and doublets.

7. Optional: Eliminate lymphocytes and neutrophils using a dump channel with CD3, CD19, and Ly6G. These contaminating cell types will have mostly been washed away in Subheading 3.7, **step 2** (*see* **Note 6**).

8. Alveolar macrophages will be side-scatter (SSC) high/CD45$^+$/CD11c$^+$/Siglec-F$^+$.

9. Interstitial macrophages will be SSC^{mid}/CD45$^+$/CD64$^+$/CD11b$^+$/CD11c$^+$

10. Monocytes will be SSC^{low}/CD45$^+$/CD11b$^+$/Ly6C$^+$.

4 Notes

1. This may allow for a more detailed characterization of the phagocytic capacity for different macrophage subpopulations in response to subsequent bacterial challenge. Pre-activation may be particularly useful in the context of studying host pathways that regulate phagocytosis but may not be appropriate for quantifying responses to specific pathogens.

2. BAL collection can be used to anatomically separate alveolar macrophages from interstitial macrophages and blood monocytes prior to the assessment of phagocytosis. However, it is important to note that BAL will not remove 100% of macrophages from the alveoli. Thus some will still be present in digested lung tissue. These alveolar macrophages can be distinguished by their high expression of CD11c and Siglec-F.

3. If desired, an independent assessment of inflammatory monocytes, interstitial macrophages, and alveolar macrophages can be achieved by purifying cells from the lung mononuclear frac-

tion using magnetic CD11b and CD11c bead selection kits and separating the individual populations using flow sorting prior to assaying for phagocytic capacity. Bead isolation serves to enrich the single-cell suspensions for macrophage subpopulations, improving the purity and yield following subsequent flow sorting.

4. We outline usage of *Streptococcus pneumoniae* as it is a relevant pulmonary pathogen, but other gram-positive organisms may be substituted as hexidium iodide selectively stains almost all gram-positive, but not gram-negative, organisms.

5. If monocyte and macrophage subpopulations were previously separated by BAL or flow sorting, pooling of cells from several mice will likely be required to achieve the desired cell density at this stage. Alternatively, fewer cells can be assayed by scaling down the size of the culture dish and media volume and adjusting the amount of *Streptococcus* added to maintain a multiplicity of infection (MOI) of 10.

6. When single-cell lung suspensions are not presorted, adherence selection serves to reduce the interference of other cell types, especially lymphocytes and neutrophils, when performing the bacterial phagocytosis assay. However, weakly adherent or non-adherent monocyte subpopulations also may be removed from the culture when performing this step.

7. Prolonged treatment or agitation with trypsin-EDTA during this step may interfere with mononuclear cell adherence, surface marker expression, and antibody binding.

References

1. Kopf M, Schneider C, Nobs SP (2015) The development and function of lung-resident macrophages and dendritic cells. Nat Immunol 16(1):36–44. https://doi.org/10.1038/ni.3052

2. Zhang X, Mosser DM (2008) Macrophage activation by endogenous danger signals. J Pathol 214(2):161–178. https://doi.org/10.1002/path.2284

3. Underhill DM, Goodridge HS (2012) Information processing during phagocytosis. Nat Rev Immunol 12(7):492–502. https://doi.org/10.1038/nri3244

4. Allen IC (2014) The utilization of oropharyngeal intratracheal PAMP administration and bronchoalveolar lavage to evaluate the host immune response in mice. J Vis Exp (86). https://doi.org/10.3791/51391

5. Van Hoecke L, Job ER, Saelens X, Roose K (2017) Bronchoalveolar lavage of murine lungs to analyze inflammatory cell infiltration. J Vis Exp (123). https://doi.org/10.3791/55398

6. Misharin AV, Morales-Nebreda L, Mutlu GM, Budinger GR, Perlman H (2013) Flow cytometric analysis of macrophages and dendritic cell subsets in the mouse lung. Am J Respir Cell Mol Biol 49(4):503–510. https://doi.org/10.1165/rcmb.2013-0086MA

7. Zaynagetdinov R, Sherrill TP, Kendall PL, Segal BH, Weller KP, Tighe RM, Blackwell TS (2013) Identification of myeloid cell subsets in murine lungs using flow cytometry. Am J Respir Cell Mol Biol 49(2):180–189. https://doi.org/10.1165/rcmb.2012-0366MA

8. Vivas-Alegre S, Fernandez-Natal I, Lopez-Fidalgo E, Rivero-Lezcano OM (2015) Preparation of inocula for experimental infection of blood with Streptococcus pneumoniae. MethodsX 2:463–468. https://doi.org/10.1016/j.mex.2015.11.003

9. Dorrington MG, Roche AM, Chauvin SE, Tu Z, Mossman KL, Weiser JN, Bowdish DM (2013) MARCO is required for TLR2- and Nod2-mediated responses to Streptococcus pneumoniae and clearance of pneumococcal colonization in the murine nasopharynx. J Immunol 190(1):250–258. https://doi.org/10.4049/jimmunol.1202113

10. Yan Q, Ahn SH, Fowler VG Jr (2015) Macrophage phagocytosis assay of *Staphylococcus aureus* by flow cytometry. Bio Protoc 5(4):e1406

11. Drevets DA, Canono BP, Campbell PA (2015) Measurement of bacterial ingestion and killing by macrophages. Curr Protoc Immunol 109:11–17. Chapter 14: Unit 16. https://doi.org/10.1002/0471142735.im1406s109

Chapter 12

Measuring Neutrophil Bactericidal Activity

Kenneth C. Malcolm

Abstract

The best-known role of neutrophils is control of pathogen growth. Neutrophils contain and kill pathogens through a variety of antimicrobial activities. Regardless of the mechanism, the ability to kill pathogens is a vital outcome. This chapter describes a method to measure the in vitro bactericidal activity of isolated neutrophils as the endpoint of converging innate immune functions.

Key words Neutrophils, Bactericidal, Bacteria killing, Innate immunity, Reactive oxygen species, Neutrophil extracellular traps, Phagocytosis

1 Introduction

Neutrophils provide the major activity against infections by bacteria and other pathogenic organisms. After migrating to sites of infection, neutrophils use several mechanisms to incapacitate and kill bacteria, including phagocytosis, reactive oxygen species (ROS) generation, neutrophil extracellular trap (NET) formation, and release of degradative enzymes [1]. Understanding how these specialized mechanisms of pathogen killing contribute to the control of bacterial infections is essential for extending our knowledge of pathogen-host interactions and for devising and testing novel therapeutic strategies against individual bacterial strains.

Phagocytosis is a highly effective means of limiting microbial spread and concentrates bactericidal processes in a tightly regulated environment. Phagocytosis is actin-dependent and can be inhibited by cytochalasins to differentiate intracellular from extracellular killing mechanisms. ROS formation is initiated with the formation of a membrane complex involving NADPH oxidase and Rho-family GTPases, which convert O_2 to superoxide ($O_2^-\cdot$) and promote subsequent formation of H_2O_2, hypochlorite (HOCl), hydroxyl radicals, and other compounds. These highly reactive species modify and inactivate biomolecules to incapacitate pathogens [2, 3]. ROS can be released extracellularly but are likely most effective inside the

Scott Alper and William J. Janssen (eds.), *Lung Innate Immunity and Inflammation: Methods and Protocols*,
Methods in Molecular Biology, vol. 1809, https://doi.org/10.1007/978-1-4939-8570-8_12,
© Springer Science+Business Media, LLC, part of Springer Nature 2018

confines of pathogen-containing phagosomes [2, 3]. NETs are released from activated neutrophils and consist of chromatin decorated with the antibacterial contents of neutrophil granules, including elastase, myeloperoxidase, lactoferrin, and MMP9/gelatinase B [4]. NETs act to confine and, under some circumstances, directly kill pathogens [4, 5]. As the name implies, NETs represent an extracellular mechanism of pathogen control. Each of these bactericidal processes can be specifically targeted by pathogens to reduce or inactivate their function, for example, by expression of antioxidants or nucleases [6, 7]. In addition to these biochemical control mechanisms, neutrophils shape subsequent immune responses through secretion of cytokines and chemokines [8].

This chapter will outline a method for measuring the bactericidal activity of neutrophils and can be adapted to killing of other pathogens. Neutrophils and bacteria of known concentrations are co-incubated for defined times, and the loss of bacterial viability is used to quantify neutrophil killing activity.

2   Materials

1. Bacterial lysogeny broth (LB): Prepare a 1 L solution of 10 g tryptone, 10 g NaCl, and 5 g yeast extract in water, and store in two 500 mL bottles. Autoclave at 121 °C for 15 min on the liquid setting. Store at room temperature or 4 °C.

2. LB agar plates: Prepare a 1 L suspension of 10 g tryptone, 10 g NaCl, 5 g yeast extract, and 1.5% agar in water in a 2 L flask. Add a Teflon stir bar, and heat to boiling with constant stirring until agar has dissolved (*see* **Note 1**). Autoclave at 121 °C for 15 min on the liquid setting. Allow to cool to 50–60 °C with stirring or with agitation in a water bath. Place warm agar solution on a 50–60 °C hot plate, and pipette 12–15 mL molten LB agar into sterile 10 cm petri dishes. Allow to solidify overnight, package in original sleeves, and store at 4 °C until day of use.

3. 0.1% Triton X-100: Add one volume of sterile-filtered 1.0% Triton X-100 to nine volumes of sterile 0.9% NaCl in a sterile tube (*see* **Note 2**). Prepare a final volume appropriate for experimental needs.

4. Complete RPMI: Supplement L-glutamine-containing RPMI with 10 mM HEPES, pH 7.4, and 2% serum or plasma (final concentrations). Prepare fresh complete RPMI sufficient for each experiment (*see* **Note 3**). Pass through a 0.2 μm filter into a sterile tube.

5. Neutrophils: Please refer to Chapter 4 for methods to isolate mouse neutrophils. Peripheral blood human neutrophils are routinely obtained using the Percoll gradient method [9].

Resuspend neutrophils to 1×10^7/mL in complete RPMI (*see* **Note 4**).

6. Bacterial stock: Bacteria of choice stored at −80 °C in LB containing 15% glycerol.

---

3 Methods

3.1 Growth of Bacteria

1. Prepare a sterile 14 mL round-bottom tube with 5 mL LB. Work under sterile conditions, either in a dedicated bacterial hood or near a flame (*see* **Note 5**).

2. Inoculate the bacteria into the 5 mL LB (*see* **Note 6**). Incubate at 37 °C, shaking at 180–225 rpm, ideally tilted at an angle, for 12–18 h.

3. On the following day, remove 1 mL of culture to a 1.6 mL microfuge tube, and wash twice in 1.5 mL sterile 0.9% NaCl, centrifuging each time for 5 min at 8000 × g (*see* **Note 7**). Resuspend by repeatedly pipetting up and down.

4. Dilute the washed culture to an optical density at 600 nm (OD_{600}) of 1.0 using 0.9% NaCl or other appropriate OD of known concentration.

3.2 Bactericidal Assay

1. Aliquot 100 μL (1×10^6) neutrophils into labeled, sterile microcentrifuge tubes for each experimental condition (e.g., time points, inhibitors, bacteria/neutrophil ratios). Prepare control tubes containing 100 μL complete RPMI without neutrophils.

2. Dilute OD_{600} = 1.0 suspension appropriately so that bacteria can be added in a reasonable volume to achieve the desired multiplicity of infection (MOI). Typical MOI is 1:1 but may vary by orders of magnitude as required to answer your specific question.

3. Add bacterial suspension in a volume of 5–10 μL to neutrophil-containing and control tubes. Incubate at 37 °C for up to 4 h (*see* **Note 8**). To promote and synchronize bacteria-neutrophil contact, one may centrifuge tubes for 1 min at 4000 × g at 22 °C, incubate without disturbing the pellet for up to 5 min at 37 °C, and resuspend the pellet in the same supernatant. Return tubes to 37 °C for intended incubation times. Incubations can be continued statically to promote settling of cells or under gentle rotation or rocking (8–15 rpm) to maintain cells in suspension.

4. Stop the killing activity by adding 900 μL 0.1% Triton X-100 at the appropriate time. This solution yields a 10^{-1} dilution (*see* **Note 9**).

3.3 Determining Bacteria Counts

1. Prepare dilution tubes containing 450 μL 0.9% NaCl (*see* **Notes 2** and **10**).

2. Incubate the 10^{-1} dilution tube for 10–30 min at room temperature or on ice, and vortex for 20 s to promote release of intracellular bacteria and disrupt bacterial aggregates and binding to cellular debris.

3. Add 50 µL from initial 10^{-1} dilution to the next tube, and continue to perform serial tenfold dilutions (*see* **Note 11**). Mix each dilution well by repeated pipetting.

4. Plate 20–50 µL of several dilutions each on LB agar plates. The plated volume can be spread on a portion of the plate using a sterile spreader or puddled and dispersed by repeated tipping of the plate. Allow the plates to absorb the liquid while upright. Place plates, agar side up, in a 37 °C incubator overnight.

5. Count colonies, and determine colony-forming units (cfu) per mL for each dilution. Bacterial concentration in cfu/mL is determined by dividing the colony counts by the volume plated (e.g., 0.02 mL) and multiplying by the dilution (e.g., 10^3).

6. Percent of bacteria killed is calculated by dividing the number of cfu that grow in the presence of neutrophils by the number of cfu that grow in the absence of neutrophils (control tube incubated for the same time) (*see* **Note 12**).

4   Notes

1. Use insulated gloves to avoid injury. Cover flask with foil and attend at all times to avoid boiling over.

2. The use of phosphate-buffered saline is also acceptable.

3. Use either human heat-inactivated serum, autologous plasma from the mouse or human neutrophil isolation, or heat-inactivated FBS. We routinely use human heat-inactivated platelet-poor plasma, pooled from five donors. Heat inactivation, by incubation in a 56 °C water bath for 30 min, is necessary to reduce complement activity, which otherwise activates immune cells.

4. Keep neutrophils at room temperature at all times before performing the experiment. Changes in temperature will activate neutrophils.

5. Typical pathogenic bacteria, including *Staphylococcus aureus* and *Pseudomonas aeruginosa*, are grown in LB or tryptic soy broth. However, other bacteria, including mycobacteria, may require specialized broth and growth conditions.

6. Bacterial stocks are transported from the −80 °C freezer to the work area using a freezer block or dry ice in a Styrofoam or other suitable container. Use dedicated sterile pipette tips for all work, and prepare pipettes by spraying with 70% ethanol and allowing to dry. Jab the tip into the frozen aliquot of

bacterial stock, and dip into the LB, being careful to avoid touching the side of the tube.

7. Bacteria can be pelleted by centrifugation at temperatures from 4 to 24 °C.

8. Unstimulated neutrophils undergo spontaneous apoptosis over this time frame; neutrophils grown in the presence of bacteria may survive longer, but this is balanced by the possibility of induced necrotic cell death. Overgrowth of bacteria can also complicate analyses at later time points.

9. The 0.1% Triton X-100 solution can be at 4 °C, but room temperature is sufficient for most situations. The detergent kills neutrophils, allows release of intracellular bacteria, and promotes dispersal of bacterial aggregates.

10. Prepare as many dilutions as you need to get to a reasonable number of plated colony-forming units (cfu); back-calculate from the inoculated (100%) cfu/mL to determine the number of theoretically achieved colonies at any given dilution. Ten to two hundred colonies are ideal to obtain reliable results—otherwise colony crowding or pipetting variability occurs. Similarly, determine the dilution necessary to achieve a countable number of colonies after killing, which may be several dilutions different from the inoculum if killing is efficient. Some bacteria will withstand overnight, or longer, incubations at 4 °C. In this case, it may be beneficial to have the lowest dilution tubes available to replate in case bacterial growth plates become contaminated.

11. The use of autoclaved microtiter tubes and a multichannel pipettor facilitates dilutions of a large number of tubes.

12. Over the course of a killing assay, typical fast-growing bacteria continue multiplying, so comparing killed bacteria to the inoculum at the beginning of the incubation, instead of to bacteria incubated for the same time in the absence of neutrophils, may give aberrant results.

Acknowledgments

This work was supported by a research grant from the Cystic Fibrosis Foundation.

References

1. Segal AW (2005) How neutrophils kill microbes. Annu Rev Immunol 23:197–223. https://doi.org/10.1146/annurev.immunol.23.021704.115653

2. Dupre-Crochet S, Erard M, Nubetae O (2013) ROS production in phagocytes: why, when, and where? J Leukoc Biol 94(4):657–670. https://doi.org/10.1189/jlb.1012544

3. Nunes P, Demaurex N, Dinauer MC (2013) Regulation of the NADPH oxidase and associated ion fluxes during phagocytosis. Traffic 14(11):1118–1131. https://doi.org/10.1111/tra.12115

4. Brinkmann V, Reichard U, Goosmann C, Fauler B, Uhlemann Y, Weiss DS, Weinrauch Y, Zychlinsky A (2004) Neutrophil extracellular

traps kill bacteria. Science 303(5663):1532–1535. https://doi.org/10.1126/science.1092385

5. Parker H, Albrett AM, Kettle AJ, Winterbourn CC (2012) Myeloperoxidase associated with neutrophil extracellular traps is active and mediates bacterial killing in the presence of hydrogen peroxide. J Leukoc Biol 91(3):369–376. https://doi.org/10.1189/jlb.0711387

6. Henningham A, Dohrmann S, Nizet V, Cole JN (2015) Mechanisms of group a Streptococcus resistance to reactive oxygen species. FEMS Microbiol Rev 39(4):488–508. https://doi.org/10.1093/femsre/fuu009

7. Storisteanu DM, Pocock JM, Cowburn AS, Juss JK, Nadesalingam A, Nizet V, Chilvers ER (2017) Evasion of neutrophil extracellular traps by respiratory pathogens. Am J Respir Cell Mol Biol 56(4):423–431. https://doi.org/10.1165/rcmb.2016-0193PS

8. Tecchio C, Cassatella MA (2016) Neutrophil-derived chemokines on the road to immunity. Semin Immunol 28(2):119–128. https://doi.org/10.1016/j.smim.2016.04.003

9. Haslett C, Guthrie LA, Kopaniak M, Johnston RB Jr, Henson PM (1985) Modulation of multiple neutrophil functions by trace amounts of bacterial LPS and by preparative methods. Am J Pathol 119:101–110

Chapter 13

Modulation of Myeloid Cell Function Using Conditional and Inducible Transgenic Approaches

Alexandra L. McCubbrey and William J. Janssen

Abstract

Transgenic mice have emerged as a central tool in the study of lung myeloid cells during homeostasis and disease. The use of Cre/Lox site-specific recombination allows for conditional deletion of a gene of interest in a spatially controlled manner. The basic Cre/Lox system can be further refined to include an inducible trigger, enabling conditional deletion of a gene of interest in a spatially and temporally controlled manner. Here we provide an overview of commercially available conditional and inducible conditional mouse strains that target lung myeloid cells and describe the appropriate breeding schemes and controls for transgenic animal systems that can be used to modulate myeloid cell function.

Key words Transgenic mice, Conditional, Inducible, Cre, CreER, rtTA, Breeding, ROSA26

1 Introduction

Genetically manipulated mouse models are a critical tool for the study of gene function. The simplest systems are gene knockouts in which a given gene is deleted from germline DNA, resulting in loss of expression in all cells. While the use of these systems has led to important discoveries, they are restricted by the lack of cellular specificity. In addition, some gene knockouts are embryonic lethal, which limits their use. Transgenic approaches represent a critical advance and overcome problems with embryonic lethality while enabling cell-specific targeting [1–3]. The objectives of this chapter are to describe commercially available resources that enable transgenic manipulation of myeloid cells and to assist investigators in selecting the best mouse lines for their studies. In addition, we describe optimal breeding schemes to generate transgenic animals and their appropriate controls. Notably, this chapter will focus on transgenic mice for loss-of-function studies. However, similar approaches can be used to generate powerful tools for gain-of-function studies.

Scott Alper and William J. Janssen (eds.), *Lung Innate Immunity and Inflammation: Methods and Protocols*,
Methods in Molecular Biology, vol. 1809, https://doi.org/10.1007/978-1-4939-8570-8_13,
© Springer Science+Business Media, LLC, part of Springer Nature 2018

Transgenic approaches differ from traditional gene knockout methods in that new transgenes are inserted into the DNA rather than deleting an existing gene from the germline [2, 4]. This allows for the creation of animals in which a gene of interest can be overexpressed or deleted in specific tissues or cells. Transgenic constructs are designed with all of the critical elements for gene expression, including promoters, introns, the protein-coding sequence for the desired gene, and a poly(A) site [5]. Importantly, the promoter sequence can grant spatial constraint of transgene expression to specific cells or tissues. This type of promoter-mediated control is termed *conditional* [2]. Ideally, to achieve specificity for a certain cell type, the promoter that is chosen represents a gene that is uniquely expressed by that cell. As an example, a transgene that uses a LysM promoter will be expressed by LysM-expressing macrophages [6]. However, since it is rare to find genes that are 100% specific for a specific cell type, many conditional systems have "off-target" effects [7, 8]. For instance, in addition to being expressed by macrophages, LysM is also expressed by neutrophils, monocytes, and a small percentage of eosinophils and dendritic cells [7, 9].

For conditional loss-of-function studies, two basic mouse lines are crossed [2]. In the first, *Cre recombinase* (Cre), a site-specific recombinase that catalyzes recombination between pairs of 34-bp DNA recognition sites named LoxP sites [10], is expressed under the control of a conditional promoter. In the second strain, the endogenous gene of interest (or key exons in the endogenous gene of interest) is flanked by LoxP sites (termed "floxed" sites) [1]. In the double-transgenic offspring, the floxed gene is excised in the specific cell subsets that express Cre. As an example, in PPARγ$^{flox/flox}$ × LysM^{cre} mice, the PPARγ gene is deleted from all cells that express LysM (i.e., monocytes, macrophages, and neutrophils).

In additional to spatial control, conditional systems can provide an element of temporal control, since specific tissues or cells may only express a gene at a certain point in development. However, temporal control is best achieved by combining a conditional approach with an *inducible* approach. Inducible transgenic systems control transgene expression through an exogenous element that regulates the activity of a transgene promoter or protein. Tamoxifen is used for loss-of-function systems in which the conditional Cre transgene is fused to a mutant estrogen receptor (CreER) [11, 12]. The fused ER prevents Cre activity until the mutated ER is engaged by its ligand, tamoxifen. As in normal conditional systems, the transgene promoter can be chosen to target expression of the fusion CreER to the tissue or cells of interest; however, Cre is only activated upon administration of tamoxifen. For loss-of-function studies, conditional CreER mice are crossed to mice where the endogenous gene of interest is floxed. For example, in PPARγ$^{flox/flox}$ × CX3CR1^{CreER} mice, the PPARγ gene will be deleted in all cells that express CX3CR1 (the fractalkine receptor) at the time tamoxifen is administered. Notably, multiple versions of the mutant CreER have been generated to enhance Cre activity; ERT2 is currently the superior ER option [13].

Doxycycline is used for loss-of-function systems where the Cre expression is controlled by a tetracycline operator sequence (tetO) in the promoter [14, 15]. Cre expression in this system is regulated by two transgenes [16]. The first encodes a reverse tetracycline-controlled transactivator (rtTA) under the control of a conditional promoter. The second encodes a tetO promoter fused to Cre (tetO-Cre). In the presence of a tetracycline analogue, such as doxycycline, rtTA binds the tetO sequence and drives Cre expression. For loss-of-function studies, these conditional rtTA and tetO-Cre mice are crossed to mice where the endogenous gene of interest is floxed [16]. An example is the cFLIP$^{flox/flox}$ × CD68-rtTA × tetO-Cre mouse [17]. In these animals, c-FLIP will be deleted in all cells that express CD68 at the time doxycycline is administered. Notably, multiple versions of rtTA have been generated; rtTA-M2 has greater affinity for tetO sequences with reduced background compared to M1 [18, 19].

Although hundreds of transgenic mouse lines are available with Cre expression driven by a variety of promoters, specific targeting to myeloid cell subpopulations remains a challenge [7, 9, 20]. The lungs possess a variety of myeloid cells during homeostasis, including embryonic-derived resident alveolar macrophages and interstitial macrophages, CD11b+ and CD103+ dendritic cells, and monocytes that survey lung tissue in steady-state (*see* Chapter 3, Isolation and Characterization of Mouse Mononuclear Phagocytes for further detail) [21–24]. Importantly, during inflammation, new myeloid cells are recruited to the lungs, including monocyte-derived recruited macrophages, neutrophils, and eosinophils [25, 26]. Here we summarize the transgenic tools available for targeting these diverse myeloid populations.

2 Materials

2.1 Mouse Genotyping

2.1.1 DNA Extraction

1. Collection tubes for tail or ear clips.

2. Sterile surgical scissors.

3. STE buffer: 100 mM Tris-HCl, 5 mM EDTA, 200 mM NaCl, 0.2% SDS in DEPC-treated water.

4. Proteinase K stock solution: PCR-grade proteinase K (10 mg/mL) in RNase-free, DNase-free water.

5. 1.5 mL microfuge tubes.

6. Molecular biology-grade isopropanol.

7. Molecular biology-grade 70% ethanol: dilute seven parts molecular biology-grade 100% ethanol with three parts RNase-free, DNase-free water.

8. RNase-free, DNase-free water (may also use DEPC-treated water).

9. Heating block capable of fitting 1.5 mL microfuge tubes.

10. Refrigerated microcentrifuge.

2.1.2   PCR

1. Thin-walled PCR tubes or plates.

2. MangoMix 2× master mix (Bioline), or equivalent Taq DNA polymerase and dNTP master mix.

3. RNase-free, DNase-free water.

4. Desalted forward and reverse primer stock solutions resuspended at 200 µM concentration in RNase-free, DNase-free water (used at 10 µM final concentration in PCR reactions).

5. Thermocycler.

2.1.3   Gel Electrophoresis

1. Agarose.

2. 50× TAE buffer (242 g Tris Base, 57.1 mL glacial acetic acid, and 37.2 g EDTA disodium salt, dehydrate, in distilled water to 1 L—pH ~8.5).

3. 250 mL Erlenmeyer flask.

4. Microwave.

5. Silicone potholder or heat-insulated glove.

6. Ethidium bromide solution.

7. Gel casting tray and comb.

8. Gel-running apparatus including power supply.

9. 100 bp DNA ladder.

10. 1× DNA gel loading dye.

11. UV light source.

2.2   Generation of Mice with Conditional Loss of Function

1. Mouse expressing conditional Cre (*see* Table 1).

2. Mouse expressing floxed target gene of interest.

2.3   Generation of Mice with Inducible Conditional Loss of Function (CreER)

1. Mouse expressing inducible conditional CreER (*see* Table 2).

2. Mouse expressing floxed target gene of interest.

3. Tamoxifen ≥99%.

4. Corn Oil, delivery vehicle for fat-soluble compounds (Sigma).

5. 50 mL conical tubes.

6. 22 µm vacuum filters.

7. 1 mL syringes.

8. 25 gauge × 5/8 needles.

9. 250 mL glass beaker.

Table 1
List of commercially available myeloid targeted conditional Cre lines[a]

Common name	Promoter	Known expression	JAX ID	Notes	Ref.
LysMcre	Mouse lysozyme 2	Macrophages, neutrophils, monocytes, some dendritic cells and eosinophils, also lung epithelial cells	004781	Cre transgene is a knock-in to the endogenous promoter resulting in **loss** of LysM expression	[6, 7, 9]
Csf1r-icre	Mouse Csf1r (cfms)	Macrophages, neutrophils, monocytes, dendritic cells, eosinophils, also lymphocytes	021024		[30]
Csfr1[Cre]	Mouse Csf1r (cfms)	Macrophages, neutrophils, monocytes, dendritic cells, eosinophils, also lymphocytes	029206		[31]
MafB-mCherry-Cre	Mouse Mafb	Macrophages	029664	Promoter simultaneously drives Cre and FLAG-tagged-mCherry expression	[32]
Cx3cr1[Cre]	Mouse CX3CR1 (fractalkine receptor)	Macrophages, CD11b + dendritic cells, some monocytes	025524	Cre transgene is a knock-in to the endogenous promoter resulting in **loss** of CX3CR1 expression	[7, 21]
Cd11c-Cre	Mouse CD11c	Macrophages, dendritic cells, some monocytes, also some lymphocytes	008068		[7, 33]
zDC-Cre	Mouse Zbtb46	Dendritic cells	028538		[34]
Mrp8Cre[Tg]	Human S100 calcium binding protein A8	Neutrophils, some macrophages, and monocytes	021614	Promoter simultaneously drives Cre and EGFP expression	[7, 35]

Common name is shown, followed by the identity of the specific conditional promoter, known expression pattern, Jackson Laboratory stock number (Jax ID), and important notes
[a]Additional conditional targeting lines targeting myeloid cells have been generated including CD11b-Cre [36], CD68-Cre [37], and Epx-Cre [38], but at the time of publishing, these lines were not commercially available

2.4 Generation of Mice with Inducible Conditional Loss of Function (rtTA)

1. Mouse expressing inducible conditional rtTA (*see* Table 2).

2. Mouse expressing tetO-Cre (JAX stock number 006224).

3. Mouse expressing floxed target gene of interest.

4. Doxycycline chow containing Doxycycline Hyclate at 625 mg/ kg (*see* **Note 1**).

Table 2
List of commercially available myeloid targeted inducible conditional Cre lines[a]

Common name	Promoter	Known expression	JAX ID	Notes	Ref.
Cx3xr1[creER]	Mouse CX3CR1 (fractalkine receptor)	IM, **not** resAM when tamoxifen given to adult mice, CD11b+ dendritic cells, some monocytes, also some lymphocytes	020940	Tamoxifen-driven	[21]
Csf1r-CreER	Mouse Csf1r (cfms)		019098	Tamoxifen-driven	[39]
CD68-rtTA	Human CD68	IM, **not** resAM when doxycycline given to adult mice, dendritic cells, monocytes, neutrophils	32044-JAX	Doxycycline-driven	[20, 40]

Common name is shown, followed by the identity of the specific conditional promoter, known expression pattern, Jackson Laboratory stock number (Jax ID), and important notes. Lung interstitial macrophages (IM), resident alveolar macrophages (resAM)

[a]Additional inducible lines targeting myeloid cells have been generated including Csf1r-rtTA [41], SRA-rtTA [42], and Mafb-CreER [43], but at the time of publishing, these lines were not commercially available

2.5 Analysis of Recombination Using Reporter Mice

1. Mouse expressing conditional or inducible conditional Cre/CreER/rtTA (*see* Tables 1 and 2).

2. Mouse expressing Cre or rtTA-driven reporter (*see* Table 3).

3 Methods

3.1 Mouse Genotyping

3.1.1 DNA Extraction

1. Prior to weaning, pups should be identified by tattoo or ear tag in accordance with institutional animal use guidelines. At the time of tattoo or ear tag, a ~2–3 mm tail or ear clip should be cut using sterile surgical scissors and placed in a labeled collection tube. If necessary, an ear clip may be collected from adult mice. Tissue samples can be kept at room temperature for up to 4 h with no detectable loss in DNA. Tail or ear tissue can be stored at −20 °C for genotyping at a later time.

2. Place the tissue in 720 μL STE buffer with 30 μL proteinase K stock solution in a microfuge tube.

3. Incubate tissue in buffer/proteinase overnight on a heating block at 55 °C or on a shaker in a 55 °C oven. After overnight incubation, tissue should be dissolved, and only fur should be visible. If tissue is still visible, vortex and incubate for an additional hour. Repeat until tissue is dissolved.

4. Inactivate proteinase K by incubating samples for 5 min at 70 °C on a heating block.

Table 3
List of commercially available reporter strains to report expression of conditional or conditional inducible Cre lines

Common name	Promoter	Reporter	JAX ID	Notes	Ref.
R26-TdTomato (Ai14)	Rosa26	TdTomato	007914	Floxed stop codon prevents TdTomato expression until Cre is expressed within the cell. All daughter cells will also express TdTomato	[44]
R26-EYFP (R26R-EYFP)	Rosa26	EYFP	006148	Floxed stop codon prevents EYFP expression until Cre is expressed within the cell. All daughter cells will also express EYFP	[45]
Rosa26-EGFP^f	Rosa26	EGFP	004077	Floxed stop codon prevents EGFP expression until Cre is expressed within the cell. All daughter cells will also express EGFP	[46]
R26-ZsGreen1 (Ai6)	Rosa26	ZsGreen1	007906	Floxed stop codon prevents ZsGreen1 expression until Cre is expressed within the cell. All daughter cells will also express ZsGreen1	[44]
pTRE-H2BGFP	tetO	GFP	005104	GFP is fused to a human histone protein; even if doxycycline is withdrawn, GFP will persist in long-lived cells	[47]

Common name is shown, followed by the identity of the specific conditional promoter, known expression pattern, Jackson Laboratory stock number (Jax ID), and notes on the mechanism of reporter action

5. Quench on ice for 5 min.

6. Centrifuge digested samples at $10,000 \times g$ at 4 °C in a microcentrifuge for 10 min.

7. Remove supernatant, avoiding the pellet containing SDS and undigested fur, and transfer to a new Eppendorf tube containing 720 μL isopropanol.

8. Precipitate the DNA by vortexing the Eppendorf tube. It will appear as thin, whitish-gray strands.

9. Centrifuge DNA at $10,000 \times g$ at 4 °C in a microcentrifuge for 10 min.

10. Discard supernatant. This can be done by inverting the tube and subsequently dabbing edge on a KimWipe or other absorbent surface to remove residual supernatant.

11. Wash pellet with 500 μL 70% ethanol; centrifuge DNA at $10,000 \times g$ at 4 °C in a microcentrifuge for 10 min.

12. Discard supernatant. This can be done by inverting the tube and subsequently dabbing the edge on a KimWipe or other absorbent surface to remove residual supernatant.

13. Leave tube inverted, and allow the DNA to dry for 1–2 min.

14. Resuspend DNA in 100 µL RNase-free, DNase-free water.

15. Incubate DNA at 55 °C for 1 h in a heating block to facilitate dissolution of DNA into water.

16. Genomic DNA can be stored at 4 °C for up to 12 months. Proceed to next step for PCR.

3.1.2 PCR

1. Clearly label thin-walled 0.65 mL PCR tubes or plates for each reaction.

2. Prepare a PCR Master Mix of 15 µL of 2× MangoMix with 13 µL of RNase-free, DNase-free water, 1 µL of forward primer stock, and 1 µL of reverse primer stock for each sample.

3. Aliquot 27 µL of the PCR Master Mix into each of the labeled PCR reaction tubes or plate wells.

4. Add 3 µL of digested template DNA.

5. Place tubes in a thermal cycler and cycle according to protocols specific to each primer set (*see* **Note 2**).

6. Repeat the steps above for every transgene (*see* **Note 3**).

3.1.3 Gel Electrophoresis

1. Prepare a 2% agarose/TAE gel by adding 1 g of agarose to 1 mL 50× TAE buffer and 49 mL distilled water in a dedicated 250 mL flask. Place the flask in the microwave, and heat on high for approximately 2 min (time required will vary across microwaves). Watch the solution as it heats. As soon as the solution begins to boil, remove the flask from the microwave using an insulated glove or potholder, and gently swirl the solution, taking care not to spill. Place the flask back in the microwave, and repeat the process of heating, removing, and swirling the solution until the solution is completely clear and the agarose is completely dissolved.

2. Place the flask on the bench to allow the solution to cool to 55 °C, the temperature when you can comfortably pick up and hold the flask with a gloved hand.

3. Add 5 µL of ethidium bromide solution to the cooled solution and swirl to mix (*see* **Note 4**).

4. Carefully pour the agarose/TAE/ethidium bromide mixture into the gel tray. Place a comb in the notches on the gel casting tray. Use a dry pipette tip to pop or drag out any bubbles that are present in the gel. Care should be taken to perform these steps before the gel solidifies.

5. When the gel has solidified (after approximately 15–20 min), carefully remove the comb to reveal the sample wells by gently and firmly pulling up on the comb.

6. Remove the gel tray containing the gel from the casting unit. Gel is not adhered to the tray, so care should be taken to prevent the gel from sliding out during transport. Place the gel tray containing the gel into the running apparatus, and cover the gel with running buffer. Running buffer should cover the gel by 1–3 mm.

7. Add 2 μL of 100 bp DNA ladder (0.1 μg/μL) mixed with 8 μL of 1× DNA loading dye in the first well of the gel.

8. Add 10 μL of each PCR product to the remaining wells (*see* **Note 5**). Use care to document the position and order of each sample as you load the gel.

9. Cover the gel box with the lid, and attach to the power supply. The black or negative lead should be at the top of the gel where DNA is loaded. The red or positive lead should be at the bottom of the gel. Run at 5–10 V/cm distance between electrodes for 20–40 min.

10. Place gel onto a UV light source for visualization of DNA bands. Positive and negative controls are important for the interpretation of PCR genotyping results (*see* **Note 6**).

3.2   Generation of Mice with Conditional Loss of Function

For conditional loss-of-function studies, two basic mouse lines are crossed. In the first, Cre recombinase (Cre) is expressed under control of a conditional promoter that enables expression of Cre in specific myeloid subsets (*see* Table 1). In the second, the endogenous gene of interest is flanked by LoxP sites ("floxed"). In double-transgenic mice that result from this cross, the floxed gene is excised in the specific myeloid subsets that express Cre, determined by the conditional promoter. An appropriate breeding scheme for conditional loss-of-function mice (Fig. 1) will generate equal numbers of experimental animals and littermate controls. However, best practice is to also include a second control group of heterozygous Cre mice that lack floxed genes to confirm that production of Cre recombinase does not explain any observed phenotype [8, 27]. Importantly, Experimental animals should be homozygous for the "floxed" gene and preferably heterozygous for the conditional Cre (*see* **Note 7**).

1. Make a breeder pair of a homozygous conditional Cre mouse (i.e., LysM-Cre$^{+/+}$) and a homozygous target-floxed mouse (i.e., PPARg$^{fl/fl}$). This is the F1 cross. These founder mice should be obtained from a commercial breeder (*see conditional lines detailed in* Table 1), reputable academic collaborator, or other trusted source.

2. Pups from the F1 cross should be identified prior to weaning by tattoo or ear tag in accordance with institutional policy. At the time of tattoo or ear tag, a tail or ear snip should be cut and

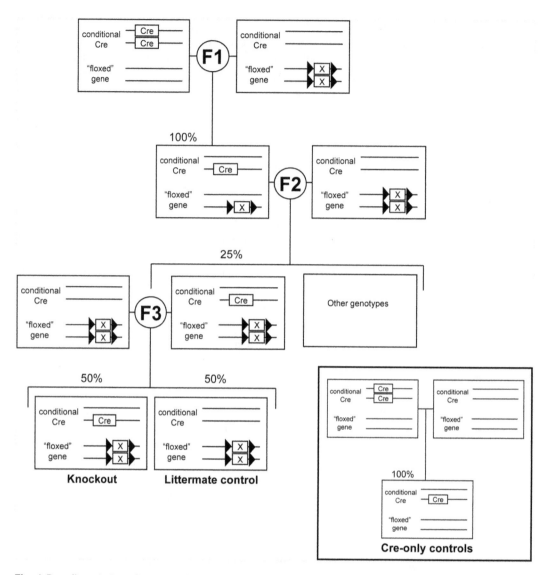

Fig. 1 Breeding strategy for generation of conditional knockout mice. F1 cross between homozygous floxed mouse and homozygous conditional Cre mouse will generate offspring that are 100% heterozygous for the floxed gene of interest and the conditional Cre. F2 cross between an F1 offspring (heterozygous for floxed gene and conditional Cre) and a homozygous floxed mouse will generate offspring that are 25% homozygous for the floxed gene of interest and heterozygous for the conditional Cre. F3 cross between these F2 offspring (homozygous for floxed gene and heterozygous for conditional Cre) will generate offspring that are 50% homozygous for the floxed gene of interest and heterozygous for conditional Cre (knockout), 50% homozygous for floxed gene without conditional Cre (littermate control). Cre-only controls for Cre toxicity (inset box) can be generated by crossing a homozygous conditional Cre mouse with a wild-type mouse

placed in a labeled collection tube for genotyping as described above (*see* Subheading 3.1). All F1 offspring are expected to be heterozygous for both genes (i.e., LysMcre$^{+/wt}$ × PPARγ$^{fl/wt}$) (*see* Fig. 1, F1).

3. Make a F2 breeder pair using one of the F1 offspring (heterozygous for Cre and target-floxed gene—e.g., LysMcre$^{+/wt}$ × PPARγ$^{fl/wt}$) and a homozygous target-floxed mouse (i.e., PPARγ$^{fl/fl}$) (*see* Fig. 1). This may be a backcross where the homozygous target-floxed mouse used is one of the parent founders, although a separate animal is recommended to reduce inbreeding.

4. Pups from the F2 cross should be identified prior to weaning by tattoo or ear tag. At the time of tattoo or ear tag, a tail or ear snip should be obtained and placed in a labeled collection tube for use in PCR as described above (*see* Subheading 3.1). Assuming Mendelian inheritance, 25% of F2 offspring will be the desired Cre$^{+/wt}$ × Target$^{fl/fl}$ genotype (i.e., LysMcre$^{+/wt}$ × PPARγ$^{fl/fl}$) needed for the F3 cross (*see* Fig. 1, F2).

5. Make a F3 breeder pair using one of the F2 offspring that is Cre$^{+/wt}$ × Target$^{fl/fl}$ and a homozygous target-floxed mouse (i.e., LysMcre$^{+/wt}$ × PPARγ$^{fl/fl}$ crossed with PPARγ$^{fl/fl}$) (*see* Fig. 1).

6. F3 pups should be identified prior to weaning by tattoo or ear tag. At the time of tattoo or ear tag, a tail or ear snip be used to determine the genotype of the generation F3 offspring (*see* Subheading 3.1). Assuming Mendelian inheritance, 50% of F3 offspring will be the desired experimental Cre$^{+/wt}$ × Target$^{fl/fl}$ genotype (i.e., LysMcre$^{+/wt}$ × PPARγ$^{fl/fl}$). Fifty percent of F3 offspring will be littermate controls with no Cre (i.e., LysMcre$^{wt/wt}$ × PPARγ$^{fl/fl}$) (*see* Fig. 1, F3).

7. Repeat **steps 5** and **6** as needed to generate and genotype experimental and littermate control mice.

8. In order to generate heterozygous Cre mice for Cre toxicity controls, make a breeder pair of a homozygous conditional Cre mouse (i.e., LysMcre$^{+/+}$) and a wild-type mouse. All F1 offspring are expected to be heterozygous for Cre (i.e., LysMcre$^{+/wt}$) (*see* Fig. 1, inset box). Tail or ear tissue can be collected from pups to confirm the expected genotype (*see* Subheading 3.1).

3.3 Generation of Mice with Inducible Conditional Loss of Function (CreER)

For temporally controlled loss-of-function studies using the tamoxifen-inducible system, conditional CreER mice (Table 2) are crossed to mice where the endogenous gene of interest is floxed. After administration of tamoxifen, the conditionally expressed CreER will become active and will recombine floxed DNA sites, deleting the gene of interest. As with simple conditional loss-of function lines, an appropriate breeding scheme will generate experimental animals and littermate controls (*see* Fig. 2). Again, best practice is to also include a second control group of heterozygous Cre mice with no floxed gene to confirm that Cre toxicity did not generate any observed phenotype. Control groups should also be treated with tamoxifen (*see* **Note 8**).

1. Make a breeder pair using a homozygous conditional Cre mouse (i.e., CX3CR1-CreER$^{+/+}$) and a homozygous target-floxed mouse (i.e., PPARγ$^{fl/fl}$). This is the F1 cross. These founder mice should be obtained from a commercial breeder (*see inducible conditional lines detailed in* Table 2), reputable academic collaborator, or other trusted source.

2. Pups from the F1 cross should be identified prior to weaning by tattoo or ear tag. At the time of tattoo or ear tag, a 1–2 mm tail or ear snip should be obtained from each pup to determine F1 genotype (*see* Subheading 3.1). All F1 offspring are expected to be heterozygous for both genes (i.e., CX3CR1-CreER$^{+/wt}$ × PPARγ$^{fl/wt}$) (*see* Fig. 2, F1).

3. Make a breeder pair using an F1 offspring that is heterozygous for both Cre and the target-floxed gene (i.e., CX3CR1-CreER$^{+/wt}$ × PPARγ$^{fl/wt}$) and a homozygous target-floxed mouse (i.e., PPARγ$^{fl/fl}$) (*see* Fig. 2). This may be a backcross where the homozygous target-floxed mouse used is one of the parent founders, but a separately procured animal is recommended to reduce inbreeding.

4. Pups from the F2 cross should be identified prior to weaning by tattooing or ear tags. Obtain tail or ear snips to determine F2 genotype (*see* Subheading 3.1). Assuming Mendelian inheritance, 25% of F2 offspring will be the desired Cre$^{+/wt}$ × Target$^{fl/fl}$ genotype (i.e., CX3CR1-CreER$^{+/wt}$ × PPARγ$^{fl/fl}$) needed for the F3 cross (*see* Fig. 2, F2).

5. Make a F3 breeder pair using an F2 offspring that is Cre$^{+/wt}$ × Target$^{fl/fl}$ (i.e., CX3CR1-CreER$^{+/wt}$ × PPARγ$^{fl/fl}$) and a homozygous target-floxed mouse (i.e., PPARγ$^{fl/fl}$) (*see* Fig. 2).

6. Pups from the F3 cross should be identified prior to weaning using tattoos or ear tags. Obtain tail or ear snips to determine F3 genotype (*see* Subheading 3.1). Assuming Mendelian inheritance, 50% of F3 offspring will be the desired experimental Cre$^{+/wt}$ × Target$^{fl/fl}$ genotype (i.e., CX3CR1-CreER$^{+/wt}$ × PPARγ$^{fl/fl}$). Fifty percent of F3 offspring will be littermate controls with no Cre (i.e., CX3CR1-CreER$^{wt/wt}$ × PPARγ$^{fl/fl}$) (*see* Fig. 2, F3).

7. Repeat **steps 5** and **6** as needed to generate and genotype experimental and littermate control mice.

8. In order to generate heterozygous Cre mice for Cre toxicity controls, make a breeder pair of a homozygous conditional Cre mouse (i.e., CX3CR1-CreER$^{+/+}$) and a wild-type mouse. All F1 offspring are expected to be heterozygous for Cre (i.e., CX3CR1-CreER$^{+/wt}$) (*see* Fig. 2, inset box). Tail or ear tissue can be collected from pups to confirm expected genotype (*see* Subheading 3.1).

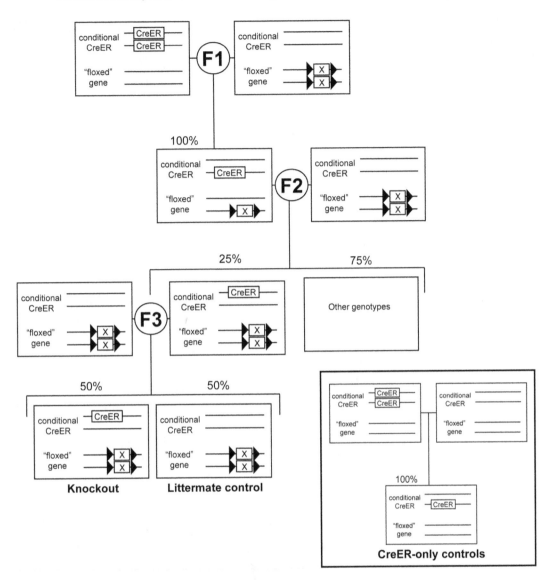

Fig. 2 Breeding strategy for generation of CreER inducible conditional knockout mice. F1 cross between homozygous floxed mouse and homozygous conditional CreER mouse will generate offspring that are 100% heterozygous for the floxed gene of interest and the conditional CreER. F2 cross between an F1 offspring (heterozygous for floxed gene and conditional CreER) and a homozygous floxed mouse will generate offspring that are 25% homozygous for the floxed gene of interest and heterozygous for the conditional CreER. F3 cross between these F2 offspring (homozygous for floxed gene and heterozygous for conditional CreER) will generate offspring that are 50% homozygous for the floxed gene of interest and heterozygous for conditional CreER (knockout) and 50% homozygous for floxed gene without conditional CreER (littermate control). CreER-only controls for Cre toxicity (inset box) can be generated by crossing a homozygous conditional CreER mouse with a wild-type mouse

3.3.2 Administration of Tamoxifen

1. Separate and weigh mice to be used for experimental studies, including controls (*see* **Note 9**).

2. Calculate amount of tamoxifen needed to inject mice with 0.25 mg tamoxifen per g body weight. Multiply by 1.25 to account for solution lost in each needle.

3. Resuspend tamoxifen in corn oil at 20 mg/mL in a 50 mL conical tube. The tube should be wrapped in foil to protect from light.

4. To dissolve tamoxifen in oil, heat the solution in a water bath at 65 °C for 1 h, swirling or vortexing every 15 min to mix.

5. Filter tamoxifen solution through a 22 μm sterile vacuum filter into a fresh 50 mL tube. Sterile filtered tamoxifen solution may be stored at 4 °C for 1 week. When using 4 °C stored tamoxifen, it should be reheated at 65 °C for 20 min, swirled, or vortexed after the first 10 min.

6. To maintain tamoxifen in suspension, keep the sterile filtered tamoxifen solution warm by placing it in a glass beaker containing hot tap water. To prevent contamination from tap water, make sure the lid is sealed and that it does not get submerged.

7. Using a fresh 1 mL syringe and 25 gauge × 5/8 needle for each mouse, inject with 20 mg/mL tamoxifen solution intraperitoneally at a dose of 0.25 mg tamoxifen/g body weight (generally 200–300 μL in an adult mouse). Do not pre-draw syringes. Syringes should be drawn fresh from warm tamoxifen solution; tamoxifen solution becomes highly viscous as it cools.

8. Mark cages of treated mice (*see* **Note 10**).

9. Repeat **steps 1–7** depending on desired duration of tamoxifen treatment. Mice should be allowed to rest 2–4 days between each tamoxifen injection. Do not inject more than three times in 1 week; twice a week is generally sufficient (*see* **Note 11**).

10. Perform experiments at desired time after beginning tamoxifen injections (*see* **Note 12**).

3.4 Generation of Mice with Inducible Conditional Loss of Function (rtTA)

Temporally controlled loss-of-function studies using the doxycycline-inducible system employ conditional rtTA mice (Table 2) that are crossed with mice expressing tetO-Cre and with mice where the endogenous gene of interest is floxed (i.e., triple-transgenic system). Administration of doxycycline or an alternate tetracycline analogue will activate the conditionally expressed rtTA. This will trigger Cre expression and recombination of floxed DNA sites and delete the gene of interest. An appropriate breeding scheme will generate experimental animals and littermate controls (*see* Fig. 3). Best practice is to also include a group of conditional rtTA mice crossed with tetO-Cre but with no

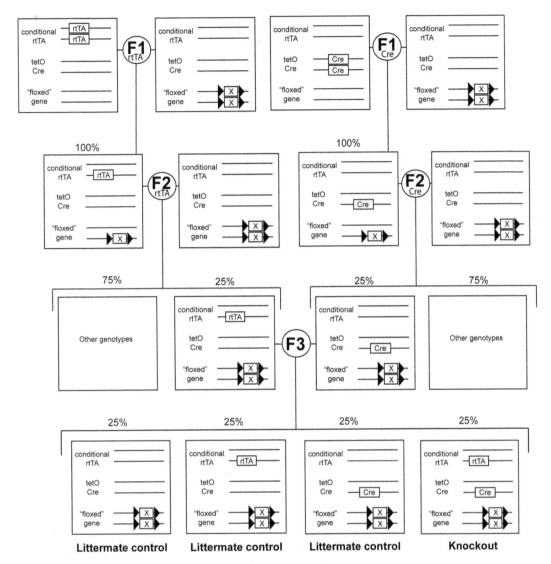

Fig. 3 Breeding strategy for generation of rtTA inducible conditional knockout mice. F1-rtTA cross between homozygous floxed mouse and homozygous conditional rtTA mouse will generate offspring that are 100% heterozygous for the floxed gene of interest and the conditional rtTA. A parallel F1-Cre cross between homozygous floxed mouse and homozygous tetO-Cre mouse will generate offspring that are 100% heterozygous for the floxed gene of interest and the tetO-Cre. F2-rtTA cross between an F1 offspring (heterozygous for floxed gene and conditional rtTA) and a homozygous floxed mouse will generate offspring that are 25% homozygous for the floxed gene of interest and heterozygous for the conditional rtTA. A parallel F2-Cre cross between an F1 offspring (heterozygous for floxed gene and tetO-Cre) and a homozygous floxed mouse will generate offspring that are 25% homozygous for the floxed gene of interest and heterozygous for the tetO-Cre. F3 cross between these F2-rtTA and F2-Cre offspring (homozygous for floxed gene and heterozygous for conditional rtTA or tetO-Cre) will generate offspring that are 25% homozygous for the floxed gene of interest and heterozygous for both conditional rtTA and tetO-Cre (knockout). The remaining 75% of F3 offspring may be used as littermate controls

floxed gene to confirm that Cre toxicity did not occur and lead to off-target effects. All control groups should be treated with doxycycline (*see* **Note 13**).

3.4.1 Breeding rtTA Inducible Conditional Mice

1. Make a breeder pair using a homozygous conditional rtTA mouse (i.e., CD68-rtTA$^{+/+}$) and a homozygous target-floxed mouse (i.e., PPARγ$^{fl/fl}$). This is the rtTA F1 cross. These founder mice should be obtained from a commercial breeder (*see inducible conditional lines detailed in* Table 2), reputable academic collaborator, or other trusted source.

2. Simultaneously, make a second breeder pair using a homozygous tetO-Cre mouse and a homozygous target-floxed mouse (i.e., PPARγ$^{fl/fl}$). This is the tetO-Cre F1 cross.

3. Pups from these F1 crosses should be identified prior to weaning by tattoo or ear tag. At the time of tattoo or ear tag, a tail or ear snip should be cut and used to determine F1 genotype (*see* Subheading 3.1). All F1 offspring are expected to be heterozygous for both genes (i.e., CD68-rtTA$^{+/wt}$ × PPARγ$^{fl/wt}$ and tetO-Cre$^{+/wt}$ × PPARγ$^{fl/wt}$) (*see* Fig. 3, F1).

4. Make a new breeding pair by crossing rtTA F1 offspring (heterozygous for rtTA and target-floxed gene) with a homozygous target-floxed mouse (i.e., PPARγ$^{fl/fl}$) (*see* Fig. 3). This may be a backcross where the homozygous target-floxed mouse used is one of the parent founders, but a separately procured animal is recommended to reduce inbreeding.

5. Simultaneously with **step 4**, make a second breeding pair by crossing tetO-Cre F1 offspring (heterozygous for tetO-Cre and target-floxed gene) with a homozygous target-floxed mouse (i.e., PPARγ$^{fl/fl}$).

6. F2 pups should be identified prior to weaning by tattoo or ear tag. At the time of tattoo or ear tag, a tail or ear snip should be obtained from each pup and used to determine F2 genotype (*see* Subheading 3.1). Assuming Mendelian inheritance, 25% of F2 offspring will be the desired rtTA$^{+/wt}$ × Target$^{fl/fl}$ or tetO$^{+/wt}$ × Target$^{fl/fl}$ genotype (i.e., CD68rtTA$^{+/wt}$ × PPARγ$^{fl/fl}$ or tetO-cre$^{+/wt}$ × PPARγ$^{fl/fl}$) needed for the F3 cross (*see* Fig. 2, F2).

7. Make a F3 breeder pair by crossing rtTA F2 offspring that are heterozygous for rtTA and homozygous for the target-floxed gene (i.e., CD68-rtTA$^{+/wt}$ × PPARγ$^{fl/fl}$) with tetO-Cre F2 offspring that are heterozygous for tetO-Cre and homozygous for the target-floxed gene (i.e., tetO-Cre$^{+/wt}$ × PPARγ$^{fl/fl}$) (*see* Fig. 3).

8. F3 pups should be identified prior to weaning by tattoo or ear tag. At the same time, determine F3 genotypes. Assuming Mendelian inheritance, 25% of F3 offspring will be the desired experimental rtTA+/wt × tetO-Cre$^{+/wt}$ × Target$^{fl/fl}$ genotype

(i.e., CD68-rtTA$^{+/wt}$ × tetO-Cre$^{+/wt}$ × PPARγ$^{fl/fl}$). Seventy-five percent of F3 offspring will be littermate controls lacking rtTA, tetO-Cre, or both (*see* Fig. 2, F3) (*see* **Note 14**).

9. Repeat **steps 7** and **8** as needed to generate and genotype experimental and littermate control mice.

10. In order to generate heterozygous Cre mice for Cre toxicity controls, make a breeder pair of a homozygous rtTA mouse (i.e., CD68-rtTA$^{+/+}$) and a homozygous tetO-Cre mouse. All F1 offspring are expected to be heterozygous for rtTA and tetO-Cre (i.e., CD68-rtTA$^{+/wt}$ × tetO-Cre$^{+/wt}$). Tail or ear tissue can be collected from pups to confirm expected genotype (*see* Subheading 3.1).

3.4.2 Administration of Doxycycline

1. Separate mice to be used for experimental studies, including controls.

2. Remove normal chow from cage and replace with doxycycline chow.

3. Mark cages of mice on doxycycline diet (*see* **Note 15**).

4. Perform experiments at desired time after beginning doxycycline chow (*see* **Note 12**)

3.5 Analysis of Recombination Using Reporter Mice

Best practice for the use of conditional and inducible conditional systems incorporates crossing the chosen system with reporter mice (*see* Table 3) to assess Cre activity. This is especially important when an injury or infection model will be used (*see* **Note 16**). Transgenic reporter mice enable investigators to understand the penetrance, specificity, and kinetics of conditional and inducible conditional lines, although this should not replace testing the knockdown of gene or protein expression in target cell populations (*see* **Note 17**). Transgenic reporter mice may also be used for lineage tracing or fate-mapping myeloid cells of interest.

1. Make a breeder pair of a homozygous conditional Cre or rtTA mouse with an appropriate homozygous reporter mouse (*see* Table 3).

2. All F1 pups will be Cre$^{+/wt}$ × Reporter$^{+/-}$ or rtTA$^{+/wt}$ × Reporter$^{+/-}$ and may be used for experiments such as flow cytometric analysis of reporter expression by cell populations (*see* **Notes 18** and **19**) (Fig. 4).

4 Notes

1. Doxycycline chow can be purchased with green coloring. This enables it to be visually distinguished from normal chow. Common formulations (such as Teklad rodent diet) aim to provide 2–3 mg of doxycycline per day. This is based

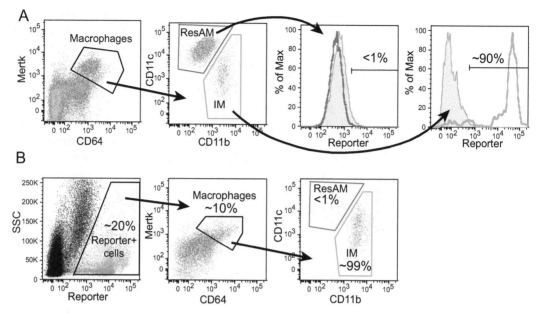

Fig. 4 Flow cytometry gating of reporter mice to assess efficiency and specificity of Cre-driver lines. CX3CR1-CreER mice were crossed with R26-TdTomato reporter mice. Offspring were administered tamoxifen for 1 week, then lungs were harvested, stained with surface antibodies, and reporter expression was assessed by flow cytometry. Efficiency and specificity of CX3CR1-CreER for lung macrophages were assessed. (**a**) Example of gating to assess reporter efficiency in target populations. Lung macrophages were identified from the lung digests by selecting CD45+ Ly6G− cells (not shown) and then selecting for co-expression of the macrophage markers CD64 and Mertk. Macrophages were separated into resident alveolar macrophages (ResAM) and interstitial macrophages (IM) by expression of CD11c and CD11b. ResAM (blue histogram) and IM (orange histogram) reporter expression was assessed using ResAM and IM from non-reporter control animals (gray histogram) to set gating. CX3CR1-CreER efficiently targets IM (~90% of cells are reporter-positive) and does not target ResAM (<1% of cells are reporter-positive). (**b**) Example of gating to assess reporter specificity to target populations. Reporter-positive cells (red dots) were distinguished from non-reporter control animals (black dots). Next, macrophages were identified from within all reporter-positive cells by co-expression of CD64 and Mertk. CX3CR1-CreER is not specific to macrophages (macrophages only account for ~10% of reporter-positive cells), although reporter-positive macrophages are IM (~99%). The identity of remaining reporter-positive cells can be interrogated using additional surface markers

on average consumption of 4–5 g of chow per day. Tamoxifen chow is optimally obtained through your institutional animal care facility. The facility may choose whether or not to irradiate chow.

2. For commercially available mice, the company will usually provide suggested PCR conditions optimal for the suggested primer sets. Generally, a brief 95 °C heat activation step followed by 25–29 cycles of 95 °C denaturation, 55–65 °C annealing, and 72 °C extension is sufficient. The time needed for extension will depend on the size of product being amplified, with 15–30 s needed per kilobase.

3. For transgenic mice, different primer sets are needed for each transgene. Some mice require multiple reactions for a single transgene to distinguish wild type from transgenic DNA. Double- and triple-transgenic mice will require two to six separate PCR reactions, depending on the mice used. This can make the genotyping of transgenic mice expensive in terms of both time and money. Paying a trusted commercial source or institutional facility to perform genotyping (generally done with automated systems) may cost less than the time and reagents needed to perform genotyping in house.

4. Ethidium bromide is a carcinogen that requires special handling and special disposal, in accordance with institutional safety regulations. Ethidium bromide for PCR gels may be replaced with more expensive but less hazardous alternatives such as GelRed (Biotium) or SYBRsafe (Life Technologies). Ethidium bromide should never be heated but rather added to the cooled gel solution as described.

5. MangoMix contains a loading dye for running PCR samples on an agarose gel, and thus the PCR product can be added straight to the gel.

6. Amplify DNA from mice with known wild-type alleles (negative control) and known transgenic alleles (positive control), as well as with no template DNA (no template control). Include these controls with experimental DNA run on each gel to ensure that primer reactions occurred as expected and to aid in proper identification of the gel bands for wild-type and transgenic alleles.

7. Cre can be toxic to cells; heterozygous Cre produces less toxicity. Further, reports have found an increase in nonspecific or background Cre expression by nontarget cells when using homozygous Cre. Additionally, some conditional Cre mice, such as LysMcre, have been made in a way that disrupts the endogenous gene activity; using these mice as heterozygotes insures that endogenous gene expression is not knocked out.

8. Tamoxifen, a chemotherapeutic, has been shown to cause intestinal damage in mice [28]; this must be accounted for by parallel treatment of control groups in order to interpret experimental data. Depending on the experimental question, experimental and control animals given vehicle may also be an important control to account for background recombination in the absence of tamoxifen.

9. Tamoxifen injection is a common way to activate CreER systems. However tamoxifen may also be administered through diet. A variety of tamoxifen diets are available (Teklad Rodent Diets) geared to dose 40–80 mg tamoxifen per kg body weight each day. Feed aversion is common with the use of tamoxifen

chow, which is why injection is often preferred. A gradual acclimation to tamoxifen chow by mixing 1:1 with normal chow for the first week may increase intake. Mice should be monitored daily while on tamoxifen diet and for at least 5 days after normal chow is reinstated. Injections of IP fluids or providing additional wet food can be used to support mice experiencing weight loss.

10. Human contact with tamoxifen should be limited, particularly in groups sensitive to chemotherapeutic agents such as women who are pregnant. In accordance with institutional regulations, bedding may need to be specially handled by animal care technicians. Cages should be clearly labeled with the dates of tamoxifen administration and appropriate hazard information. In addition, mice should be observed daily for at least 5 days following tamoxifen injection. IP fluids or wet food can be provided to support mice experiencing weight loss in response to tamoxifen.

11. Limiting the number of injections is recommended to reduce side effects. Two injections given over 1 week are sufficient to induce high-efficiency Cre recombination. While this is a standard starting protocol, the number of injections and even the dose of tamoxifen used for a study should be determined empirically by the investigator depending on the specific CreER strain and experimental design. To limit tamoxifen toxicity, investigators should strive to provide the lowest dose of tamoxifen necessary to reliably activate CreER.

12. The inducible nature of the CreER and rtTA systems allows investigators to induce DNA recombination at any point. Pregnant mothers may be dosed with tamoxifen or doxycycline to induce recombination in utero. More commonly, tamoxifen or doxycycline is not provided until mice are 6–8 weeks old, allowing for normal development. After recombination, the DNA is permanently altered. For the study of long-lived cells, continual administration is not required; to maintain gene deletion in cells with rapid turnover, continual administration is required.

13. Doxycycline is an anti-inflammatory antimicrobial drug and may be protective in some models of infection or injury; this must be accounted for by parallel treatment of control groups in order to interpret experimental data. Depending on the experimental question, experimental and control animals given vehicle may also be an important control to account for background recombination in the absence of doxycycline.

14. The choice of littermate control is at the investigators discretion; all three may provide critical information about the system, and all three may be used. The rtTA-floxed gene littermate control accounts for potential rtTA toxicity, which

has been reported in specific strains [29]. Because Cre is not expressed without rtTA, the Cre-floxed gene littermate control is not an accurate control for Cre toxicity, although it does serve as a control to assess non-specific Cre leak. Finally, floxed gene only littermates serve as the controls against which rtTA toxicity or Cre leak can be assessed. In all cases, the use of true sibling or littermate controls accounts for potential strain variation introduced by crossing transgenic mice that are frequently generated on mixed strain backgrounds.

15. Unlike tamoxifen, doxycycline is not hazardous to humans or mice. Where in accordance with institutional animal use guidelines, biohazard marking and daily monitoring should not be required. However cages should be marked such that animal care technicians are aware that mice are being fed a special diet.

16. In some cases, inflammation has been found to alter Cre recombinase expression. Additionally, new populations of leukocytes are recruited to the lung during inflammation including CD11b[hi] macrophages in the airspaces termed recruited macrophages, as well as monocytes and neutrophils. Very few studies describing conditional Cre expression through the use of reporter mice have done so during inflammation.

17. Reporter mice are a surrogate for Cre expression that enables critical understanding of the chosen transgenic system, but deletion of the target gene in cells of interest should be confirmed at a mRNA or protein level by isolating cells of interest (discussed elsewhere in this volume) and performing RT-PCR or western blot (*see* Chapter 14 Modulation of lung epithelial cell function using conditional and inducible transgenic approaches for further detail). Cre recombination rarely occurs in a true 100% of target cells; even in very efficient systems, recombination may peak at 99%. If the floxed target gene is critical for cell function, selection pressure will drive any non-recombined cells to fill the niche, leaving the investigator with relatively normal rather than conditional knockout cells.

18. R26-driven reporters provide a highly accurate representation of Cre or CreER activity. These systems use a floxed stop codon to prevent reporter expression until the floxed DNA is recombined by Cre; the reporter will subsequently be expressed for the life of the cell and all daughter cells. This is equivalent to the Cre-mediated recombination of a floxed target gene of interest. The pTRE-H2BGFP reporter used to study rtTA systems is accurate in systems where doxycycline is continually given to adult mice. Further, the GFP-histone fusion protein is very long-lived. However, if the investigator is interested in doxycycline withdrawal, it is recommended to cross the conditional rtTA with a tetO-Cre mouse and then a R26-driven

reporter system, which would be achieved through the breeding strategy discussed in Subheading 3.4. This will provide the most accurate information regarding rtTA-driven Cre activity.

19. Surface markers assessed by flow cytometric analysis can identify myeloid subpopulations in the lung (*see* Chapter 3, Isolation and Characterization of Mouse Mononuclear Phagocytes for further detail). Using reporter strains, the efficiency and specificity of conditional lines for given cell populations be determined by (A) identifying specific cell populations and then measuring the level of reporter expression in target cell populations (efficiency) or (B) gating first on reporter-positive cells and assessing how many are the target cell population(s) versus other "off-target" cell types (specificity). Examples of flow cytometry gating to assess efficiency or specificity are shown in Fig. 4.

References

1. Orban PC, Chui D, Marth JD (1992) Tissue- and site-specific DNA recombination in transgenic mice. Proc Natl Acad Sci U S A 89(15):6861–6865

2. Rajewsky K, Gu H, Kuhn R, Betz UA, Muller W, Roes J, Schwenk F (1996) Conditional gene targeting. J Clin Invest 98(3):600–603. https://doi.org/10.1172/jci118828

3. Wang X (2009) Cre transgenic mouse lines. Methods Mol Biol 561:265–273. https://doi.org/10.1007/978-1-60327-019-9_17

4. Miller RL (2011) Transgenic mice: beyond the knockout. Am J Physiol Renal Physiol 300(2):F291–F300. https://doi.org/10.1152/ajprenal.00082.2010

5. Haruyama N, Cho A, Kulkarni AB (2009) Overview: engineering transgenic constructs and mice. Curr Protoc Cell Biol Chapter 19:Unit 19.10. doi:https://doi.org/10.1002/0471143030.cb1910s42

6. Clausen BE, Burkhardt C, Reith W, Renkawitz R, Forster I (1999) Conditional gene targeting in macrophages and granulocytes using LysMcre mice. Transgenic Res 8(4):265–277

7. Abram CL, Roberge GL, Hu Y, Lowell CA (2014) Comparative analysis of the efficiency and specificity of myeloid-Cre deleting strains using ROSA-EYFP reporter mice. J Immunol Methods 408:89–100. https://doi.org/10.1016/j.jim.2014.05.009

8. Schmidt-Supprian M, Rajewsky K (2007) Vagaries of conditional gene targeting. Nat Immunol 8(7):665–668. https://doi.org/10.1038/ni0707-665

9. Jakubzick C, Bogunovic M, Bonito AJ, Kuan EL, Merad M, Randolph GJ (2008) Lymph-migrating, tissue-derived dendritic cells are minor constituents within steady-state lymph nodes. J Exp Med 205(12):2839–2850. https://doi.org/10.1084/jem.20081430

10. Sauer B, Henderson N (1988) Site-specific DNA recombination in mammalian cells by the Cre recombinase of bacteriophage P1. Proc Natl Acad Sci U S A 85(14):5166–5170

11. Feil S, Valtcheva N, Feil R (2009) Inducible Cre mice. Methods Mol Biol 530:343–363. https://doi.org/10.1007/978-1-59745-471-1_18

12. Belteki G, Haigh J, Kabacs N, Haigh K, Sison K, Costantini F, Whitsett J, Quaggin SE, Nagy A (2005) Conditional and inducible transgene expression in mice through the combinatorial use of Cre-mediated recombination and tetracycline induction. Nucleic Acids Res 33(5):e51. https://doi.org/10.1093/nar/gni051

13. Casanova E, Fehsenfeld S, Lemberger T, Shimshek DR, Sprengel R, Mantamadiotis T (2002) ER-based double iCre fusion protein allows partial recombination in forebrain. Genesis 34(3):208–214. https://doi.org/10.1002/gene.10153

14. Gossen M, Freundlieb S, Bender G, Muller G, Hillen W, Bujard H (1995) Transcriptional activation by tetracyclines in mammalian cells. Science 268(5218):1766–1769

15. Kistner A, Gossen M, Zimmermann F, Jerecic J, Ullmer C, Lubbert H, Bujard H (1996) Doxycycline-mediated quantitative and tissue-

specific control of gene expression in transgenic mice. Proc Natl Acad Sci U S A 93(20):10933–10938

16. Zhu Z, Zheng T, Lee CG, Homer RJ, Elias JA (2002) Tetracycline-controlled transcriptional regulation systems: advances and application in transgenic animal modeling. Semin Cell Dev Biol 13(2):121–128

17. McCubbrey AL, Barthel L, Mohning MP, Redente EF, Mould KJ, Thomas SM, Leach SM, Danhorn T, Gibbings SL, Jakubzick CV, Henson PM, Janssen WJ (2017) Deletion of c-FLIP from CD11bhi macrophages prevents development of Bleomycin-induced lung fibrosis. Am J Respir Cell Mol Biol 58:66–78. https://doi.org/10.1165/rcmb.2017-0154OC

18. Kamper MR, Gohla G, Schluter G (2002) A novel positive tetracycline-dependent transactivator (rtTA) variant with reduced background activity and enhanced activation potential. FEBS Lett 517(1–3):115–120

19. Koponen JK, Kankkonen H, Kannasto J, Wirth T, Hillen W, Bujard H, Yla-Herttuala S (2003) Doxycycline-regulated lentiviral vector system with a novel reverse transactivator rtTA2S-M2 shows a tight control of gene expression in vitro and in vivo. Gene Ther 10(6):459–466. https://doi.org/10.1038/sj.gt.3301889

20. McCubbrey AL, Barthel L, Mould KJ, Mohning MP, Redente EF, Janssen WJ (2016) Selective and inducible targeting of CD11b+ mononuclear phagocytes in the murine lung with hCD68-rtTA transgenic systems. Am J Physiol Lung Cell Mol Physiol 311(1):L87–l100. https://doi.org/10.1152/ajplung.00141.2016

21. Yona S, Kim KW, Wolf Y, Mildner A, Varol D, Breker M, Strauss-Ayali D, Viukov S, Guilliams M, Misharin A, Hume DA, Perlman H, Malissen B, Zelzer E, Jung S (2013) Fate mapping reveals origins and dynamics of monocytes and tissue macrophages under homeostasis. Immunity 38(1):79–91. https://doi.org/10.1016/j.immuni.2012.12.001

22. Zaynagetdinov R, Sherrill TP, Kendall PL, Segal BH, Weller KP, Tighe RM, Blackwell TS (2013) Identification of myeloid cell subsets in murine lungs using flow cytometry. Am J Respir Cell Mol Biol 49(2):180–189. https://doi.org/10.1165/rcmb.2012-0366MA

23. Misharin AV, Morales-Nebreda L, Mutlu GM, Budinger GR, Perlman H (2013) Flow cytometric analysis of macrophages and dendritic cell subsets in the mouse lung. Am J Respir Cell Mol Biol 49(4):503–510. https://doi.org/10.1165/rcmb.2013-0086MA

24. Gibbings SL, Thomas SM, Atif SM, McCubbrey AL, Desch AN, Danhorn T, Leach SM, Bratton DL, Henson PM, Janssen WJ, Jakubzick CV (2017) Three unique interstitial macrophages in the murine lung at steady state. Am J Respir Cell Mol Biol 57(1):66–76. https://doi.org/10.1165/rcmb.2016-0361OC

25. Janssen WJ, Barthel L, Muldrow A, Oberley-Deegan RE, Kearns MT, Jakubzick C, Henson PM (2011) Fas determines differential fates of resident and recruited macrophages during resolution of acute lung injury. Am J Respir Crit Care Med 184(5):547–560. https://doi.org/10.1164/rccm.201011-1891OC

26. Gibbings SL, Goyal R, Desch AN, Leach SM, Prabagar M, Atif SM, Bratton DL, Janssen W, Jakubzick CV (2015) Transcriptome analysis highlights the conserved difference between embryonic and postnatal-derived alveolar macrophages. Blood 126(11):1357–1366. https://doi.org/10.1182/blood-2015-01-624809

27. Lee JY, Ristow M, Lin X, White MF, Magnuson MA, Hennighausen L (2006) RIP-Cre revisited, evidence for impairments of pancreatic beta-cell function. J Biol Chem 281(5):2649–2653. https://doi.org/10.1074/jbc.M512373200

28. Huh WJ, Khurana SS, Geahlen JH, Kohli K, Waller RA, Mills JC (2012) Tamoxifen induces rapid, reversible atrophy, and metaplasia in mouse stomach. Gastroenterology 142(1):21–24.e27. https://doi.org/10.1053/j.gastro.2011.09.050

29. Morimoto M, Kopan R (2009) rtTA toxicity limits the usefulness of the SP-C-rtTA transgenic mouse. Dev Biol 325(1):171–178. https://doi.org/10.1016/j.ydbio.2008.10.013

30. Deng L, Zhou JF, Sellers RS, Li JF, Nguyen AV, Wang Y, Orlofsky A, Liu Q, Hume DA, Pollard JW, Augenlicht L, Lin EY (2010) A novel mouse model of inflammatory bowel disease links mammalian target of rapamycin-dependent hyperproliferation of colonic epithelium to inflammation-associated tumorigenesis. Am J Pathol 176(2):952–967. https://doi.org/10.2353/ajpath.2010.090622

31. Loschko J, Rieke GJ, Schreiber HA, Meredith MM, Yao KH, Guermonprez P, Nussenzweig MC (2016) Inducible targeting of cDCs and their subsets in vivo. J Immunol Methods 434:32–38. https://doi.org/10.1016/j.jim.2016.04.004

32. Wu X, Briseno CG, Durai V, Albring JC, Haldar M, Bagadia P, Kim KW, Randolph GJ,

Murphy TL, Murphy KM (2016) Mafb lineage tracing to distinguish macrophages from other immune lineages reveals dual identity of Langerhans cells. J Exp Med 213(12):2553–2565. https://doi.org/10.1084/jem.20160600

33. Caton ML, Smith-Raska MR, Reizis B (2007) Notch-RBP-J signaling controls the homeostasis of CD8- dendritic cells in the spleen. J Exp Med 204(7):1653–1664. https://doi.org/10.1084/jem.20062648

34. Loschko J, Schreiber HA, Rieke GJ, Esterhazy D, Meredith MM, Pedicord VA, Yao KH, Caballero S, Pamer EG, Mucida D, Nussenzweig MC (2016) Absence of MHC class II on cDCs results in microbial-dependent intestinal inflammation. J Exp Med 213(4):517–534. https://doi.org/10.1084/jem.20160062

35. Passegue E, Wagner EF, Weissman IL (2004) JunB deficiency leads to a myeloproliferative disorder arising from hematopoietic stem cells. Cell 119(3):431–443. https://doi.org/10.1016/j.cell.2004.10.010

36. Ferron M, Vacher J (2005) Targeted expression of Cre recombinase in macrophages and osteoclasts in transgenic mice. Genesis 41(3):138–145. https://doi.org/10.1002/gene.20108

37. Franke K, Kalucka J, Mamlouk S, Singh RP, Muschter A, Weidemann A, Iyengar V, Jahn S, Wieczorek K, Geiger K, Muders M, Sykes AM, Poitz DM, Ripich T, Otto T, Bergmann S, Breier G, Baretton G, Fong GH, Greaves DR, Bornstein S, Chavakis T, Fandrey J, Gassmann M, Wielockx B (2013) HIF-1alpha is a protective factor in conditional PHD2-deficient mice suffering from severe HIF-2alpha-induced excessive erythropoiesis. Blood 121(8):1436–1445. https://doi.org/10.1182/blood-2012-08-449181

38. Doyle AD, Jacobsen EA, Ochkur SI, Willetts L, Shim K, Neely J, Kloeber J, Lesuer WE, Pero RS, Lacy P, Moqbel R, Lee NA, Lee JJ (2013) Homologous recombination into the eosinophil peroxidase locus generates a strain of mice expressing Cre recombinase exclusively in eosinophils. J Leukoc Biol 94(1):17–24. https://doi.org/10.1189/jlb.0213089

39. Qian BZ, Li J, Zhang H, Kitamura T, Zhang J, Campion LR, Kaiser EA, Snyder LA, Pollard JW (2011) CCL2 recruits inflammatory monocytes to facilitate breast-tumour metastasis. Nature 475(7355):222–225. https://doi.org/10.1038/nature10138

40. Pillai MM, Hayes B, Torok-Storb B (2009) Inducible transgenes under the control of the hCD68 promoter identifies mouse macrophages with a distribution that differs from the F4/80- and CSF-1R-expressing populations. Exp Hematol 37(12):1387–1392. https://doi.org/10.1016/j.exphem.2009.09.003

41. Yan C, Lian X, Li Y, Dai Y, White A, Qin Y, Li H, Hume DA, Du H (2006) Macrophage-specific expression of human lysosomal acid lipase corrects inflammation and pathogenic phenotypes in lal−/− mice. Am J Pathol 169(3):916–926. https://doi.org/10.2353/ajpath.2006.051327

42. Pan H, Mostoslavsky G, Eruslanov E, Kotton DN, Kramnik I (2008) Dual-promoter lentiviral system allows inducible expression of noxious proteins in macrophages. J Immunol Methods 329(1–2):31–44. https://doi.org/10.1016/j.jim.2007.09.009

43. Di Meglio T, Kratochwil CF, Vilain N, Loche A, Vitobello A, Yonehara K, Hrycaj SM, Roska B, Peters AH, Eichmann A, Wellik D, Ducret S, Rijli FM (2013) Ezh2 orchestrates topographic migration and connectivity of mouse precerebellar neurons. Science (New York, NY) 339(6116):204–207. https://doi.org/10.1126/science.1229326

44. Madisen L, Zwingman TA, Sunkin SM, Oh SW, Zariwala HA, Gu H, Ng LL, Palmiter RD, Hawrylycz MJ, Jones AR, Lein ES, Zeng H (2010) A robust and high-throughput Cre reporting and characterization system for the whole mouse brain. Nat Neurosci 13(1):133–140. https://doi.org/10.1038/nn.2467

45. Srinivas S, Watanabe T, Lin CS, William CM, Tanabe Y, Jessell TM, Costantini F (2001) Cre reporter strains produced by targeted insertion of EYFP and ECFP into the ROSA26 locus. BMC Dev Biol 1:4

46. Mao X, Fujiwara Y, Chapdelaine A, Yang H, Orkin SH (2001) Activation of EGFP expression by Cre-mediated excision in a new ROSA26 reporter mouse strain. Blood 97(1):324–326

47. Tumbar T, Guasch G, Greco V, Blanpain C, Lowry WE, Rendl M, Fuchs E (2004) Defining the epithelial stem cell niche in skin. Science (New York, NY) 303(5656):359–363. https://doi.org/10.1126/science.1092436

Chapter 14

Modulation of Lung Epithelial Cell Function Using Conditional and Inducible Transgenic Approaches

Adrianne L. Stefanski, Dorota S. Raclawska, and Christopher M. Evans

Abstract

In the lungs, the epithelium is a first line of innate defense. In acute settings, such as infection or particulate exposure, the epithelium is protective. Protection is conferred by the epithelium's role as a physical barrier and by its ability to synthesize proteins that promote defense directly through physical interactions (e.g., mucins and anti-microbial peptides) and indirectly through the production of proteins that regulate inflammation (e.g., cytokines and chemokines). Despite its importance as a first line of host defense, the epithelium is also a significant target and an effector in lung pathologies. Accordingly, to determine the significance and biological mechanisms of genes involved in pulmonary defense, it is important to be able to interrogate the lung epithelium. In mice, this presents challenges related to the cellular location and timing of interventions. Effective genetic strategies for targeting the lung epithelium using tissue-/cell-specific and inducible control have been developed over the past decade. Methods for spatiotemporal targeting of gene expression are described here.

Key words Lungs, Airways, Cilia, Goblet cell, Mucus, Mucous, Mucin

1 Introduction

With a surface area of 50–100 m² and a daily exposure to approximately 10,000 L of air containing hundreds of billions of particles, it is crucial for the lungs to defend against a plethora of exposures. It is equally important that protection occurs with minimal physiological disruption or injury caused by inflammation. Accordingly, the lungs employ defenses that are noninflammatory under homeostatic conditions. On the other hand, when inflammation is needed, defenses are restrained and transient.

The lungs are lined by an epithelial layer that serves as a first line of innate defense. In healthy individuals, epithelial cells lining the conducting airways (the trachea, bronchi, and bronchioles) are equipped with the means to physically prevent the accumulation of particles that could cause injury to the airways or damage to delicate gas-exchange surfaces (alveoli). Airway epithelial cells form a

Scott Alper and William J. Janssen (eds.), *Lung Innate Immunity and Inflammation: Methods and Protocols*,
Methods in Molecular Biology, vol. 1809, https://doi.org/10.1007/978-1-4939-8570-8_14,
© Springer Science+Business Media, LLC, part of Springer Nature 2018

physical barrier that is further aided by a dynamic mucociliary clearance mechanism. Two epithelial cell types are crucial for mucociliary clearance: secretory cells and ciliated cells. Secretory cells include surface and glandular mucin-producing cells. Traditionally, these are referred to as goblet and mucous cells, respectively. Both cell types synthesize polymeric mucin glycoproteins that are the chief macromolecules in airway mucus, comprising MUC5AC and MUC5B mucin isoforms predominantly. Inhaled particles entering the airways adhere to mucus, become trapped, and are then removed by mucociliary clearance. Removal is controlled by ciliated cells, whose apical motile cilia engage in coordinated beating that propels mucus and its contents cranially for eventual elimination by expectoration or swallowing. As a result, healthy mucociliary clearance allows for the efficient removal of particles (and pathogens) before they are able to stimulate a significant inflammatory response.

In some cases, the need for additional lines of defense is unavoidable. For example, very small particles (<1 μm diameter) may not impact on conducting airway surfaces. Instead, they reach alveolar surfaces. Alveoli are lined by type I and type II epithelial cells. Type I epithelia are the thin cells that normally allow for gas exchange, but in some cases extremely small particles, as well as vapor phase agents, may pass through them. Type II epithelia secrete surfactant to maintain alveolar patency, and they also secrete defensive molecules such as collectins, anti-microbial peptides, and cytokines, which are anti-microbial directly or via facilitation of leukocyte responses [1]. Notably, the expression of these anti-microbial factors is not exclusively limited to alveolar type II cells, as surface and glandular secretory cells of the conducting airways are also sources [2, 3]. Altogether, these define a second layer of epithelial protection, the production of secreted defense factors.

The third defense strategy is the ability of lung epithelia in both the airways and alveoli to coordinate inflammation. The epithelium is an important site for the initiation of pro-inflammatory signal transduction, such as through the activation of pattern recognition receptors [4]. Epithelia also produce cytokines, chemokines, and adhesion molecules that mediate generalized pro-inflammatory signals, stimulate the recruitment of leukocytes into the lungs, and provide a location for them to dock and elicit defensive functions [5, 6].

In a healthy or temporarily challenged state, the barrier, microbicidal, and inflammatory factors described above are crucial for maintaining effective defense. However, acute and chronic diseases are associated with mucociliary dysfunction, impaired pathogen defense, and severe or persistent inflammation [7–9]. Therefore, determining the factors that are critical for epithelial defense, the regulation of their activation, and the control of their effector functions is an important undertaking. While this can be done in vitro using both primary cells and cell lines, to develop an understanding of physiological and pathophysiological mechanisms in vivo, animal models are often used. Mice are employed widely in pulmonary research, and while mice differ substantially from humans and other mammals with respect to normal lung functions and to respiratory disease processes, they still provide a justified means for discovering mechanisms that regulate lung homeostasis and pathobiology.

This chapter does not focus on a specific disease or challenge model per se, as these vary widely and should be tailored for use by individual investigators. Instead, the emphasis is on the ability to deploy mouse genetic tools within specific user-defined challenge models. In this case, reagents and methods are described that utilize elements of lung innate defense and inflammation that are common to mouse and human species. For example, secretory and ciliated cells are present in both mice and humans, and the airway mucins MUC5AC/Muc5ac and MUC5B/Muc5b are also abundant in both species. For these reasons, many of the methods detailed below will focus on the use of Muc5ac and Muc5b in mice.

To provide an appropriate context for the methods detailed below, it is helpful to define the language used when describing genetically engineered mice. Terms used in the field and within this chapter are as follows:

1.1 Transgenic Mice

Mice whose genomes have been altered via insertion of foreign genomic material, deletion of endogenous genetic material, substitution of genomic sequence by exchange of nucleotides, or combinations of these. Examples of transgenes that can be inserted into mice include sequences that can be expressed to produce proteins including direct reporters such as green fluorescent protein (GFP) or enzymes such as beta-galactosidase (β-Gal) or Cre recombinase. GFP and β-Gal can be used to identify the timing and location of expressed genes. Cre recombinase is used to cut DNA at specific 34-base pair target sequences called loxP sites. The Cre-mediated removal of loxP flanked DNA ("floxed" DNA) can be used to disrupt gene expression or in some cases to induce expression of reporter genes. Gene disrupted "knockout" mice are transgenic animals in whom a gene is made defective by Cre-mediated excision of DNA or the introduction of enzymes that disrupt or replace DNA for encoded genetic material.

1.2 Targeted Versus Non-targeted

These terms refer to the directed insertion of genes into known vs. unknown loci. With a targeted approach, a mutation is directed at known genetic locations by use of homologous recombination to introduce material that is flanked with sequences that match specifically mapped genetic targets. Specific targeting thus utilizes endogenous cellular machinery to direct the introduced material to loci that match homologous targeting "arms" and recruit DNA repair enzymes. A non-targeted approach is one in which a mutation is introduced without homologous targeting as a goal. Rather, introduced material is incorporated into random sites in the genome. Targeted approaches generally confer specificity in terms of where and when a transgene is expressed, but this can come at the cost of the amount of expression that occurs. Non-targeted approaches can confer higher levels of expression since exogenous promoter fragments with higher activity than endogenous promoters can be used, and non-targeted DNAs usually integrate into genomic loci as multi-copy inserts. Targeted and non-targeted approaches can both be used to generate conditional and non-conditional transgenics.

1.3 Conditional Versus Non-conditional

Conditional transgenesis refers to the use of targeted or non-targeted approaches that allow for control over the location of expression (e.g., the respiratory epithelium), the timing of expression, or both. Non-conditional targeting results in constitutive, non-specific, and/or widespread effects of mutations.

1.4 Tissue Specific Versus Widespread

Tissue specificity is usually achieved by inserting fragments of DNA containing promoters for genes specifically expressed within a given tissue or cell type or achieved by inserting DNA into endogenous loci to drive expression (e.g., knock-in of an enzyme or reporter gene). Widespread transgenesis is achieved by inserting the DNA into loci that are not specific for a given cell type (e.g., knock-in to the Rosa26 locus), or by inserting DNA containing non-cell-specific promoters (e.g., non-targeted insertion using a beta actin or a viral promoter) driving a recombinant enzyme or reporter, or by causing germline mutation of DNA from endogenous loci.

1.5 Inducible Versus Constitutive

For constitutive expression, widespread or tissue-/cell-specific promoters can be used to directly drive expression of transgenes encoding cDNAs without means for turning expression on or off. Controlling when a gene is expressed (inducible) can be achieved by several means. The most common methods are to use viral transduction or pharmacologic regulation. Viral vectors encoding recombinases such as Cre that can excise DNA can be introduced directly into the lungs. Pharmacologic control is most often gained using doxycycline (dox) or tamoxifen (tmx) to regulate tansgene expression or activity, respectively. For dox-regulated transgenes, a system is employed in which expression is controlled by two transgenes. First, a transgene driven by a 5'-flanking sequence containing a repeated tetracycline operator fused with a cytomegalovirus promoter (tetO) is introduced to drive expression of a target gene upon dox treatment. Second, a reverse tetracycline transactivator (rtTA) transgene driven by a 5'-flanking sequence containing DNA that drives either widespread or tissue-specific expression is introduced. For tmx-regulated transgenes, mice express a mutant form of Cre recombinase fused to a fragment of the estrogen receptor (CreER). In CreER mice, Cre remains in the cytoplasm and lacks activity in the absence of tmx treatment. For inducible systems, in addition to the timing of expression, the location can also be controlled by using tissue-/cell-specific promoters to drive rtTA or CreER expression.

These methods have been used in a variety of settings to explore the effects of full gene deficiency, tissue-specific deficiency and overexpression, and inducible regulation of gene expression. For spatial control, the use of promoters that are specific for genes expressed within epithelial cell subtypes in the conducting and respiratory zones are deployed (Fig. 1). Examples of mice employing these temporal and spatial controller strategies are shown in Fig. 2 and summarized in Table 1. Chapter 13 outlines complementary conditional and inducible transgenic approaches to modulate gene function in myeloid cells.

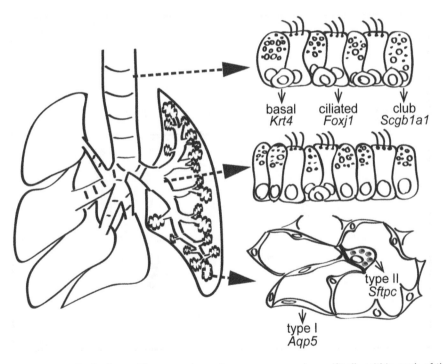

Fig. 1 Epithelial targets in the lungs. The promoters of genes expressed specifically within each of these cell types are commonly used for transgenic targeting. These include *keratin 4* (*Krt4*) for basal cells, *forkhead box J1* (*Foxj1*) for ciliated cells, *secretaglobin 1a1* (*Scgb1a1*) for club cells, *aquaporin 5* (*Aqp5*) for type I alveolar epithelial cells, and *surfactant protein C* (*Sftpc*) for type II alveolar epithelial cells

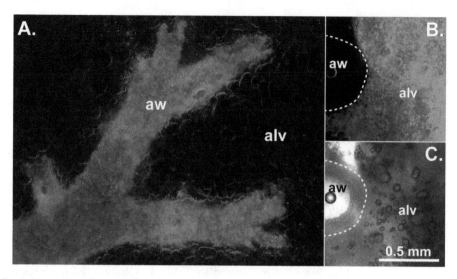

Fig. 2 Spatial targeting of conducting airways and alveoli in adult mice. Two lines of Muc5b overexpressing mice were made using cloned mouse *Scgb1a1* or human *SFTPC* gene promoter fragments to drive conditional expression of Muc5b. In each line, the Muc5b allele was also linked to a fluorescent reporter gene using an internal ribosomal entrance sequence (IRES). Lungs were isolated and prepared for imaging as whole mount tissues. (**a**) Club cells in the conducting airways (aw) of Scgb1a1-Muc5b mice express green fluorescent protein, and there is little to no labeling in surrounding alveolar tissues. (**b** and **c**). Alveolar (alv.) type 2 cells of SFTPC-Muc5b Tg mice express the monomeric cherry variant of red fluorescent protein, which is restricted to alveolar tissues and is absent from conducting airways (outlined with dashed lines in **b** and **c**). Fluorescence of mCherry localized to alveoli is shown in **b**; bright-field image shows airway and alveolar structures in **c**. Strong signals are seen in parenchymal tissues surrounding a bronchial airway. Scale bar = 0.5 mm. Mouse lung samples were generously provided by Dr. David Schwartz

Table 1
Whole body and lung-targeted transgenic and transgene regulator lines

Common name	Promoter sp.	Expression traits	Jax no.	Ref.
Widespread ("global")				
Constitutive				
CMV-Cre	Virus	Targets all cells, including germline (both sexes)	006054	[10]
ACTB-Cre	Human	Targets all cells, including germline (both sexes)	019099	[11]
Zp3-Cre	Mouse	Targets premeiotic oocytes. F1 females used as founders for germline heterozygous transmission	003651	[12, 13]
Inducible				
UBC-CreER^T2	Human	Targets tamoxifen-regulated Cre throughout the body in adults, thus avoiding effects in early life. Tamoxifen can also be given to pregnant dams, with additional precautions to reduce embryonic/fetal toxicities. (*see* **Notes 1** and **2**)	007001	[14]

(continued)

Table 1
(continued)

Common name	Promoter sp.	Expression traits	Jax no.	Ref.
Lung targeted				
Constitutive				
Shh^{Cre/GFP}	Mouse	Knock-in that disrupts *Shh* and drives both Cre and GFP. Not specific to lung only. Targets all epithelia from trachea to alveoli via precocious *Shh* expression in embryonic foregut endoderm	005622	[15]
Nkx2-5^{Cre}	Mouse	Knock-in that disrupts *Nkx2-5* and drives Cre. Targets all epithelia via precocious *Nkx2-5* expression in embryonic foregut endoderm	030047	[16, 17]
Nkx2-5^{IRES/Cre}	Mouse	Knock-in that drives Cre. Targets all epithelia via *Nkx2.1* expression in embryonic foregut endoderm	024637	[18]
Nkx2.1-Cre	Mouse	Bacterial artificial chromosome expressing Cre. Targets all epithelia via constitutive expression in club and type II alveolar cells	008661	[19]
SFTPC-Cre	Human	Targets all epithelia from trachea to alveoli via precocious transgene expression in embryonic foregut endoderm. Leaky in male germline	n/a	[20]
KRT5-Cre	Human	Targets basal and basal-like cells. Not specific to lung	n/a	[21, 22]
KRT5-Cre	Cattle	Targets basal and basal-like cells. Not specific to lung	n/a	[23]
Krt14^{Cre}	Mouse	Targets basal and basal-like cells. Not specific to lung	n/a	[24]
KRT14-Cre	Human	Targets basal and basal-like cells. Not specific to lung	018964	[25, 26]
Scgb1a1^{Cre}	Mouse	Knock-in that disrupts *Scgb1a1* and drives Cre. Targets club cells and cells that differentiate from them	n/a	[27]
Scgb1a1-Cre	Mouse	Targets club cells and cells that differentiate from them	n/a	[28]
Scgb1a1-Cre	Rat	Targets club cells and cells that differentiate from them. Multiple lines have been made	n/a	[29–32]
FOXJ1-Cre	Human	Targets ciliated cells. Not specific to lung only	n/a	[33]
Aqp5^{Cre}	Mouse	Knock-in that disrupts *Aqp5* and drives both Cre and dsRed. Not specific to lung only	n/a	[34]
Pdpn^{Cre/dsRed}	Mouse	Knock-in that disrupts *Pdpn* (also called T1α) drives Cre. Not specific to lung only	n/a	[35]
Inducible				
KRT5-rtTA	Cattle	Targets dox-regulated expression in basal cells and cells that differentiate from them (*see* **Note 3**)	n/a	[36, 37]

(continued)

Table 1
(continued)

Common name	Promoter sp.	Expression traits	Jax no.	Ref.
KRT5-CreER	Cattle	Targets tamoxifen-regulated Cre expression in basal cells and cells that differentiate from them	n/a	[38–40]
KRT5-CreER[T2]	Human	Inducibly targets basal cells and cells that differentiate from them in developing and adult lungs. Only affects ~10% of cells in the trachea	n/a	[41]
KRT5[IRESCreERT2]	Mouse	Knock-in that inducibly targets basal cells and cells that differentiate from them	029155	[42, 43]
KRT14-rtTA	Human	Targets dox-regulated expression in basal cells and cells that differentiate from them. Multiple lines have been made	n/a	[44, 45]
KRT14-Cre[ER]	Human	Inducibly targets basal cells and cells that differentiate from them in developing and adult lungs. Partial efficiency	n/a	[46–49]
Nkx2.1[CreERT2]	Mouse	Targets tamoxifen-regulated Cre expression in club and type II alveolar cells and cells that differentiate from them	014552	[50]
Scgb1a1-rtTA (line 1)	Rat	Inducibly targets club cells and cells that differentiate from them in developing and adult lungs	006232	[51]
Scgb1a1-rtTA (line 2)	Rat	Inducibly targets club cells and cells that differentiate from them in developing and adult lungs. This line has less expression of rtTA than Scgb1a1-rtTA (line 1) in the absence of dox	016145	[52]
Scgb1a1-rtTA2(s)-M2	Rat	Inducibly targets club cells and cells that differentiate from them in developing and adult lungs. There is less transgene expression than Scgb1a1-rtTA lines above in the absence of dox. Dox induces expression as early as 4 h that is maximal by 24 h. Approximately 10% of alveolar type II cells are also targeted with both embryonic and adult dox treatments	n/a	[53]
Scgb1a1[IRES/CreER]	Mouse	Knock-in that inducibly targets club cells and cells that differentiate from them in developing and adult lungs	016225	[54]
Scgb1a1[EGFP/Cre/ERT2]	Mouse	Knock-in that inducibly targets club cells and cells that differentiate from them in developing and adult lungs	n/a	[55]
FOXJ1-CreER[T2]	Human	Inducibly targets ciliated cells. Not specific to lung only	n/a	[56]
FOXJ1-CreER[T2]/ IRES/eGFP	Human	Inducibly targets ciliated cells. Not specific to lung only	n/a	[57]
Foxj1[CreERT2/GFP]	Mouse	Knock-in that inducibly targets both Cre and GFP to ciliated cells in developing and adult lungs. Not specific to lung only	027012	[56]

(continued)

Table 1
(continued)

Common name	Promoter sp.	Expression traits	Jax no.	Ref.
$Sftpc^{CreERT2}$	Mouse	Knock-in that inducibly targets alveolar type II cells and cells that differentiate from them. Multiple lines have been made	028054	[58, 59]
$Sftpc^{CreERT2/rtTA}$	Mouse	Knock-in that disrupts $Sftpc$ and drives Cre. Inducibly targets alveolar type II cells and cells that differentiate from them in developing and adult lungs. Can be used in both tamoxifen and dox-regulated systems	n/a	[60]
Sftpc-rtTA;Nudt18-Cre	Mouse	Inducibly targets alveolar type II cells and cells that differentiate from them in developing and adult lungs. Leak observed in adult lungs in the absence of dox	n/a	[59]
Sftpc-rtTA (line 1)	Rat	Inducibly targets alveolar type II cells and cells that differentiate from them in developing and adult lungs. Leak observed in adult lungs in the absence of dox	006235	[51]
Sftpc-rtTA (line 2)	Rat	Inducibly targets alveolar type II cells and cells that differentiate from them in developing and adult lungs. Less leak, but lower rtTA levels, than line 1	006235	[51]

Not all lines are shown. Priority is placed on lines that are available, with notes on relative specificity to the lung, leakiness of expression, and maintenance on a C57BL/6 lineage

2   Materials

The generation and use of experimental mice should be conducted in accordance with all institutional, governmental, and other regulations or guidelines required by your facility. Due to variations in these guidelines between institutions, exact procedures may need to be adapted to meet the requirements of a given institution. Nonetheless, the materials below are suitable for almost any laboratory. These include materials that are important for breeding and identifying experimental animals, determining their genotypes, and validating the efficiency of transgene expression (overexpression and gene disruption). To determine the transcript levels of an overexpressed or disrupted gene, quantitative RT-PCR and immunoblotting procedures will be discussed. Additional methods, such as reporter gene expression can also be used (*see* Fig. 2), and these are described in detail elsewhere [61].

2.1 Mice,
Identification,
and Gene Induction

1. Obtain 1–2 breeding pairs of transgenic mice to initiate a colony. Whole-body, lung-targeted transgenic, or transgene regulator mouse lines most appropriate for an experimental

study design should be obtained from a reputable academic collaborator, commercial breeder, or other trusted source (*see* Table 1). For mice bred on-site, animals with appropriate genotypes should be selected and mated at 8–10 weeks of age and no older than 6 months. It is possible to breed aged mice, but fertility may be low. In this vein, breeder pairs should be replaced every 6–12 months in order to prevent colony depletion associated with breeding aging mice.

2. Tattoo ink: USDA approved tattoo ink for use in laboratory animals (Stone Manufacturing).

3. Needles: Sterile 25 gauge × 5/8 (0.5 mm × 16 mm) needles.

4. Syringes: Sterile 1 mL syringes for delivering tattoo ink.

5. Scissors: Sharp surgical scissors for obtaining tail biopsies to be used for genetic analysis.

6. Microcentrifuge tube for storage of tail biopsies at −20 °C. Sizes can vary from 200 to 650 μL to allow for placement of samples into a thermocycler for digestion.

7. If using an inducible system, a drug (e.g., tamoxifen or doxycycline) may be required and is available through multiple delivery routes (e.g., injectable, in diet) depending on the longevity of transgene expression/induction required for a given study. ENVIGO provides tmx and dox supplemented chow (with or without irradiation) at concentrations that are suitable for use in most in vivo systems (400 ppm tmx and 625 ppm dox). Dox and tmx can also be purchased and administered to mice directly. Dox can be delivered in drinking water, and tmx can be dissolved in corn oil and delivered by intraperitoneal injection. As a general rule, while dox can be administered at any time, the use of tmx during prenatal gestation requires additional steps (*see* **Notes 1–3**).

2.2 DNA Extraction

1. Nuclease-free water.

2. Alkaline lysis reagent: Add 62.5 μL of 10 N NaOH and 10 μL 0.5 M disodium EDTA to a 50 mL conical tube. Add nuclease-free water to final volume of 25 ml. The final solution contains 25 mM NaOH and 0.2 mM disodium EDTA (pH ~12). Store at room temperature and prepare fresh every 1–2 months.

3. Neutralization reagent: Weigh 325 mg Tris–HCl. Add nuclease-free water to a final volume of 50 mL. The final solution is 40 mM Tris–HCl in water at a pH of approximately 5. Store solution at room temperature.

2.3 PCR and Gel Electrophoresis

1. Reaction containers: Depending on the number and sizes of reactions used for genotyping and RT-PCR, as well as the heat blocks and thermocyclers available, appropriate thin-walled vials and 96- or 384-well plates are needed.

2. Template DNA: Non-purified DNA is taken directly from the mouse tissue in extraction buffer. Use positive control DNA obtained from a founder parent of the mouse colony. Also use negative control DNA from known wild-type or nonmutant mice, and use nuclease-free water as a no-template control.

3. Taq DNA polymerase.

4. PCR reaction buffer.

5. Targeted primers: Prepare lyophilized sense and antisense primers to a final concentration of 100 mM in nuclease-free water (*see* **Note 4**).

6. Working stock of mixed PCR primers: In a 1.5 mL microcentrifuge tube, combine 25 μL sense and 25 μL antisense primer (from **Step 5**–100 mM) with 450 μL ultrapure, nuclease-free water.

7. Thin-walled PCR tubes (1 per sample). Tubes can be substituted with 96 or 384 well plates depending on thermocycler(s) used.

8. A 500 mL glass Erlenmeyer flask.

9. Microwave able to accept a 500 mL flask.

10. Agarose: LE agarose.

11. 5× TBE stock solution: Weigh 54 g of Tris base and 27.5 g boric acid and add to a 1 L glass beaker or bottle. Add 20 mL of 0.5 M EDTA at pH 8.0. Add ultrapure water to a volume of 800 mL and mix. Adjust the pH to 8.3 and bring the solution up to 1 L with ultrapure water.

12. 0.5× TBE working solution: Add 100 mL 5× TBE and 900 mL ultrapure Milli-Q water to a 1 L glass bottle or beaker.

13. Ethidium bromide solution (10 mg/mL).

14. Gel electrophoresis system: A unit with casting tray, well combs (1.5 mm think), buffer chamber, and lid. These vary in size and sample capacities.

15. Power supply. A unit capable of generating constant voltage or constant current power at ranges of 10–30 V and 4–400 mA (75 W maximum) is sufficient. A PowerPac™ basic power supply (Bio-Rad, Hercules, CA) was used for the examples described here.

16. Silicone pot holder.

17. Gel documentation system: A UV light source with shielding to prevent skin and eye exposures is required to visualize DNA. Digital camera or polaroid film systems are both suitable for recording data. UV protective eyewear is also recommended.

18. DNA loading dye: Dissolve 0.6 g of blue dextran in 30 mL of ultrapure water. Mix well. Add 20 mL of glycerol. Mix well (*see* **Note 5**).

19. DNA molecular weight ladder: 100 bp DNA ladder.

2.4 Validation of Transgene Expression and Target Gene Disruption

2.4.1 Sample Procurement Materials

1. Euthanasia: Appropriate means are determined by institutional regulations. The most common procedures are anesthetic overdose and CO_2 inhalation.

2. Ethanol: Prepare a solution of 70% ethanol in water and store in a spray bottle. This will be used to wet and disinfect mouse tissues. It is also convenient for limiting fur spreading on bench surfaces and into samples during tissues extraction.

3. Surgical tools: Three pairs of sharp scissors are needed. Two pairs of 4–5-inch light dissecting scissors (straight and curved, blunt tipped) are used for surface and underlying tissue dissections. One pair of 3-inch curved micro-dissecting spring scissors (0.1–0.15 mm cutting edge) is used to perform tracheostomies. In addition, three pairs of rounded serrated forceps are needed for manipulating skin and underlying tissues. One pair of "heavy" 4–6-inch forceps with curved 1–2 mm tips is used for grasping external tissues. Two pairs of 4–4.5-inch Moloney forceps with 0.8 mm partial or full curve tips are used for dissection of underlying tissues during surgery. For experiments shown here, surgical instruments were purchased from Roboz (Gaithersburg, MD).

4. Surgical thread: Gütermann top-stitch heavy duty polyester sewing thread (Joann Fabrics, Hudson, OH).

5. RNAse decontamination: RNase AWAY (Molecular Bio-Products, San Diego, CA) is available as liquid in spray and pour-off containers.

6. Plastic dishes for tissue collection: Disposable 100 mm petri dishes.

7. Collection vials: Plastic nuclease and pyrogen-free 0.5–2 mL microcentrifuge tubes and 1–2 mL screw-top cryovials.

2.4.2 Detecting Transgenic Manipulation at the Gene Expression Level

1. RNA purification: RNeasy and AllPrep DNA/RNA kits are recommended (Qiagen, Germantown, MD). These have nuclease-free water, extraction buffers, purification columns, wash buffers, and collection vials. Molecular biology grade 100% ethanol must be supplied separately. In addition, RNase-free DNase will also be required (Qiagen, Germantown, MD).

2. RNA quantification: A spectrophotometer capable of measuring absorbance at 260 and 280 nm wavelengths is suitable for assessing nucleic acid quantities (*see* **Note 6**).

3. Reverse transcription: Reactions can be carried out using reagents from numerous vendors. The examples provided in this chapter use Superscript™ III reverse transcriptase enzyme mix (Thermo Fisher Scientific, Waltham, MA), which is supplied with a 2× reaction buffer containing oligo(dT)$_{20}$, random hexamers, magnesium chloride (MgCl$_2$), and deoxynucleotide triphosphates (dNTPs). These can be purchased separately or in a kit with the reverse transcriptase from Thermo Fisher Scientific (the supplier used here), as well as other vendors. Examples shown in this chapter use random hexamers as reverse transcription primers, but oligonucleotides for poly-A tail or target gene-specific antisense primers can also be used (*see* **Note 7**).

4. Quantitative PCR: Quantitative PCR (qPCR) is performed with specific primers for selected targets (the transcripts of disrupted or transgenically expressed genes) and for reference genes. qPCR amplification can be carried out using reagents from numerous vendors. The examples shown here use TaqMan™ Gene Expression Assay 20× and TaqMan™ Fast Advanced Master Mix 2× (Thermo Fisher Scientific). Reactions of 10 μL were performed on a ViiA 7 real-time PCR system in 384-well plate format (ThermoFisher Scientific, Applied Biosystems).

2.4.3 Detecting Transgenic Manipulation at the Protein Level

1. Protease inhibition: Protease inhibitor cocktail (Sigma, St. Louis, MO) containing [4-(2-aminoethyl)benzenesulfonyl fluoride] (AEBSF), aprotinin, bestatin HCl, [*N*-(trans-Epoxysuccinyl)-L-leucine 4-guanidinobutylamide] (E-64), leupeptin hemisulfate salt, and pepstatin A can be added to samples at 1/10 the volume of extracted material. This is suitable for inhibition of serine, cysteine, and acid proteases. It should be included in all protein isolation preparations.

2. Protein extraction: For detection of proteins in lung tissues, extraction is required. Protein extraction can be carried out using mechanical- or detergent-based procedures. Mechanical methods include physical extrusion with a mortar and pestle, glass tissue grinder, bead-based disrupter, mechanical homogenizer, or sonicator. These are available from numerous vendors and are often shared across laboratories. During mechanical disruption, tissues should be kept in an appropriate liquid. For mortar-and-pestle disruption, which is performed on frozen tissue, liquid nitrogen is suitable. For other forms of mechanical disruption, solutions contain standard pH buffers and vary in ionic strengths and concentrations that must be determined by the user. For detergent-based disruption, simple solutions containing 1–2% (w/v) sodium dodecyl sulfate (SDS) in 0.01 M phosphate-buffered saline (PBS; 0.138 M

NaCl, 0.0027 M KCl, pH 7.4) or tris(hydroxymethyl)amino-methane hydrochloride (Tris–HCl) can be used (Sigma). Alternative detergent lysis buffers can be found in the following references [62–66].

3. Lung lavage buffer: For examples shown here, samples of lung lavage fluid are used. Lavage is performed using PBS containing 0.001 M ethylenediaminetetraacetic acid (EDTA).

4. Solubilization buffer: Depending on the protein and tissue of interest, additional solubilization reagents may also be needed (e.g., guanidinium or urea buffers). These can be purchased from numerous vendors, including Sigma. In the examples shown here, lung lavage fluid is dissolved in solubilization buffer (3 M urea/0.05 M DTT) for membrane-based dot-blot enzyme-linked immunosorbent assay (ELISA).

5. Total protein quantification: Use a bicinchoninic acid (BCA) assay kit (Thermo Scientific/Pierce Biotechnology, Rockford, IL).

6. Protein separation by SDS-polyacrylamide gel electrophoresis (SDS-PAGE): SDS-polyacrylamide gels of varying concentrations or gradients should be determined based on the size and nature of proteins being analyzed. For ease, a 4–20% gradient is sufficient for detecting proteins in a broad 2–400 kDa range. Pre-cast gels of single polyacrylamide concentration and gradient SDS-polyacrylamide gels are available from various suppliers. In most cases, this will dictate the form and format of the gel and unit used for electrophoresis. For separating proteins such as mucins that are larger than 500 kDa, agarose gel electrophoresis may be necessary; please *see* ref. [67] for a detailed protocol.

7. Sample loading buffer (10×): For reducing buffer, dissolve 1.54 g DTT, 25 mg bromophenol blue, and 100 mg of SDS in 10 mL of deionized water. For non-reducing buffer, omit DTT.

8. Protein blotting: For immunoblot detection, proteins are transferred from gels onto nitrocellulose or polyvinylidene fluoride (PVDF) membranes. For SDS-PAGE, this is performed using a wet electroblot apparatus with Tris-glycine transfer buffer (3.3 g Tris, 14.4 g glycine, 200 mL methanol, mixed in a final volume of 1 L with deionized water). In the immunoblot examples shown here, mucins separated by agarose gel electrophoresis are transferred by vacuum blotting; please *see* ref. [67] for a detailed protocol. Immunohistochemical and ELISA methods for protein detection can also be employed depending on user need and reagent availability (*see* **Notes 8** and **9**).

3 Methods

3.1 Mice, Identification, and Gene Induction

1. To maintain strain purity, cross targeted and non-targeted transgenic mice with wild-type animals to maintain as heterozygotes and hemizygotes, respectively (*see* **Note 10**).

2. Cross transgenic animals together as appropriate for specific compound transgenic targeting systems. For example, cross Scgb1a1-rtTA mice with a tetO-driven target gene for expression in club cells or FOXJ1-Cre mice with a "floxed" target containing two loxP sites for gene disruption in ciliated cells. Methods for the creation of a de novo floxed mouse are outside the scope of this chapter. Please refer to ref. [68] or contact a service that will engineer lines (*see* **Note 11**).

3. Observe mice for signs of pregnancy, and separate pregnant dams if necessary. Most one-to-one mating pairs can be maintained together during gestation through weaning. Follow institutional requirements (*see* **Note 12**).

4. Between postnatal days 5–9, mice are suitable for identification by tattoo labeling. Fill a 25 gauge 1 mL syringe with 0.4 mL of a USDA approved tattoo ink for use in laboratory animals.

5. Restrain mouse pup so that the ventral side of the animal is visible.

6. Use the footpad tattooing chart to identify which paws should be tattooed to generate the appropriate identification number when added together (Fig. 3).

7. Once the correct paws have been identified, gently slide needle under skin of foot, bevel side up, injecting a small amount under skin of pad (~20–50 µL). Designate the first mouse with a single

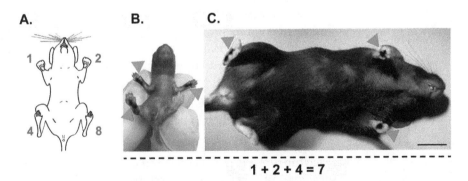

Fig. 3 Tattoo identification of mutant mice. Mice are identified using a numerical encoding system based on label placed on the front and rear footpads. (**a**) Digits are assigned to the right-front, left-front, right-rear, and left-rear footpads. (**b**) Pups are tattoo-labeled between 5 and 9 days of age. (**c**) Adult mice display retained labeling. In the example shown, the right-front (1), left-front (2), and right-rear (4) footpads were marked. The sum of these (7) becomes the ones place value digit for the animal's identification. Tens, hundreds, and thousands place values are tracked separately in notes by investigative and animal care personnel

digit "1." Numbers 1, 2, 4, and 8 are marked with single tattoos; numbers 3, 5, 6, 7, and 9 are marked with additive combinations. Mouse number 10 would receive no tattoos, which indicated zero in the one's place. This same system is then used for higher numbers, with notes placed on cages to indicate place values in the 10s, 100s, etc. Thus, in this system each mouse receives a tattoo which corresponds with the final digit of their identification number. For example, mouse 11 would receive a tattoo of 1, so a notation is made on a cage card and notebook to identify 11 as its unique identifier (*see* **Note 13**).

8. Cut ~1 mm of tissue from the tip of the tail and place in a collection tube. Samples can be kept at room temperature for up to 4 h with no detectable loss in DNA needed for genotyping. If DNA is not extracted within this time frame, the tails can be stored at −20 °C.

3.2 DNA Extraction for Mouse Genotyping

1. Add 100 μL alkaline lysis reagent to tails. Make sure that the tissue is completely submerged

2. Incubate at 95 °C for 45 min and then store at 4 °C (10–60 min). Shake gently to begin digestion.

3. Add 100 μL neutralization reagent for each sample. Shake gently again for completely digestion.

4. Samples may then be put into freezer or used immediately for PCR.

3.3 PCR and Gel Electrophoresis for Mouse Genotyping

1. Identify and clearly label thin-walled tubes or plates for each reaction.

2. Prepare reaction Master Mix: For each reaction add 17.7 μL nuclease-free water, 6 μL MyTaq™ reaction buffer, 3 μL working stock of mixed primers, and 0.3 μL MyTaq™ DNA polymerase. Increase the total number of reactions by 1 to account for loss of volume during pipetting (e.g., for 29 DNA samples, make 30 times the recipe above). The PCR Master Mix should be prepared for all reactions together, and mixed thoroughly by gentle agitation of the tube.

3. Aliquot 27 μL of the PCR Master Mix into each of the labeled 0.65 mL PCR reaction tubes.

4. Add 3 μL of digested template DNA.

5. Place tubes in a thermal cycler and cycle according to protocols specific to primers' annealing temperatures. Appropriate times and temperature for the denaturation, annealing, and extension steps of the cycle will be determined based on the nucleotide composition and length of the primers. Generally, a 2-min 95 °C heat activation step followed by 25–29 cycles of 95 °C denaturation, 55–65 °C annealing, and 72 °C extension is sufficient (*see* **Note 14**).

6. Prepare a 2% agarose/TBE gel by adding 2 g of agarose per 100 mL 0.5× TBE in a dedicated 500 mL flask. Cover the mouth of the flask with a lab wipe (Kimwipe or similar). Place the flask in the microwave and heat on high for approximately 3 min(time required will vary across microwaves). Watch the solution as it heats. When the solution begins to boil, remove the flask from the microwave using a silicone pot holder, and gently swirl the solution. Use care when swirling the solution to minimize the introduction of air bubbles and avoid spilling. Place the flask back in the microwave until it begins to boil again. Repeat this process of heating, removing, and swirling the solution until the solution is completely clear and the agarose is completely dissolved (*see* **Notes 15** and **16**).

7. Allow the solution to cool at room temperature by placing the flask on the benchtop. Allow the solution to cool until you can comfortably pick up and hold the flask with a gloved hand.

8. Add 2 μL of ethidium bromide solution (10 mg/mL) per 100 mL to the flask and swirl to mix (*see* **Note 17**).

9. Place the gel tray into the gel casting apparatus. Most units have gaskets on the gel running trays that allow for them to be used to create tight seals when placed orthogonally on the gel running unit. Other options include separate casting trays, or removable dams that fit either end of the gel tray. Carefully pour the agarose/TBE/ethidium bromide mixture into the gel tray. Place a 20-well 1.5 mm comb in the notches on the gel casting tray. Use a dry pipette tip to pop or drag out any bubbles that are present in the gel. Care should be taken to perform these steps before the gel solidifies (*see* **Note 18**).

10. When the gel has solidified, remove the gel tray from the casting unit. Use care to ensure that the gel stays in the tray. Place the gel tray into the tank oriented for electrophoresis, and cover the gel with 0.5× TBE.

11. Carefully remove the comb to reveal the sample wells by gently and firmly pulling up on the comb.

12. Add 10 μL of 100 bp DNA ladder (0.1 μg/μL) mixed with 3 μL of DNA loading dye in the first well of the gel.

13. Add 6 μL of DNA loading dye to the PCR product. Centrifuge briefly to bring contents to bottom of tube.

14. Add 30 μL of PCR product/loading dye mixture to wells 2–20. Use care to document the position and order of each sample as you load the gel.

15. Cover the gel box with the lid and attach to the power supply. Run at 5–10 V/cm distance between electrodes for 45–60 min.

16. Place gel onto a Bio-Rad ChemiDoc or other UV light source for visualization of DNA bands (Fig. 4).

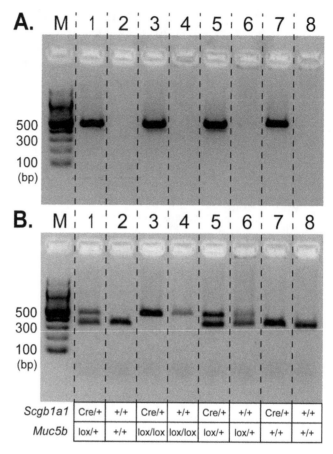

	Scgb1a1	Muc5b
1	Cre/+	lox/+
2	+/+	+/+
3	Cre/+	lox/lox
4	+/+	lox/lox
5	Cre/+	lox/+
6	+/+	lox/+
7	Cre/+	+/+
8	+/+	+/+

Fig. 4 Genotyping transgenes. PCR was performed on mouse tail lysates to identify the presence of two transgenes: the lung-targeted IRES-CreER transgene inserted into the 3′UTR of the mouse *Scgb1a1* locus (**a**) and the "floxed" *Muc5b* gene with loxP sites inserted 5′ and 3′ relative to exon 1 (**b**). The sequences of PCR primers used to amplify these are as follows: 5′-TCA TGC AAG CTG GTG GCT GG (CreER sense), 5′-GGG CTC AGC ATC CAA CAA GG (CreER antisense), 5′-AGG ACC TGA CTT TGA TGG AAG AGG (*Muc5b* sense), and 5′-GGG TAT CTT ACC GGT CTG TTC TGG (*Muc5b* antisense). PCR reactions were carried out using the following parameters: 95 °C for 30 s, 60 °C for 30 s, and 72 °C for 30 s, repeated for 27 cycles. Amplified products are analyzed by gel electrophoresis and compared to a molecular weight marker. Eight samples were genotyped for the two transgenes. M: 100 bp DNA ladder; 1: positive control; 2: (**a**) negative control for Cre transgene; (**b**) negative control for "floxed" Muc5b gene, but positive control for wild-type allele. 3–8: DNA samples PCR screened. In **a**, samples that yield PCR product are positive for the Cre transgene; samples yielding no PCR product are negative for the transgene. In **b**, two sizes of PCR products are observed. A wild-type allele and a slightly higher molecular weight allele that is amplified by the same primers in the same reaction. The higher molecular weight is due to loxP sites and cloning sites. Samples with only a higher molecular weight band are homozygous floxed (*Muc5b*$^{loxP/loxP}$). Sample with high and low molecular weight bands are heterozygous floxed (*Muc5b*$^{loxP/+}$). Samples with only a low molecular weight band are homozygous wild type (*Muc5b*$^{+/+}$)

17. Appropriate positive and negative controls are critical for the interpretation of PCR genotyping results (*see* **Note 19**).

18. Choose mice with appropriate genotypes for breeding and experimentation. Breeders can be homozygous for targeted alleles (knock-out or knock-in mutations), but only hemizygous mice should be used for non-targeted alleles (those generated by random insertion). For experiments, heavily backcrossed congenic mice can be compared to commercially available wild-type inbred strains. However, due to (1) potential genomic carry over from the creation of the mouse, (2) random variation within sub-strains (genetic drift), and (3) environmental issues related to local versus commercial housing facilities, it is recommended to use littermate controls instead of commercially available wild-type mice. Experimental mice should be used at ages appropriate for a given study (*see* **Note 20**).

3.4 Validation of Transgene Expression and Target Gene Disruption

3.4.1 Sample Procurement

For non-conditional and constitutive tissue-specific conditional transgenes, mice should be studied in settings that allow for comparison to non-transgenic controls that are as closely related as possible (littermates and age-/sex-matched controls). Using inducible systems, it is best to control for both transgenes and inducing agents. When using viral induction, empty vector controls or vectors expressing an irrelevant reporter gene serve as controls. Under dox and tmx driven systems, mice should be compared under the most stringent and comprehensive controls as possible. As a best practice for controls, a cohort of transgene negative mice will receive dox or tmx, and a cohort of transgene positive mice will also be studied in the absence of dox or tmx. Together, these controls allow for analysis of off-target effects of tmx or dox, as well as leakiness of transgene expression. In experimental lines, the timing of induction will depend on the nature of the question being asked (e.g., prenatal lung development or acute vs. chronic diseases in adults). Other important factors include the efficiencies of the promoters being used to drive transgene expression at given mouse ages, as well as the turnover of recombined cells and their replacement by self-renewal vs. progenitor cell activation. These should be tested by users using the methods below. For Cre-mediated gene disruption, efficiency may be increased using extended periods of dox or tmx treatment and multiple copies of Cre (if using a targeted transgenic Cre line) (Fig. 5).

Lung tissue extracts can be used to assess both transcript and protein expression. Alternatively, proteins secreted into airspaces may be detected in lung lavage fluid. In some cases (e.g., upon introduction of a cytokine or chemokine expressing transgene), the effects of transgenic manipulation can result in changes in inflammation that can also be detected in lung lavage fluid.

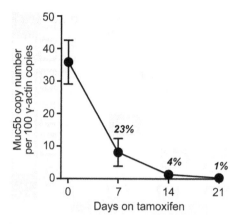

Fig. 5 Determining the effects of Cre-mediated gene disruption on target gene expression. Quantitative RT-PCR was performed on cDNA extracted from the lungs of mice targeted for disruption of the airway mucin Muc5b (*Muc5b*^loxP/loxP^;*Scgb1a1*^CreER/CreER^ mice). Mice were placed on a standard diet supplemented with tmx (400 ppm) for the times indicated. Expression of *Muc5b* was measured relative to the reference gene γ-actin 1 (*Actg1*). TaqMan primer/probe sets for *Muc5b* (Mm00466391_m1) and *Actg1* (Mm01963702_s1) were used. Template cDNA in reactions for *Actg1* was diluted 1:100 to ensure that observations were made in the linear range of detection. To provide specific copy number quantities, standard curves were generated using known quantities of previously amplified and purified *Muc5b* and *Actg1* target amplicons (10^0–10^8 copies per reaction)

Therefore, this protocol includes steps for performing lung lavage prior to excising tissues (**Steps 2–7** below in this section). These lavage steps can be omitted if only lung tissue samples will be investigated.

1. For surgical preparation, mice are first euthanized in accordance with institutional animal use guidelines and then sprayed with 70% ethanol to soak the fur and disinfect surfaces. Mice are then placed on a benchtop or a surgical platform and secured using pins, tape, or thread.

2. To tracheostomize mice, first cut a ~1 cm diameter patch of skin from the neck, and expose the trachea by removing deep tissues (connective tissue, glands, and muscle).

3. Using rounded forceps, separate the trachea from the esophagus, and pass a folded 30–40 cm length of surgical thread underneath the trachea.

4. Cut the thread at the center point releasing two 15–20 cm segments that will be tied to stabilize a tracheal cannula.

5. Using spring scissors, make a partial incision across the ventral half of the trachea.

6. Insert a blunt Luer-Stub adapter (20 gauge), and secure it by tying the surgical threads across the steel portion of the stub

both cranial and caudal to the insertion site. This will secure the cannula, preventing it from slipping out and from pivoting laterally during subsequent manipulations.

7. Lung lavage is then performed by instilling 1 ml of PBS/ EDTA lung lavage buffer using a 1 cc syringe and then withdrawing the PBS/EDTA buffer by pulling back on the plunger. To maintain high concentrations of secreted proteins, it is recommended that the same 1 mL solution be instilled and withdrawn 3–5 times to ensure even sampling. To maximize cell returns, the process can be repeated 2–4 more times using fresh PBS/EDTA for each sample.

8. Collect lavage fluid in a 1.5 mL microcentrifuge tube (1 mL lavage) or in a 15 ml conical tube (2–5 mL lavage).

9. Store fluid unspun ("neat") or centrifuge at $2000 \times g$ for 5 min at 4 °C or room temperature to remove leukocytes and cellular debris. Neat lavage and supernatants can be used to test levels of secreted factors that may be affected by genetic manipulations. In the examples described below, secreted mucin proteins in neat lavage are assessed. Inflammatory cell numbers can also be quantified using microscopy and/or flow cytometry with methods described in other chapters in this volume and elsewhere [69–72] (*see* **Notes 21** and **22**).

10. Add 1/10 volume of protease inhibitor cocktail to neat lavage and/or lavage supernatants.

11. Snap freeze samples in liquid nitrogen or dry ice.

12. To expose thoracic tissues, make a transverse incision across the abdomen below the diaphragm. Then, make a second incision to remove the skin from the thorax. This superficial incision is made longitudinally over the sternum and from the xyphoid process to the site where neck incision had been made previously.

13. To expose lung tissues, use clean RNAse-free surgical forceps and scissors. If using the same surgical tools as in the steps above, clean all surgical tools that have touched external skin surfaces and fur with RNAse AWAY. Cut the diaphragm away from the rib cage, and then cut the rib cage along the sternum up through the manubrium. Cut away the ribs in a caudal-to-cranial direction along the dorsal angle of the ribcage.

14. Excise the lungs, and place them in a petri dish. As a best practice, the petri dish should be pre-cleaned with RNAse AWAY and kept cold on wet ice or ice packs.

15. Remove extraneous portions of the diaphragm, esophagus, heart, and thymus from the lungs. Depending on the study, removing tracheal segments may also be recommended.

16. Rapidly mince lungs to 2–4 mm pieces and mix with surgical scissors. Keeping the Petri dish on ice and mincing for <10 s is recommended.

17. Place tissues into two separate cryovials (one for RNA, one for protein), and snap freeze in liquid nitrogen or on dry ice. Frozen samples can be stored long-term in liquid nitrogen or in a freezer at −20 °C to −80 °C (*see* **Note 23**).

3.4.2 Analysis of Transgenic Manipulation at the Gene Expression Level

1. Thaw frozen lung samples on wet ice (*see* **Note 24**).

2. Prepare RLT lysis buffer. Add β-mercaptoethanol (BME) to buffer RLT (10 μL BME/ml RLT) using an appropriate amount of RLT lysis buffer for the tissues being used. The manufacturer's protocol recommends using 600 μL of RLT lysis buffer per 20–30 mg tissue. The total mass of intact adult mouse lungs is approximately 100–150 mg.

3. Disrupt tissues. Transfer thawed lung samples into disposable plastic tissue grinders (VWR). Add the appropriate volume or RLT/BME lysis buffer. Manually grind for 10–20 s, and place sample back into wet ice.

4. Homogenize tissues. Using a 3–10 mL syringe, pass the tissue lysates from **Step 3** above through a 1.5-inch 20-gauge needle 5–10 times. Distribute equal volumes of homogenates into microcentrifuge tubes.

5. Centrifuge homogenates for 3 min at >12,000 × g at room temperature.

6. Carefully remove supernatants (lysates), and place into new microcentrifuge tube.

7. Add an equal volume of 70% ethanol to each lysate, and mix by pipetting.

8. Transfer up to 700 μL of lysate/ethanol mixture to an RNeasy spin column fitted to a 2 mL collection tube (provided in kit).

9. Centrifuge for 15 s at ≥8000 × g at room temperature, discard flow-through. Keep the collection tube for reuse in subsequent steps.

10. Transfer any additional lysate/ethanol mixture (up to 700 μL) to the RNeasy spin column, and repeat **Step 9** until the entire lysate/ethanol mixture is consumed.

11. Add 350 μL of buffer RW1 to the RNeasy spin column fitted to the 2 mL collection tube used in the steps above.

12. Centrifuge for 15 s at ≥8000 × g at room temperature, discard flow-through.

13. Perform on-column DNase digestion. Prepare a solution containing 10 μL DNase I stock solution and 70 μL buffer RDD (Qiagen) for each sample plus 10% additional volume. Mix

gently by inverting. Add 80 μL of the DNase/RDD solution directly to the RNeasy column membrane and incubate at room temperature for 15 min.

14. Add 350 μL of buffer RW1 to the RNeasy spin column fitted to the 2 mL collection tube used in the steps above.

15. Centrifuge for 15 s at ≥8000 × g at room temperature, discard flow-through. Keep the collection tube for reuse in subsequent steps.

16. Add 500 μL of buffer RPE to the RNeasy spin column fitted to the 2 mL collection tube used in the steps above.

17. Centrifuge for 15 s at ≥8000 × g at room temperature, discard flow-through. Keep the collection tube for reuse in subsequent steps.

18. Repeat **Steps 14** and **15**, with the exception that centrifugation should be performed for 2 min. The longer centrifugation time allows for complete removal of residual ethanol from the RNeasy column.

19. Place the RNeasy spin column into a new RNase-free 1.5 mL microcentrifuge tube (provided in kit).

20. Add 50 μL of nuclease-free water directly to the RNeasy spin column membrane.

21. Centrifuge for 1 min at ≥8000 × g at room temperature. Keep the flow-through, and either store frozen at −80 °C, or proceed to subsequent analysis steps.

22. Quantify total RNA using a UV spectrophotometer by measuring absorbance at 260-and 280-nm wavelengths. An absorbance unit of 1.0 is approximates a total RNA concentration of approximately 44 μg/mL (*see* **Note 25**).

23. For reverse transcription, prepare a working stock of total RNA by adding nuclease-free water to an aliquot of total RNA to obtain RNA at a concentration of 10 ng/μL.

24. For each reaction, combine 10 μL of 2× SuperScript III First Strand Synthesis SuperMix, 8 μL of RNA (10 ng/μL), and 2 μL SuperScript III reverse transcriptase enzyme in a 0.65 mL thin-walled PCR tube.

25. Mix tube contents gently and centrifuge briefly in a bench top centrifuge to collect all liquid. Perform this step for each RNA sample you wish to reverse transcribe into cDNA.

26. Place all tubes into a thermocycler and incubate as follows: 25 °C for 10 min, 50 °C for 30 min, and 85 °C for 5 min.

27. Remove tubes from thermocycler, and place on ice.

28. Add 1 μL RNase H from *E. coli* (supplied in SuperScript III kit) to each 20 μL reaction, and heat to 37 °C for 20 min in a

thermocycler. The cDNA can be used immediately for quantitative PCR or stored between −20 °C and −80 °C.

29. For qPCR using 384-well plates, identify which wells will be used for PCR reactions, and add 3.5 μL of nuclease-free water to each defined well.

30. Combine 1 μL of reverse transcribed cDNA with 3.5 μL nuclease-free water for each well of each sample. (e.g., for two gene expression assays run in technical triplicate, combine 6.6 μL cDNA template with 23.1 μL nuclease-free water).

31. Add 4.5 μL of this diluted cDNA to each of the appropriate wells in the 384-well plate.

32. Prepare PCR reaction master mixes for target and reference genes by mixing 5 μL TaqMan™ Fast Advanced Master Mix 2× with 0.5 μL TaqMan™ Gene Expression Assay 20×. It is critical to account for the number of biological and technical replicates, as well as overage.

33. Mix tube contents gently (*see* **Note 26**).

34. Add 5.5 μL of the PCR reaction master mix (**Step 32**) to the appropriate wells of the 384-well plate.

35. Perform these steps for each desired combination of cDNA template and Taqman™ probe.

36. Seal the plate and briefly centrifuge using a plate adapter.

37. Cycling parameters will be determined based on the real-time PCR system available to you.

38. Evaluate the levels of expression of target genes relative to reference to determine the efficiency of transgene expression.

3.4.3 Analysis of Transgenic Manipulation at the Protein Level

The methods below will provide a supplement to measuring gene expression above. In some cases, transcripts may only partly reflect levels of functional protein produced. In addition, depending on the ease of procurement and sensitivity of an assay, measuring protein may be the preferred analysis.

1. Thaw lavage or tissue sample to prepare for protein extraction. To extract proteins from lung lavage or lysates, use a procedure and buffer that is best-suited for the targets being manipulated by transgenic overexpression and/or disruption. For examples shown here, samples of lung lavage fluid are used. Samples are obtained by lavage using 1.0 mL PBS/EDTA, which is preserved "neat" in 50-100 μL aliquots.

2. Add extraction buffer supplemented with protease inhibitor cocktail diluted to a suitable working concentration. A twofold dilution with extraction buffer supplemented with a 100× protease cocktail that is diluted 1:50 (thus also present at 2× before mixing with sample) is usually sufficient for achieving protein concentrations that are useful for immunoblot and ELISA analyses.

3. Quantify total protein concentrations using a BCA assay. It is important to carefully follow the manufacturer's protocol, and it is also important to note that many extraction buffer components can interfere with the BCA reaction. For example, SDS must be kept at <5% w/v, dithiothreitol must be kept at <1 mM, and BME must be kept at <0.01% v/v (*see* **Note 27**).

4. Determine the levels of transgenically regulated protein production by immunolabeling methods. The "best" methods must be determined by the user based on the specific localization of the protein among lung epithelial cells and other cell types in the lungs (e.g., fibroblasts, smooth muscle, or macrophages), as well as the specific localization of the protein within epithelial cell types (e.g., club, ciliated, or alveolar type II cells). The examples shown here (Fig. 6) use two methods to detect the airway mucin Muc5b by ELISA and immunoblot. Samples are compared using mice that were generated on constitutive lung-specific SFPTC-Cre and tamoxifen-inducible club cell-specific $Scgb1a1^{CreER}$ drivers. Please see figure legends for details relating to each protocol (*see* **Notes 28–30**).

4   Notes

1. To counteract the mixed estrogen agonist effects of tamoxifen injections, which can result in late fetal abortions in pregnant mice, progesterone may be co-administered.

2. CreER is Cre recombinase fused to a G525R mutant form of the mouse estrogen receptor. $CreER^{T2}$ depicts Cre fused to a G400V/M543A/L544A triple mutant form of the human estrogen receptor.

3. To generate knockout mice using the doxycycline regulated system, animals must also be crossed with a tetracycline operator-controlled Cre recombinase expression line. A useful non-targeted tetO-Cre transgenic line is available from the Jackson laboratories.

4. PCR primers should be ~20–22 bp long, with 50–55% GC content. Primer design software can be used to test for GC content, melting point, and dimer and hairpin formation. A BLAST analysis should also be performed to ensure unique specificity of primers. For convenience, the National Center for Biotechnology Information has a useful site that allows for both analyses to be conducted simultaneously (https://www.ncbi.nlm.nih.gov/tools/primer-blast/).

5. Standard loading dyes use 0.25% (w/v) bromophenol blue (BB) or xylene cyanol (XC), which migrate at low (BB) and high (XC) relative molecular weights based on size and charge

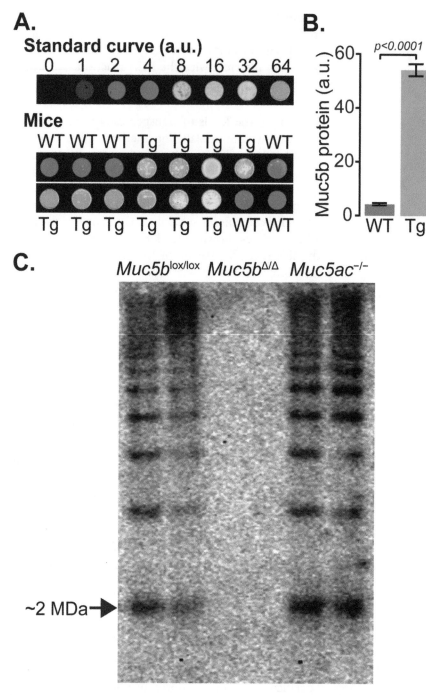

Fig. 6 Evaluating the effects of transgene expression and gene disruption at the protein level. (**a**) A dot-blot ELISA was performed on mouse lung lavage fluid applied to a PVDF membrane using a 96-well vacuum manifold (Whatman Minifold I, GE Healthcare Bio-Sciences). A standard curve was generated by applying lavage fluid from a mouse with lipopolysaccharide-induced lung inflammation. Serial twofold dilutions (1:1 vol/vol in 3 M urea/100 mM DTT) were applied to a PVDF membrane. Membranes were blotted with rabbit-anti-Muc5b antisera (1:2000 dilution). Signal detection was performed using an Odyssey CLx Imager (LI-COR Biotechnology); anti-Muc5b-labeled samples were detected using IRDye 680RD goat anti-rabbit IgG, (1:5000 diluted in TBS-

of the dye molecules. These are mixed in a solution of 30% glycerol diluted in water, for a final concentration of 6×. We also recommend using 2.0% (w/v) blue dextran (BD), which will not migrate out of wells and is thus convenient for running gels of varying agarose content containing PCR products of different sizes. Care must be taken not to overrun gels since BD will not move through the gel.

6. Benchtop Beckman UV/VIS or NanoDrop instruments are well suited. Alternative methods such as Qbit or PicoGreen are also useful for small quantities but require additional instrument availability.

7. For target gene-specific priming, antisense oligonucleotides that are the same as or 3′ to the antisense primers used in quantitative PCR are suitable.

8. Depending on the methods used for protein probing (e.g., types and sources of primary antibodies used, detection chemiluminescence vs. fluorescence), the specific reagents and methods used will vary.

9. In many cases, immunohistochemical and enzyme-linked immunosorbent assays (ELISAs) are also useful as additional or alternative assays that should be determined by the user. Immunohistochemistry may be especially useful if a gene of interest is expressed within multiple cell types normally but only transgenically overproduced or disrupted within a specific subpopulation.

10. Crossing non-targeted transgenic mice of the same genotype together to create double hemizygotes is not recommended. Data could be confounded if unknown loci are disrupted.

11. There are numerous resources available to assist in finding and generating mice with floxed alleles in genes of interest. The Jackson Labs' Mouse Genome Informatics database (www.informatics.jax.org) is useful for finding published lines as well as those in development from outside resources. The European Conditional Mouse Mutagenesis (EUCOMM) program has

---

Fig. 6 (continued) formulated LI-COR Odyssey Blocking Buffer). Accurate amounts of Muc5b are not known, but relative levels can be precisely estimated and assigned arbitrary units (a.u.). In the example shown, Scgb1a1-Muc5b transgenic (Tg) and wild-type (WT) littermate controls were compared on dot-blot membranes. (**b**) Enumerated results of Muc5b dot-blot ELISA are shown in a bar graph plot. Differences were determined by Mann-Whitney U-test comparison of a.u. Muc5b signal intensities. (**c**) Western blot analysis was used to examine mouse lung lavage samples from *Muc5b* conditional knockout mice (*Muc5b*$^{\Delta/\Delta}$), which were compared to SFTPC-Cre negative littermate (*Muc5b*$^{lox/lox}$) and *Muc5ac* knockout controls. Neat lung lavage samples (25 μL) were diluted in non-reducing 6× loading buffer (5 μL) and separated by electrophoresis in an SDS-agarose gel (42 V, 4 °C, overnight). Samples were transferred to a PVDF membrane, blotted with rabbit-anti-Muc5b antisera (1:2000 dilution) and detected by chemiluminescence using biotinylated goat-anti-rabbit IgG and streptavidin-horseradish peroxidase. Blots were imaged using a ChemiDoc Imaging System (Bio-Rad)

produced conditionally targeted genes in C57BL/6N embryonic stem (ES) cells that are available for purchase from the International Knockout Mouse Consortium (IKMC; http://www.mousephenotype.org). At present, the IKMC collection currently has coverage for over 20,000 genes.

12. Pairs may need 2–3 weeks upon first breeding. More experienced breeders may only need 1–4 days for successful breeding. Breeding can be tracked by observing mucus plugging, but this is only a sign of mating and not necessarily successful fertilization.

13. Care needs to be taken to avoid confusion if combining animals in a single cage after tattooing to avoid redundancies in tattooing. Using the single last digit works well in most instances. However, if two or more mice have the same one's place digit, an additional identification method may be required. A small excision of tissue on the right or left ear is useful.

14. To avoid false positives from PCR contamination, using less than 30 cycles is strongly recommend.

15. Significant water loss can result in changes in gel density that negatively impact interpretation of DNA band sizes. Best practice is to weigh the flask containing agarose and TBE buffer before and after microwaving. Water evaporation during boiling can result in changes in mass that should be compensated for by adding back water (1 mL of water per gram of mass lost).

16. If PCR results in amplicons that are >500 bp, a 1–1.5% (w/v) agarose gel is recommended.

17. Use caution when handling ethidium bromide and dispose of any generated waste according to institutional and local regulations. It is useful to dedicate one flask for use with ethidium bromide to avoid contamination of multiple pieces of glassware.

18. To increase the number of samples run on a single gel, multiple combs can be used when casting (dependent upon the capabilities of the casting unit). When doing so, caution should be used interpreting the bottom half of the gel. During electrophoresis, ethidium bromide will migrate in the opposite direction of DNA, and signal strength will be diminished on lower portions of gels. To prevent misinterpretation, load molecular weight ladders into at least one well of each row using the same volume/mass of DNA ladder. Differences in intensity of PCR products can be accounted for relative to the intensities of DNA bands on the ladders.

19. Use known wild-type and known mutant templates as positive and negative controls for PCR.

20. Postnatal day 36–38 is considered to be the end of normal mouse lung development.

21. Other isotonic buffers can be substituted here based on user-defined needs.

22. Centrifugation at settings as low as $300 \times g$ for 1–2 min can remove 50% or more of high molecular weight non-cellular material (e.g., mucins).

23. More vials may be used as back-ups and/or for additional analytic endpoints. This may depend on expected concentrations of extracted materials. Typically 2–5 mg each of total RNA and total protein can be extracted from homogenates of whole mouse lungs (combined right and left lobes). Aliquots of samples should be made to allow for adequate sample loading while avoiding repeated freeze-thaw cycles for samples used in more than one assay.

24. To avoid sample degradation, thaw only a few samples at a time unless using automated machinery. Keeping the time from initial thawing to extraction to 10 min or less is ordinarily sufficient.

25. RNA purity can also be assessed by measuring spectrophotometric light absorbance at 280 nm wavelength. Contaminants such as protein absorb 280 nm light preferentially in comparison with nucleic acids. If division of absorbance values obtained at 260 nm by values obtained at 280 nm yields a ratio whose value is 2.0 or greater, this is generally considered to indicate pure RNA.

26. Quantification using other methods such as SYBR green can be substituted. In addition, non-quantitative methods using ethidium bromide detection can also be used to determine expression in "yes" vs. "no" expression scenarios.

27. A complete list of compatible substances is given in the BCA manufacturer's manual.

28. Proteins to be separated are diluted to 10–100 µg of total protein of tissue lysate fluid and then should be heated to 55 °C for 30 min, 70 °C for 10 min, or 99 °C for 5 min depending on the need for denaturation and/or reduction. Under non-reducing conditions, no heating is need.

29. Gels and membranes are assembled in a transfer sandwich and immersed in compatible transfer buffer and tanks. Transfer conditions may vary, but standard protocols use a constant current (0.1–1.0 A) or constant voltage (5–30 V) for 1 h on the benchtop with ice or overnight in a refrigerator.

30. As an additional or alternative method, immunohistochemistry can be used. Methods are specific to the users' needs and antibody characteristics.

References

1. Mason RJ (2006) Biology of alveolar type II cells. Respirology 11(Suppl):S12–S15. https://doi.org/10.1111/j.1440-1843.2006.00800.x

2. Singh PK, Jia HP, Wiles K, Hesselberth J, Liu L, Conway BA, Greenberg EP, Valore EV, Welsh MJ, Ganz T, Tack BF, McCray PB Jr (1998) Production of beta-defensins by human airway epithelia. Proc Natl Acad Sci U S A 95(25):14961–14966

3. Wine JJ, Joo NS (2004) Submucosal glands and airway defense. Proc Am Thorac Soc 1(1):47–53. https://doi.org/10.1513/pats.2306015

4. Basu S, Fenton MJ (2004) Toll-like receptors: function and roles in lung disease. Am J Physiol Lung Cell Mol Physiol 286(5):L887–L892. https://doi.org/10.1152/ajplung.00323.2003

5. Evans SE, Xu Y, Tuvim MJ, Dickey BF (2010) Inducible innate resistance of lung epithelium to infection. Annu Rev Physiol 72:413–435. https://doi.org/10.1146/annurev-physiol-021909-135909

6. Georas SN, Rezaee F (2014) Epithelial barrier function: at the front line of asthma immunology and allergic airway inflammation. J Allergy Clin Immunol 134(3):509–520. https://doi.org/10.1016/j.jaci.2014.05.049

7. Janssen WJ, Stefanski AL, Bochner BS, Evans CM (2016) Control of lung defence by mucins and macrophages: ancient defence mechanisms with modern functions. Eur Respir J 48(4):1201–1214. https://doi.org/10.1183/13993003.00120-2015

8. Williams OW, Sharafkhaneh A, Kim V, Dickey BF, Evans CM (2006) Airway mucus: from production to secretion. Am J Respir Cell Mol Biol 34(5):527–536. https://doi.org/10.1165/rcmb.2005-0436SF

9. Fahy JV, Dickey BF (2010) Airway mucus function and dysfunction. N Engl J Med 363(23):2233–2247. https://doi.org/10.1056/NEJMra0910061

10. Schwenk F, Baron U, Rajewsky K (1995) A cre-transgenic mouse strain for the ubiquitous deletion of loxP-flanked gene segments including deletion in germ cells. Nucleic Acids Res 23(24):5080–5081

11. Lewandoski M, Meyers EN, Martin GR (1997) Analysis of Fgf8 gene function in vertebrate development. Cold Spring Harb Symp Quant Biol 62:159–168

12. de Vries WN, Binns LT, Fancher KS, Dean J, Moore R, Kemler R, Knowles BB (2000) Expression of Cre recombinase in mouse oocytes: a means to study maternal effect genes. Genesis 26(2):110–112

13. Lewandoski M, Wassarman KM, Martin GR (1997) Zp3-cre, a transgenic mouse line for the activation or inactivation of loxP-flanked target genes specifically in the female germ line. Curr Biol 7(2):148–151

14. Ruzankina Y, Pinzon-Guzman C, Asare A, Ong T, Pontano L, Cotsarelis G, Zediak VP, Velez M, Bhandoola A, Brown EJ (2007) Deletion of the developmentally essential gene ATR in adult mice leads to age-related phenotypes and stem cell loss. Cell Stem Cell 1(1):113–126. https://doi.org/10.1016/j.stem.2007.03.002

15. Harfe BD, Scherz PJ, Nissim S, Tian H, McMahon AP, Tabin CJ (2004) Evidence for an expansion-based temporal Shh gradient in specifying vertebrate digit identities. Cell 118(4):517–528. https://doi.org/10.1016/j.cell.2004.07.024

16. Que J, Luo X, Schwartz RJ, Hogan BL (2009) Multiple roles for Sox2 in the developing and adult mouse trachea. Development 136(11):1899–1907. https://doi.org/10.1242/dev.034629

17. Moses KA, DeMayo F, Braun RM, Reecy JL, Schwartz RJ (2001) Embryonic expression of an Nkx2-5/Cre gene using ROSA26 reporter mice. Genesis 31(4):176–180

18. Stanley EG, Biben C, Elefanty A, Barnett L, Koentgen F, Robb L, Harvey RP (2002) Efficient Cre-mediated deletion in cardiac progenitor cells conferred by a 3'UTR-ires-Cre allele of the homeobox gene Nkx2-5. Int J Dev Biol 46(4):431–439

19. Xu Q, Tam M, Anderson SA (2008) Fate mapping Nkx2.1-lineage cells in the mouse telencephalon. J Comp Neurol 506(1):16–29. https://doi.org/10.1002/cne.21529

20. Okubo T, Knoepfler PS, Eisenman RN, Hogan BL (2005) Nmyc plays an essential role during lung development as a dosage-sensitive regulator of progenitor cell proliferation and differentiation. Development 132(6):1363–1374. https://doi.org/10.1242/dev.01678

21. Mao CM, Yang X, Cheng X, Lu YX, Zhou J, Huang CF (2003) Establishment of keratinocyte-specific Cre recombinase transgenic mice. Yi Chuan Xue Bao 30(5):407–413

22. Tarutani M, Itami S, Okabe M, Ikawa M, Tezuka T, Yoshikawa K, Kinoshita T, Takeda J (1997) Tissue-specific knockout of the mouse pig-a gene reveals important roles for GPI-

anchored proteins in skin development. Proc Natl Acad Sci U S A 94(14):7400–7405

23. Berton TR, Matsumoto T, Page A, Conti CJ, Deng CX, Jorcano JL, Johnson DG (2003) Tumor formation in mice with conditional inactivation of Brca1 in epithelial tissues. Oncogene 22(35):5415–5426. https://doi.org/10.1038/sj.onc.1206825

24. Huelsken J, Vogel R, Erdmann B, Cotsarelis G, Birchmeier W (2001) beta-Catenin controls hair follicle morphogenesis and stem cell differentiation in the skin. Cell 105(4):533–545

25. Dassule HR, Lewis P, Bei M, Maas R, McMahon AP (2000) Sonic hedgehog regulates growth and morphogenesis of the tooth. Development 127(22):4775–4785

26. Li M, Chiba H, Warot X, Messaddeq N, Gerard C, Chambon P, Metzger D (2001) RXR-alpha ablation in skin keratinocytes results in alopecia and epidermal alterations. Development 128(5):675–688

27. Li H, Cho SN, Evans CM, Dickey BF, Jeong JW, DeMayo FJ (2008) Cre-mediated recombination in mouse Clara cells. Genesis 46(6):300–307. https://doi.org/10.1002/dvg.20396

28. Bertin G, Poujeol C, Rubera I, Poujeol P, Tauc M (2005) In vivo Cre/loxP mediated recombination in mouse Clara cells. Transgenic Res 14(5):645–654. https://doi.org/10.1007/s11248-005-7214-0

29. Simon DM, Arikan MC, Srisuma S, Bhattacharya S, Tsai LW, Ingenito EP, Gonzalez F, Shapiro SD, Mariani TJ (2006) Epithelial cell PPAR[gamma] contributes to normal lung maturation. FASEB J 20(9):1507–1509. https://doi.org/10.1096/fj.05-5410fje

30. Ji H, Houghton AM, Mariani TJ, Perera S, Kim CB, Padera R, Tonon G, McNamara K, Marconcini LA, Hezel A, El-Bardeesy N, Bronson RT, Sugarbaker D, Maser RS, Shapiro SD, Wong KK (2006) K-ras activation generates an inflammatory response in lung tumors. Oncogene 25(14):2105–2112. https://doi.org/10.1038/sj.onc.1209237

31. Tanaka T, Rabbitts TH (2010) Interfering with RAS-effector protein interactions prevent RAS-dependent tumour initiation and causes stop-start control of cancer growth. Oncogene 29(45):6064–6070. https://doi.org/10.1038/onc.2010.346

32. Oikonomou N, Mouratis MA, Tzouvelekis A, Kaffe E, Valavanis C, Vilaras G, Karameris A, Prestwich GD, Bouros D, Aidinis V (2012) Pulmonary autotaxin expression contributes to the pathogenesis of pulmonary fibrosis. Am J Respir Cell Mol Biol 47(5):566–574. https://doi.org/10.1165/rcmb.2012-0004OC

33. Zhang Y, Huang G, Shornick LP, Roswit WT, Shipley JM, Brody SL, Holtzman MJ (2007) A transgenic FOXJ1-Cre system for gene inactivation in ciliated epithelial cells. Am J Respir Cell Mol Biol 36(5):515–519. https://doi.org/10.1165/rcmb.2006-0475RC

34. Flodby P, Borok Z, Banfalvi A, Zhou B, Gao D, Minoo P, Ann DK, Morrisey EE, Crandall ED (2010) Directed expression of Cre in alveolar epithelial type 1 cells. Am J Respir Cell Mol Biol 43(2):173–178. https://doi.org/10.1165/rcmb.2009-0226OC

35. Bertozzi CC, Schmaier AA, Mericko P, Hess PR, Zou Z, Chen M, Chen CY, Xu B, Lu MM, Zhou D, Sebzda E, Santore MT, Merianos DJ, Stadtfeld M, Flake AW, Graf T, Skoda R, Maltzman JS, Koretzky GA, Kahn ML (2010) Platelets regulate lymphatic vascular development through CLEC-2-SLP-76 signaling. Blood 116(4):661–670. https://doi.org/10.1182/blood-2010-02-270876

36. Vitale-Cross L, Amornphimoltham P, Fisher G, Molinolo AA, Gutkind JS (2004) Conditional expression of K-ras in an epithelial compartment that includes the stem cells is sufficient to promote squamous cell carcinogenesis. Cancer Res 64(24):8804–8807. https://doi.org/10.1158/0008-5472.CAN-04-2623

37. Diamond I, Owolabi T, Marco M, Lam C, Glick A (2000) Conditional gene expression in the epidermis of transgenic mice using the tetracycline-regulated transactivators tTA and rTA linked to the keratin 5 promoter. J Invest Dermatol 115(5):788–794. https://doi.org/10.1046/j.1523-1747.2000.00144.x

38. Indra AK, Warot X, Brocard J, Bornert JM, Xiao JH, Chambon P, Metzger D (1999) Temporally-controlled site-specific mutagenesis in the basal layer of the epidermis: comparison of the recombinase activity of the tamoxifen-inducible Cre-ER(T) and Cre-ER(T2) recombinases. Nucleic Acids Res 27(22):4324–4327

39. Kataoka K, Kim DJ, Carbajal S, Clifford JL, DiGiovanni J (2008) Stage-specific disruption of Stat3 demonstrates a direct requirement during both the initiation and promotion stages of mouse skin tumorigenesis. Carcinogenesis 29(6):1108–1114. https://doi.org/10.1093/carcin/bgn061

40. Liang CC, You LR, Chang JL, Tsai TF, Chen CM (2009) Transgenic mice exhibiting inducible and spontaneous Cre activities driven by a bovine keratin 5 promoter that can be used for

the conditional analysis of basal epithelial cells in multiple organs. J Biomed Sci 16:2. https://doi.org/10.1186/1423-0127-16-2

41. Rock JR, Onaitis MW, Rawlins EL, Lu Y, Clark CP, Xue Y, Randell SH, Hogan BL (2009) Basal cells as stem cells of the mouse trachea and human airway epithelium. Proc Natl Acad Sci U S A 106(31):12771–12775. https://doi.org/10.1073/pnas.0906850106

42. Van Keymeulen A, Rocha AS, Ousset M, Beck B, Bouvencourt G, Rock J, Sharma N, Dekoninck S, Blanpain C (2011) Distinct stem cells contribute to mammary gland development and maintenance. Nature 479(7372):189–193. https://doi.org/10.1038/nature10573

43. Tadokoro T, Wang Y, Barak LS, Bai Y, Randell SH, Hogan BL (2014) IL-6/STAT3 promotes regeneration of airway ciliated cells from basal stem cells. Proc Natl Acad Sci U S A 111(35):E3641–E3649. https://doi.org/10.1073/pnas.1409781111

44. Foster KW, Liu Z, Nail CD, Li X, Fitzgerald TJ, Bailey SK, Frost AR, Louro ID, Townes TM, Paterson AJ, Kudlow JE, Lobo-Ruppert SM, Ruppert JM (2005) Induction of KLF4 in basal keratinocytes blocks the proliferation-differentiation switch and initiates squamous epithelial dysplasia. Oncogene 24(9):1491–1500. https://doi.org/10.1038/sj.onc.1208307

45. Nguyen H, Rendl M, Fuchs E (2006) Tcf3 governs stem cell features and represses cell fate determination in skin. Cell 127(1):171–183. https://doi.org/10.1016/j.cell.2006.07.036

46. Vasioukhin V, Degenstein L, Wise B, Fuchs E (1999) The magical touch: genome targeting in epidermal stem cells induced by tamoxifen application to mouse skin. Proc Natl Acad Sci U S A 96(15):8551–8556

47. Hong KU, Reynolds SD, Watkins S, Fuchs E, Stripp BR (2004) Basal cells are a multipotent progenitor capable of renewing the bronchial epithelium. Am J Pathol 164(2):577–588. https://doi.org/10.1016/S0002-9440(10)63147-1

48. Hong KU, Reynolds SD, Watkins S, Fuchs E, Stripp BR (2004) In vivo differentiation potential of tracheal basal cells: evidence for multipotent and unipotent subpopulations. Am J Physiol Lung Cell Mol Physiol 286(4):L643–L649. https://doi.org/10.1152/ajplung.00155.2003

49. Li M, Indra AK, Warot X, Brocard J, Messaddeq N, Kato S, Metzger D, Chambon P (2000) Skin abnormalities generated by temporally controlled RXRalpha mutations in mouse epidermis. Nature 407(6804):633–636. https://doi.org/10.1038/35036595

50. Taniguchi H, He M, Wu P, Kim S, Paik R, Sugino K, Kvitsiani D, Fu Y, Lu J, Lin Y, Miyoshi G, Shima Y, Fishell G, Nelson SB, Huang ZJ (2011) A resource of Cre driver lines for genetic targeting of GABAergic neurons in cerebral cortex. Neuron 71(6):995–1013. https://doi.org/10.1016/j.neuron.2011.07.026

51. Tichelaar JW, Lu W, Whitsett JA (2000) Conditional expression of fibroblast growth factor-7 in the developing and mature lung. J Biol Chem 275(16):11858–11864

52. Perl AK, Zhang L, Whitsett JA (2009) Conditional expression of genes in the respiratory epithelium in transgenic mice: cautionary notes and toward building a better mouse trap. Am J Respir Cell Mol Biol 40(1):1–3. https://doi.org/10.1165/rcmb.2008-0011ED

53. Duerr J, Gruner M, Schubert SC, Haberkorn U, Bujard H, Mall MA (2011) Use of a new-generation reverse tetracycline transactivator system for quantitative control of conditional gene expression in the murine lung. Am J Respir Cell Mol Biol 44(2):244–254. https://doi.org/10.1165/rcmb.2009-0115OC

54. Rawlins EL, Okubo T, Xue Y, Brass DM, Auten RL, Hasegawa H, Wang F, Hogan BL (2009) The role of Scgb1a1+ Clara cells in the long-term maintenance and repair of lung airway, but not alveolar, epithelium. Cell Stem Cell 4(6):525–534. https://doi.org/10.1016/j.stem.2009.04.002

55. Blake JA, Eppig JT, Kadin JA, Richardson JE, Smith CL, Bult CJ, The Mouse Genome Database Group (2017) Mouse Genome Database (MGD)-2017: community knowledge resource for the laboratory mouse. Nucleic Acids Res 45(D1):D723–D729. https://doi.org/10.1093/nar/gkw1040

56. Rawlins EL, Ostrowski LE, Randell SH, Hogan BL (2007) Lung development and repair: contribution of the ciliated lineage. Proc Natl Acad Sci U S A 104(2):410–417. https://doi.org/10.1073/pnas.0610770104

57. Meletis K, Barnabe-Heider F, Carlen M, Evergren E, Tomilin N, Shupliakov O, Frisen J (2008) Spinal cord injury reveals multilineage differentiation of ependymal cells. PLoS Biol 6(7):e182. https://doi.org/10.1371/journal.pbio.0060182

58. Rock JR, Barkauskas CE, Cronce MJ, Xue Y, Harris JR, Liang J, Noble PW, Hogan BL (2011) Multiple stromal populations contribute to pulmonary fibrosis without evidence for

epithelial to mesenchymal transition. Proc Natl Acad Sci U S A 108(52):E1475–E1483. https://doi.org/10.1073/pnas.1117988108

59. Lin C, Song H, Huang C, Yao E, Gacayan R, Xu SM, Chuang PT (2012) Alveolar type II cells possess the capability of initiating lung tumor development. PLoS One 7(12):e53817. https://doi.org/10.1371/journal.pone.0053817

60. Chapman HA, Li X, Alexander JP, Brumwell A, Lorizio W, Tan K, Sonnenberg A, Wei Y, Vu TH (2011) Integrin alpha6beta4 identifies an adult distal lung epithelial population with regenerative potential in mice. J Clin Invest 121(7):2855–2862. https://doi.org/10.1172/JCI57673

61. Fu Y, Xiao W (2006) Study of transcriptional regulation using a reporter gene assay. Methods Mol Biol 313:257–264. https://doi.org/10.1385/1-59259-958-3:257

62. Linke D (2009) Detergents: an overview. Methods Enzymol 463:603–617. https://doi.org/10.1016/S0076-6879(09)63034-2

63. Peach M, Marsh N, Miskiewicz EI, MacPhee DJ (2015) Solubilization of proteins: the importance of lysis buffer choice. Methods Mol Biol 1312:49–60. https://doi.org/10.1007/978-1-4939-2694-7_8

64. Peach M, Marsh N, Macphee DJ (2012) Protein solubilization: attend to the choice of lysis buffer. Methods Mol Biol 869:37–47. https://doi.org/10.1007/978-1-61779-821-4_4

65. Komatsu S (2007) Extraction of nuclear proteins. Methods Mol Biol 355:73–77. https://doi.org/10.1385/1-59745-227-0:73

66. Blancher C, Jones A (2001) SDS-PAGE and western blotting techniques. Methods Mol Med 57:145–162. https://doi.org/10.1385/1-59259-136-1:145

67. Piccotti L, Dickey BF, Evans CM (2012) Assessment of intracellular mucin content in vivo. Methods Mol Biol 842:279–295. https://doi.org/10.1007/978-1-61779-513-8_17

68. Friedel RH, Wurst W, Wefers B, Kuhn R (2011) Generating conditional knockout mice. Methods Mol Biol 693:205–231. https://doi.org/10.1007/978-1-60761-974-1_12

69. Van Hoecke L, Job ER, Saelens X, Roose K (2017) Bronchoalveolar lavage of murine lungs to analyze inflammatory cell infiltration. J Vis Exp (123). https://doi.org/10.3791/55398

70. Misharin AV, Morales-Nebreda L, Mutlu GM, Budinger GR, Perlman H (2013) Flow cytometric analysis of macrophages and dendritic cell subsets in the mouse lung. Am J Respir Cell Mol Biol 49(4):503–510. https://doi.org/10.1165/rcmb.2013-0086MA

71. Mauderly JL (1977) Bronchopulmonary lavage of small laboratory animals. Lab Anim Sci 27(2):255–261

72. Medin NI, Osebold JW, Zee YC (1976) A procedure for pulmonary lavage in mice. Am J Vet Res 37(2):237–238

Chapter 15

Computational Analysis of RNA-Seq Data from Airway Epithelial Cells for Studying Lung Disease

Nathan D. Jackson, Lando Ringel, and Max A. Seibold

Abstract

Airway epithelial cells (AECs) play a central role in the pathogenesis of many lung diseases. Consequently, advancements in our understanding of the underlying causes of lung diseases, and the development of novel treatments, depend on continued detailed study of these cells. Generation and analysis of high-throughput gene expression data provide an indispensable tool for carrying out the type of broad-scale investigations needed to identify the key genes and molecular pathways that regulate, distinguish, and predict distinct pulmonary pathologies. Of the available technologies for generating genome-wide expression data, RNA sequencing (RNA-seq) has emerged as the most powerful. Hence many researchers are turning to this approach in their studies of lung disease. For the relatively uninitiated, computational analysis of RNA-seq data can be daunting, given the large number of methods and software packages currently available. The aim of this chapter is to provide a broad overview of the major steps involved in processing and analyzing RNA-seq data, with a special focus on methods optimized for data generated from AECs. We take the reader from the point of obtaining sequence reads from the lab to the point of making biological inferences with expression data. Along the way, we discuss the statistical and computational considerations one typically confronts during different phases of analysis and point to key methods, software packages, papers, online guides, and other resources that can facilitate successful RNA-seq analysis.

Key words Clustering, Data normalization, Differential expression, Functional enrichment, Gene mapping, Gene quantification, Pathway analysis, Transcript quantification, Transcriptome alignment, WGCNA

---

1 Introduction

The epithelial cells that line the human airway play a central role in respiratory health and homeostasis. They form a barrier to particulates, toxins, and pathogens inhaled into the lungs. Specifically, ciliated epithelial cells beating in concert move a mucus layer generated by secretory epithelial cells, which allows the airway epithelium to capture, neutralize, and remove foreign particles from the lung. In addition, the interaction and signaling between airway epithelial cells (AECs) and airway immune cells are critical in the initiation and modulation of immune responses to infiltrating microorganisms [1].

Scott Alper and William J. Janssen (eds.), *Lung Innate Immunity and Inflammation: Methods and Protocols*,
Methods in Molecular Biology, vol. 1809, https://doi.org/10.1007/978-1-4939-8570-8_15,
© Springer Science+Business Media, LLC, part of Springer Nature 2018

A corollary of the protective and homeostatic role of AECs is that dysfunction of these cells is central to the pathogenesis of many chronic airway diseases, such as chronic obstructive pulmonary disease [COPD; e.g., 2, 3], cystic fibrosis [CF; e.g., 4, 5], asthma [e.g., 6, 7], and idiopathic pulmonary fibrosis [IPF; e.g., 8, 9]. Therefore, detailed study of the genes and molecular pathways that are activated in the AECs of individuals with and without pulmonary disease is essential, both for characterizing the pathophysiology of chronic airway diseases and for elucidating potential therapeutic targets.

Over the past decade or so, high-throughput gene expression platforms have emerged as some of the most powerful and unbiased tools for exploring the potential underlying mechanisms of disease [10]. Moreover, in comparison with previous approaches, RNA sequencing (RNA-seq) technology has resulted in substantial improvements to the quality of gene expression data, particularly owing to its lack of reliance on prespecified sets of genes, its sensitivity to detecting genes across a wide range of expression levels, and its reproducibility and precision in sequence identification [11]. The past few years have seen RNA-seq-based studies that have elucidated effects of smoking on the distal airway epithelium [12], discovered nasal airway epithelial genes for distinguishing sub-phenotypes of childhood asthma [13], revealed mechanistic pathways involved in epithelial-to-mesenchymal transition in the small airways [14], described differential gene splicing in IPF [15], and identified respiratory virus-infected subjects and host responses to viral infection of the airway [16], to name a few. RNA-seq data will undoubtedly continue to play important roles in deepening our understanding of airway biology in human health and disease.

Taking full advantage of RNA-seq technology, however, requires a certain level of statistical and computational expertise. The performance of a large number of analytical steps is required to proceed from raw sequencing data to the point of producing biological insights. Since many methods and software packages exist for carrying out these steps, it can be arduous to sufficiently familiarize oneself with the available resources and to choose the most appropriate suite of methods for one's dataset and questions of interest. Detailed manuals and guides exist for the various software packages and specific steps involved in analysis of RNA-seq data. However, when beginning a project, we believe it would be helpful to have a roadmap that provides a broad overview of all the major steps involved in processing and analyzing RNA-seq data. This guide would suggest specific methods to use that have been successfully applied to data from one's particular system of interest.

The goal of this chapter is to provide such a guide for the major steps involved in analyzing RNA-seq data—with a special focus on data generated from human airway epithelial cell samples—to make biologically meaningful inferences in studies of lung disease. In doing so, we aim to outline the discrete steps of analysis, focusing on the bioinformatic and statistical considerations one will confront throughout the process. We anticipate that this chapter will be most useful for airway biologists with some computational skills that would like to apply RNA-seq to their own biological questions. This chapter may also be a useful reference to bioinformaticians who are new to analyzing human airway epithelial data, as our workflow is appropriate for these cells. In this guide, for each step in the analytical process we describe the overarching goal, discuss potential software packages, and highlight major parameters to take into account. We also note useful papers, online guides, and other resources throughout, making this a good launching pad for further exploration.

2   Materials

We begin this chapter at the point of having FASTQ files in hand, which contain cDNA sequences obtained from a sequencing platform. Thus, the materials relevant to this chapter are strictly computational. Although the methods outlined here can technically be performed on any modern-day computer, for most real-world RNA-seq datasets, it would be laborious and slow to perform some steps of analysis without access to a high-performance computing cluster. Also, many modern bioinformatics methods are geared toward a Linux or Mac environment, and thus those are the operating systems that we recommend. Happily, most (though not all) of the methods we discuss here are open source and easily available through internet download. Many methods are implemented using command line-based software, particularly R, and thus some experience with R and the terminal shell are recommended prior to beginning.

Below is a list of all the software packages we discuss in the chapter, organized by section:

2.1 Data Processing: FASTQ Quality Control (Subheading 3.1.2)

FastQC [17].

2.2 Data Processing: Removing Poor Quality Reads and Bases (Subheading 3.1.3)

Skewer [18].
Trimmomatic [19].
Cutadapt [20].

2.3 Data Processing: Aligning Reads to a Reference Genome (Subheading 3.1.4)

GSNAP [21].
TopHat [22].
MapSplice [23].
STAR [24].
HISAT [25].

	kallisto [26].
	Samtools [27].
2.4 Data Processing: Measuring Transcript or Gene Abundance (Subheading 3.1.5)	HTSeq [28]. Cufflinks [29].
2.5 Making Biological Inferences: Initial Data Checking (Subheading 3.2.2)	featureCounts [30]. Rsubread [31].
2.6 Making Biological Inferences: Identifying Relevant Genes (Subheading 3.2.3)	DESeq2 [32]. limma [33].
2.7 Making Biological Inferences: Characterizing Heterogeneity Among Samples (Subheading 3.2.4)	heatmap3 [34]. Rtsne [35].
2.8 Making Biological Inferences: Characterizing Heterogeneity Among Genes (Subheading 3.2.5)	DAVID [36]. Enrichr [37]. GOstats [38]. Ingenuity Pathway Analysis [39]. iRegulon [40]. Cytoscape [41].

3   Methods

There are two major phases of RNA-seq data analysis. The first of these involves generation of a gene expression matrix (expression levels of all genes by all subjects) from raw sequence reads. This part of the process is often referred to as a "pipeline" given that, once one has initially chosen the analytical tools and parameter settings to use, this step can usually be automated with scripting tools. We discuss the various components of this first part of analysis in Subheading 3.1. The second major phase of analysis—making biological inferences using the gene expression matrix—is a bit more open-ended, and the specific methods and tests applied will be more dependent on the nature of the dataset and project goals. We

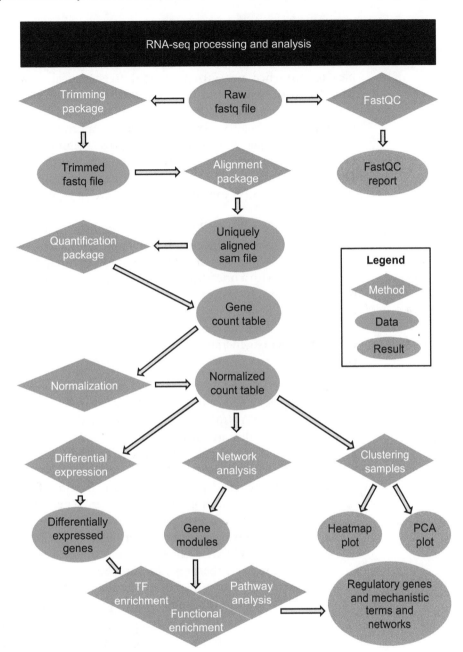

Fig. 1 A graphic overview of the major components of a typical RNA-seq workflow

discuss the elements involved in this second phase of analysis in Subheading 3.2. Figure 1 depicts a graphic overview of the major steps involved in a typical RNA-seq workflow.

In this chapter, we focus on analysis of RNA-seq data that have already been generated. However, everything that happens prior the downloading of FASTQ files from one's server can of course have a big impact on downstream computational analyses and inference. There are several experimental design issues that must be managed prior to sequencing that are critical for successful RNA-seq analysis, such as (1) the sample sequencing depth (i.e., number of sequences generated/sample), (2) the number of samples to sequence per group (related to statistical power), (3) whether to do paired-end or single-end sequencing, (4) how to barcode samples, and (5) how to best allocate samples to discrete sequencing batches to avoid batch effects. Appropriate attention and planning regarding these issues can save a significant amount of time and money and are thus definitely worth the effort. Discussion of these and other issues along with design planning suggestions and tools can be found elsewhere [e.g., 42, 43–45].

3.1 Data Processing: Generating Gene Expression Levels from RNA-Seq Data

The process of analyzing RNA-seq data begins with a FASTQ file containing cDNA sequence reads generated from a set of samples. However, to provide proper context for understanding the analysis steps, we start with a brief discussion of the nature of these sequence reads, how they are generated, and why they reflect levels of gene expression.

3.1.1 Raw Sequence Data: Nature and Generation of FASTQ Files

RNA-seq data are produced from sequencing a library of cDNA fragments generated from the RNA of a sample. The general steps in library preparation and sequencing are as follows:

1. *Library preparation.* First, total RNA is purified from subject cell samples. Then, either total RNA (for total RNA-seq) or purified mRNA (mRNA-seq) is converted to cDNA. The cDNA molecules are mechanically or enzymatically cut into fragments of a particular size range (typically 300–700 base pairs). This is followed by ligation of short DNA segments (called sequencing adapters) to the 5′ and 3′ ends of the cDNA fragments. The collection of these adapter-ligated cDNA fragments generated from a sample is known as the sample sequencing "library." Barring experimental problems, this cDNA library should contain cDNA fragments of all expressed genes in the sample. Moreover, the presence and abundance of fragments for a particular gene in the library should reflect the expression level of that gene in the sample from which it was prepared.

2. *Library sequencing.* For RNA-seq performed with an Illumina HiSeq instrument, the libraries are loaded onto a flow cell to which the library molecules attach by the annealing of adapter sequences to complementary oligonucleotides that cover the flow cell surface. These fragments are then amplified and sequenced in massively parallel reactions. Illumina sequencing is carried out using cyclic reversible termination, in which new complementary fragments are synthesized by incorporating one fluorescently labeled base at a time. Between each sequencing cycle, a computer takes a picture of the flow cell and then makes a base call by measuring the wavelength of the fluorescent tag captured in the image. These base calls for each library

fragment, along with quality scores, are recorded in a binary base call (BCL) file. The set of base calls for each position on the flow cell (and thus each cDNA fragment) forms the corresponding sequence read.

3. *FASTQ files.* The Illumina BCL file must then be converted into a FASTQ file. FASTQ is a simple text-based file format that has become the standard for storing high-throughput sequencing data. This BCL to FASTQ conversion can be performed using the bcl2fastq tool available on Illumina's proprietary cloud platform, BaseSpace.

   FASTQ files contain four lines for each sequence:

   - A unique sequence identifier beginning with the "@" symbol (*see* **Note 1** for more details)

   - The raw nucleotide sequence

   - The "+" symbol

   - A Phred quality score assigned to each nucleotide base

   Phred quality scores (Q) are encoded using ASCII characters and provide an estimate of the probability of a base being called in error (*see* **Note 2** for how Phred scores are decoded and interpreted). These scores can be used to systematically evaluate the quality of a dataset (discussed in the next section).

4. *Gene expression information contained in a FASTQ file.* To reiterate from above, sequence reads within FASTQ files can reveal those genes that are expressed in a given sample, once the gene from which the sequence read was derived has been determined. Furthermore, the abundance of reads originating from a particular gene will reflect the expression level of that gene (although there are caveats to this related to variation in gene length, which we discuss below in Subheading 3.2.1). Because the full nucleotide sequence of reads is known, other sequence-related phenomena, such as the existence of transcript variants of a gene, can be identified and studied using these data.

   It is important to note that since only a subsampling of the RNA molecules that exist within a given sample are sequenced, the expression values from RNA-seq can only provide estimates of gene expression levels. Additionally, gene transcripts that are relatively rare in a sample can be missed altogether, depending on the sequencing depth. In general, the deeper the sequencing for a particular sample the more accurately and precisely (but also expensively) one can measure gene expression [e.g., 46, 47]. Although dependent on many factors, when performing mRNA-seq with airway epithelial cells, we have found that sequencing somewhere in the neighborhood of 20–40 million, 100 base pair, paired-end reads is sufficient to accurately detect most expressed genes.

Before beginning the process of generating gene expression data from raw sequences stored in FASTQ files, it is important to carry out quality checking (*see* **Note 3**). This will allow you to familiarize yourself with important characteristics of the data as well as to verify that there are no significant errors or biases present that may reduce the accuracy of downstream analyses. The method we use for doing this is FastQC [17], a common, easy-to-use quality control (QC) software package for high-throughput sequence data. Upon inputting a FASTQ file, this Java-based program can generate quick HTML reports with graphical summaries of several important QC metrics. For each set of metrics, an overall color-coded assessment is provided that indicates whether a result appears to be normal (green), moderately abnormal (yellow), or highly abnormal (red).

When inspecting FastQC output, of initial interest is the Basic Statistics section, which includes both the total number of sequences in the file and the observed range of sequence lengths (a plot of the distribution of sequence lengths is also provided elsewhere in the report). If these values are not as expected (e.g., if the total number of sequences is considerably below the specifications for one's sequencing platform or if there is a large number of sequences flagged as poor quality), this can indicate a significant problem with the RNA sequencing performance or library generation.

The degree and distribution of sequence nucleotide base quality are also of primary interest. Summary graphics reporting sequence quality in the FastQC report are based on the nucleotide base-specific Phred quality scores embedded in the FASTQ file (discussed above). The Per Base Sequence Quality section shows the distribution of quality scores across sequences for each base in the sequence read. Ideally, the median quality score should be above ~28 (in the green section of the plot) for all bases or, minimally, above ~20 (in the orange section of the plot). It should be noted that it is typical to observe a degradation of nucleotide quality toward the 3′ end of reads. If a dip in sequence quality is due to a problem with a particular physical location (or "tile" location) on the flow cell (*see* **Note 4**), this can be identified from the per tile sequence quality heat map, which shows variation in sequence quality across tiles.

Another important QC feature to examine relates to the existence and nature of overrepresented sequences. In the Overrepresented Sequences section, FastQC reports sequences that comprise more than 0.1% of the entire dataset. The program also checks whether any overrepresented sequences match known common contaminants. Although catching possible contaminants is one of the main stated purposes of this section of the report, in our experience, the most commonly observed overrepresented sequences are untrimmed adapter sequences, mRNA poly (A) tails, and highly abundant ribosomal (rRNA) and mitochondrial (mtRNA) genes. Because of sequencing errors, which are particularly common in long reads, many overrepresented sequences or motifs may be missed when looking for exact duplicates of entire reads. Thus, FastQC also looks for overrepresentation of short sequences that are K bases long (called Kmers) and then plots the enrichment of these sequences across reads.

There are many additional features that FastQC monitors that may be of interest in particular contexts, such as the distributions of the four nucleotide bases (per base sequence content), GC content (per sequence GC content), or uncalled bases (per base N content). A relatively small amount of time spent checking these QC readouts can catch errors to be corrected that will save one from drawing spurious conclusions from one's dataset and wasting time rerunning analysis.

3.1.3 Removing Poor Quality Reads and Bases

Once you have become familiar with the overall quality of your dataset, you can proceed to remove poor quality or undesired sequences from the dataset. Many different software packages (discussed below) have been developed for parsing RNA-seq FASTQ files in this fashion. There are two major features that these software packages can remove from RNA-seq data:

Low Quality Reads and Bases

Removal of poor quality sequence data is based on the Phred quality scores contained within FASTQ files (discussed previously). First, one can trim poor quality bases found near the ends of otherwise high quality reads. As discussed above, the quality of base calls on Illumina platforms will naturally degrade as sequencing progresses, and thus these sequence ends will often need to be removed. The remaining high quality, but shorter, read will typically align better to the genome than will an untrimmed, longer read with poor quality ends. However, at times, quality trimming may reduce the length of a read considerably. Reads that are too short tend either to be mapped to multiple regions of the genome or to be mapped to the wrong region of the genome and therefore should be removed prior to downstream analysis. Trimming software allows the user to remove reads that fall below a specified length. For both read trimming and short read removal, we use the stand-alone package, Skewer [18]. Other common trimming methods include Trimmomatic [19] and Cutadapt [20].

In addition to the reads that need to be trimmed, there will be other reads that exhibit poor quality throughout the length of the read. In this case, it is appropriate to remove these reads altogether from the FASTQ file. The program Skewer [18] has the capability of filtering out entire reads based on a user-specified mean quality score cutoff.

Adapter Sequences

If the original RNA fragment is shorter than the number of bases sequenced, then a portion of the adapter will also be sequenced. Adapter contamination can lead to alignment errors since the adapter sequences do not occur in the reference genome. We use Skewer to remove these sequencing adapters by specifying the exact adapter we used in library preparation.

3.1.4   Aligning Reads to a Reference Genome

The number of sequences in the FASTQ file originating from a particular gene will be used to calculate that gene's expression levels across samples. Before this can be done, one must first identify the gene origin of all sequence reads. This is accomplished by first aligning, or "mapping," the RNA-seq reads to a reference human genome sequence. It is important to keep in mind three characteristics of RNA-seq reads that influence their alignment: (1) "short" read sequencers (e.g., Illumina) produce 50–150 bp reads, which do not span entire mRNA molecules; (2) the random shearing of library cDNA molecules means that RNA-seq reads will align randomly to different positions in the genes from which they were expressed; and (3) many reads span exon-exon junctions, especially with read lengths >100 bp, and thus will be discontiguous with respect to the genome sequence, since introns have been spliced out of the mRNA (and thus cDNA library) reads. Therefore, RNA-seq aligners must employ strategies to map millions of short sequences, with random genetic positions, many of which are genomically discontiguous. Methods exist to complete this alignment without a priori gene or exon positioning information. These methods have the advantage of allowing for the discovery of new gene transcripts and splice sites. However, these methods are also generally more computationally intensive, are susceptible to spurious results, and have limited benefit in organisms with highly annotated transcriptomes. Decades of intense study across a range of primary cell types and tissues have succeeded in generating very comprehensive transcriptome annotation files for the human genome containing known positions of exon boundaries. Most aligners use one of these transcriptome annotation files to perform a "splice-aware" alignment that is fast, accurate, and, by definition, comparable to reported gene transcripts. We recommend doing a splice-aware alignment in most situations.

The aligner we currently use is the Genomic Short-Read Nucleotide Alignment Program (GSNAP) [21]. This choice is based on our own and other published data showing that GSNAP generates one of the highest percentages of uniquely mapped reads when compared with other aligners [48]. Additionally, GSNAP has the capability of performing alignment in a way that can accommodate genetic variants (discussed further below). Other excellent RNA-seq alignment tools include TopHat [22], MapSplice [23], STAR [24], and HISAT [25]. Below, we briefly summarize the inputs, outputs, and critical issues of RNA-seq alignment analysis when using the GSNAP aligner, which will be similar and relevant to other common aligners.

Inputs

There are three major inputs when performing a splice-aware alignment of RNA-seq data to a reference genome: (1) the trimmed FASTQ file containing one's RNA sequence reads (required), (2) a reference genome (required), and (3) a transcriptome annotation file, called a gene transfer format (GTF) file, which contains the location of genes and their exon boundaries throughout the genome, from which the location of splice junctions can be inferred. If a goal of the analysis is to potentially discover new splice sites, GSNAP has the capability of carrying out splice junction-aware alignment using the provided GTF file (by invoking the "—splicesites" flag), while simultaneously allowing for the detection of novel splice sites that are not contained in the GTF file, using probabilistic modeling (by invoking the "—novelsplicing" flag).

Both reference genome assembly and GTF files for humans can be accessed from one of several online databases. Two of the major databases for accessing these files are Ensembl and the UCSC Genome Browser. Because these files are based on the same genome assembly, they will only differ in sequence annotation. The GTF files available from these databases share the same highly validated primary gene transcripts but differ with respect to the number of alternative transcripts, exons, and splice sites included. We typically use either the latest version of the iGenomes GTF file and assembly available from Illumina (which is derived from the UCSC database) or the Ensembl GTF file and assembly (*see* **Note 5** for where to download these files). The UCSC GTF file is a streamlined but reasonably comprehensive catalogue of common human coding gene transcripts that works well for most mRNA-seq experiments. For total RNA-seq experiments assessing non-coding and small RNAs, and/or if a more comprehensive assay of gene transcript variants in mRNA-seq experiments is desired, the Ensembl GTF file is preferred.

Outputs

GSNAP, like most other aligners, outputs mapping results in Sequence Alignment Map (SAM) format, which is a standard tab-delimited, text-based way to store high-throughput nucleotide alignments (*see* **Note 6**). SAM files are usually very large (many gigabytes each) but can be compressed into binary versions (BAM files) for storage. SAM to BAM conversion can be carried out using Samtools [27]. GSNAP can be configured to split alignment output into different SAM files based on alignment results, such as whether sequences were unmapped ("nomapping"), mapped to a single location in the genome ("uniq"), or mapped to multiple places ("mult"). Typically, one will want to focus on those reads within the uniquely mapped files for downstream quantification and analysis. In our experience, for high quality human samples, we would expect the percentage of reads that are uniquely mapped to be between 75% and 90%.

Critical Issues: Computation Time and Pseudoalignment

Before you can begin mapping your trimmed RNA-seq files, you should ensure that you have sufficient computational resources. Alignment run time is largely dependent upon the size of the input FASTQ file(s) as well as the available number of parallelizable processors. For most modern-day RNA-seq datasets, the alignment step is one of those for which a high-performance computing (HPC) environment is highly recommended. The queuing system on an HPC cluster allows for many samples to be mapped in parallel. If computational time is a concern, STAR [24] and HISAT [25] are promising alternative mappers that exhibit much shorter runtime than GSNAP [25, 48], although we note that STAR requires a lot of memory (~30 GB of RAM).

If you don't have access to a large computer cluster, one option is to "align" reads to transcripts using a "pseudoalignment" approach, which is much less computationally intensive than traditional alignment. Currently one of the most promising methods for implementing pseudoalignment is kallisto [26] (*see* **Note 7**). The main goal of this method is to measure transcript abundance. Thus, in contrast to traditional alignment, which must identify the base pair position within a transcript on a reference genome/transcriptome to which a target read most likely aligns, pseudoalignment matches reads directly to transcripts, ignoring information about genome coordinates, which are largely unnecessary for accurate transcript identification and quantification. This method has been shown to result in transcript quantification speeds that are two orders of magnitude faster than traditional mapping approaches, without a measurable decrease in accuracy [26], and is thus suitable for running on a standard laptop. Because the output of kallisto is transcript abundances, this method is performed in place of both mapping and transcript quantification steps (discussed in the next section) (*see* **Note 8**).

Critical Issues: Accounting for Genetic Variation During Alignment

The human genome contains a high degree of genetic variation, such as single nucleotide polymorphisms (SNPs) and structural variants (e.g., indels). The reference genome used for read alignment contains one of the two alleles (for biallelic variants) for these variants. The RNA-seq data for any particular subject will mismatch alleles for many of these variants. These mismatches reduce the precision and accuracy of the read alignment [21]. GSNAP has the capability of carrying out variant-aware alignment based on user-inputted prior information concerning the location and alleles for known genetic variants databased by dbSNP. In addition to increasing alignment accuracy, variant-aware alignment usually results in a higher percentage of mapped reads. Human genome assemblies from both UCSC and Ensembl come packaged with an annotation file documenting these known genetic variants, which can be used to create a database that GSNAP can use to carry out variant-tolerant mapping.

Critical Issues: Mapping Paired-End Versus Single-End Reads

When carrying out alignment, one must specify whether one has paired or single-end sequence reads. Paired-end sequencing allows one to sequence both 5′ and 3′ ends of a library fragment. Paired-end sequencing is more expensive but has the advantage of producing better mapping for fragments. In addition, the mapping of paired ends for a fragment contributes considerably more information regarding gene splicing. Single-end sequencing involves sequencing library molecules from only one end and is considerably cheaper to conduct since half the data is generated compared to paired-end sequencing. The majority of human RNA-seq experiments are conducted with paired-end sequencing. Our lab uses 100 bp paired-end sequencing for airway epithelial cell experiments. That being said, in humans, for whom a very well-annotated genome is available, the proportion of uniquely mapped genes will likely be similar between shorter, single-end reads and longer, paired-end reads, as long as read length is greater than 75 base pairs [e.g., 49] (*see* **Note 9**).

3.1.5 Measuring Transcript or Gene Abundance

GSNAP supplies you with the unique mapping coordinates for each read within a sample, stored within SAM files. With this information, you can add up the total number of reads aligning to a particular gene. The total count of reads aligned to a particular gene represents the expression level for that gene. We typically perform this counting (often referred to as gene expression quantification) using the python package, HTSeq [28]. Application of HTSeq requires the SAM aligned read files and the same GTF file used to align your data. Expression counts are generated at the gene level by summing the number of reads aligned to any reported exons for the gene. Generation of this gene level expression is the default option for HTSeq. You will need to choose to apply one of several available counting rules (using the –m flag), which specify how reads that align across two or more genes should be treated (*see* **Note 10** for a summary of these counting rules). Depending on the dataset and your goals, you may want to count reads covering introns of the measured gene. This can be accomplished by setting the "-t" flag to "gene" rather than to the default "exon." When successfully run, HTSeq generates vectors of counts across genes for each sample, which can easily be assembled into a single count matrix (*see* **Note 11**).

Quantification of transcripts is more difficult to accomplish accurately than the quantification of genes, given that many transcripts contain the same exons. Thus, an encountered read may reasonably be assigned to any number of that gene's transcripts. Methods such as Cufflinks [29] incorporate additional modeling to better infer transcript abundances using aligned RNA-seq data. As mentioned above, pseudoalignment approaches, such as kallisto, can also carry out transcript-level quantification. These methods will be more appropriate if your research questions require expression quantification of specific gene transcript variants.

At completion of quantification analysis, you will have an $i \times j$ gene or gene transcript-level expression matrix, where the raw number of reads (i.e., counts) are given for a particular gene or gene transcript in the ith row and a particular sample in the jth column. This data matrix will be the input for all downstream analyses.

3.2 Making Biological Inferences Using RNA-Seq Expression Data

3.2.1 Background on the Statistical Nature of RNA-Seq Count Data

With the expression count matrix in hand, it is tempting to simply jump right in to statistical analysis and visualization using common statistical approaches, such as t-tests for comparing expression among samples. However, it is important to be aware of several characteristics of RNA-seq data that require the application of specific statistical models and/or normalization of the dataset prior to analysis. Some unique attributes of most RNA-seq count data are as follows:

1. RNA-seq gene count data tend to be overdispersed (i.e., exhibit unexpectedly high variance) relative to normal or Poisson expectations [e.g., see 50].

2. The variance in RNA-seq count data tends to increase with mean expression (is heteroskedastic), which violates the assumptions of homoscedasticity implicit in many downstream statistical tests such as linear modeling, clustering, network, and functional enrichment analysis [e.g., see 51].

3. Given the shotgun approach used to sequence fragmented cDNA in RNA-seq protocols, longer genes are more likely to be sequenced than shorter genes, given the same degree of expression in a sample [e.g., see 52]. Thus, raw counts *between* genes are not directly comparable.

4. Sequencing depths will to a greater or lesser extent also differ among samples, making it difficult to compare raw, non-normalized, counts between samples [e.g., see 53].

5. Error around estimated gene expression within a sample can be high for lowly expressed genes due to the fact that random chance becomes an important determinant of the number of times a gene is sequenced when available copies are rare [i.e., sampling error; e.g., see 54]. This phenomenon can be exacerbated by insufficiently deep sequencing or in samples that tend to be dominated by the expression of a handful of extremely

abundant genes. These both tend to decrease the relative abundance of reads for lowly expressed genes in a dataset.

In light of these RNA-seq data characteristics, best practices have been determined for analyzing RNA-seq data that incorporate appropriate normalization and statistical models/tests during analysis. Below, when discussing various analytical steps, we highlight how the methods discussed accommodate these statistical features of RNA-seq data.

3.2.2 Initial Data Checking: Characterizing Broad Level Heterogeneity Among Samples and Examination of Marker Genes

Prior to carrying out more directed analyses of your expression data, it is important to perform some preliminary data exploration as a sort of data sanity check. For example, you should inspect whether groups of samples that are known a priori to be highly similar and/or highly dissimilar cluster accordingly based on their gene expression profiles. Secondly, you can verify that positive control samples or biomarker genes known to define your samples are expressed as expected. Unanticipated results from these initial analyses could signal that an error has occurred somewhere upstream. This could be anything from a simple fix, such as swapped labels in the sample table, to something more serious, such as contamination or batch effects. Either way, it is important to catch such errors early on, before you've spent a lot of time on data analysis or descended into a rabbit hole trying to make sense of what are actually spurious patterns.

Initial Visualization of Sample Groups and Outliers: Multidimensional Scaling

One quick visualization approach that our lab uses for initial inspection of expression data is the multidimensional scaling (MDS) plot. Similar to principle component analysis (PCA), MDS is a dimensionality reduction approach that allows one to visualize major sources of independent variation in a dataset. A plot of top MDS dimensions enables one to visualize which samples group together in expression space and can also help to identify outliers. There are three major steps to carrying out MDS with expression data: (1) normalizing the data, (2) selecting genes to use as input, and (3) performing the MDS analysis and plotting. We discuss each of these below.

Multidimensional Scaling (Step 1): Variance Stabilizing Transformation (VST) Normalization

For the reasons mentioned in Subheading 3.2.1 above, normalization of the count data is needed to correctly conduct MDS analysis. The normalization technique we typically use is called the variance stabilizing transformation (VST). As its name suggests, VST lognormalizes count data in a way that minimizes mean-dependent variance in the dataset, while also normalizing data with respect to library size. These transformed counts are more suitable for most downstream linear modeling, clustering, network, and enrichment analyses (*see* **Note 12**). VST transformation can be carried out using the R package, DESeq2 [32], by implementing the *varian-*

ceStabilizingTransformation function. For very large datasets, the VST transformation can be computationally intensive. If time is a factor, a similar transformation that is much quicker to calculate is the voom transformation implemented in the R package limma [33]. The voom and VST transformations perform similarly in the context of differential expression analysis [55].

Multidimensional Scaling (Step 2): Selecting Top Variant Genes

RNA-seq expression data typically consist of counts from many thousands of genes. A subset of these genes will exhibit highly divergent levels of expression among samples. It is these genes that will enable us to identify groups of samples with unique expression profiles. Moreover, it is these genes that are most likely involved in the cellular processes that distinguish sample groups of interest. However, many of the genes in our expression matrix will exhibit overall low expression and/or low variability in expression among samples. Expression patterns among such genes will often yield nothing but statistical noise. At best, including these "low-information" genes in ordination or clustering analyses will simply render these tasks more computationally taxing. At worst, including these genes may obscure patterns of interest within genes that *are* informative. Thus, it is a good idea to perform clustering and visualization analyses on a subset of genes that show the highest variability among samples. We typically sort the genes within the VST-normalized expression matrix by their variance (e.g., calculated using the *var* function in R) across samples and then select the 50–500 genes with the highest variance. It can be informative to perform MDS on increasing numbers of genes and observe how your analysis results change.

Multidimensional Scaling (Step 3): Carrying Out MDS Analysis

Unlike PCA, MDS attempts to maintain the original relative distances among samples when assigning sample coordinates for a given dimension and thus uses a distance matrix rather than the VST-normalized count data as input. One should thus first convert the VST matrix into distances (e.g., Euclidean distances). We then typically use the R function *cmdscale* to carry out the MDS. The coordinates (i.e., eigenvectors, which are stored in the *points* element of the *cmdscale* output) can then be plotted for the top dimensions (usually MDS1 and MDS2) to visualize natural groupings in distance space (there are as many columns in *points* as there are MDS dimensions). As an easy alternative to *cmdscale* that is tailored for expression data, the differential expression R package, limma, contains a function called *plotMDS* that wraps the *cmdscale* function. This function inputs the full VST matrix and then proceeds to do the gene selection, distance matrix conversion, MDS, and plotting, all in a single step.

Initial Inspection of the Most Highly Expressed Genes

Researchers usually possess prior knowledge about many of the genes that are commonly expressed within a tissue or cell type being sequenced. As a second sanity check, one can thus also inspect whether genes that are expected to be highly abundant in the dataset are present and whether genes that are known to be strongly associated with a particular cell type or treatment (i.e., biomarkers) are highly expressed in samples known to represent those cell types or treatments. For example, in airway epithelial tissue, highly expressed genes we expect to see include *SLPI*, *PIGR*, and *SCGB1A1*. If these genes are not abundantly present, this signals that something is wrong. Samples enriched in epithelial cilia cells should exhibit high expression for genes such as *CETN2*, *DNAH5*, and *DNAH9* relative to samples not enriched in cilia. Samples enriched in mucus-secreting cells should exhibit high relative expression for genes such as *MUC5AC*, *MUC5B*, and *TFF3*.

Although taking a preliminary look at highly expressed genes in a sample to ensure that expected genes or biomarkers are present can be done using raw or VST-normalized count data, we note that for any formal comparison of relative expression among genes, as mentioned in the previous section, one must also normalize the data in respect to gene length. This is commonly done by transforming counts into scaled values such as RPKM (reads per kilobase per million), FPKM (fragments per kilobase per million; a variation of RPKM for paired-end reads), or TPM (transcripts per million), which normalize reads or fragments to account for both differences in sequencing depth among samples and differences in length among genes or transcripts within a sample (*see* **Note 13**). To calculate these metrics, one must obtain length information for transcripts or genes in the dataset. The method, featureCounts [30], which is a gene count quantification function available within the R package, Rsubread [31], can output gene length information calculated directly from the GTF file. These gene lengths can then be used to calculate the above normalization metrics. Our lab typically uses kallisto (discussed in the previous section on gene alignment) to calculate TPMs that can be used to compare expression among gene isoforms. This method estimates "effective lengths," which modify transcript lengths to account for various characteristics of the fragments observed within one's particular dataset [56]. One can obtain gene-level TPMs from transcript-specific TPMs by simply summing the TPMs from across all the isoforms for each gene [e.g., *see* 57].

3.2.3 Identifying Relevant Genes: Single Gene Differential Expression Analysis

One of the key questions we can ask with RNA-seq data concerns whether the expression of individual genes is up- or downregulated between groups, treatments, time points, etc. One popular method for single gene differential expression (DE) analysis with RNA-seq data is DESeq2 [32], which is the one we primarily use in our lab. Although this is the method we will focus on here, there are many others, some of which may be better suited for particular needs [for comparisons of these methods, *see* 55, 58, 59]. As mentioned earlier, the R package, limma [33], is of particular interest when in possession of very large datasets, as it has generally been found to match DESeq2 in performance under most conditions (e.g., regarding sensitivity, false detection rate) while running considerably faster [59].

Explanation of the DESeq2 Method

DESeq2, which is an R package available from Bioconductor, carries out differential expression analysis that has been tailored to RNA-seq data and thus accounts for the statistical characteristics of RNA-seq discussed in Subheading 3.2.1. The method assumes that the expression data follow a negative binomial distribution, which can accommodate variance that exceeds the mean by way of an extra parameter, helping to mitigate the problem of overdispersion observed when using simpler models such as Poisson. To account for both the problem of different library sizes among samples and the problem of skewed counts due to the presence of highly abundant/variant genes, DESeq2 calculates "size factors," which normalize the counts with respect to these two issues, such that they are more comparable across samples. DESeq2 also automatically removes count outliers that exceed a given distance threshold from the observed distribution (*see* **Note 14**), as these outliers may exert undue influence on inferred fold changes and p-values. Furthermore, the method can automatically filter out lowly expressed genes, which tend to reduce statistical power while having very little chance of being detectably differentially expressed. We note that it is not necessary to account for variation in sequence length in differential expression analysis, as the method is comparing expression among samples using the same set of genes.

Running DESeq2

When running differential expression analysis with DESeq2, one should first create a data object using the function *DESeqDataSetFromMatrix*, which reads in the count matrix, the experimental design (a table that assigns experimental/population groups to each sample), and the formula that defines the relationship between the counts and relevant independent variables included in the design (e.g., "~subject + treatment"; *see* **Notes 15 and 16**). Any known variable expected to influence patterns of gene expression should be accounted for in this way. Once the data object is set up, one can run differential expression analysis using the *DESeq* function, which can take a while, depending on the size of the dataset and complexity of the design. The results table summarizing log fold changes and levels of significance can be obtained using *results*. A very comprehensive guide to using DESeq2 for differential expression analysis can be found on the Bioconductor website (*see* **Note 17**). A similar guide for using limma is also available (*see* **Note 18**).

3.2.4 Characterizing Heterogeneity Among Samples: Further Clustering and Visualization Analyses

Implicit to differential expression analysis is the existence of known groups of samples among which patterns of gene expression can be compared. Often, these groups and comparisons are defined a priori, such as when you have sampled from both healthy and diseased subjects or from both treated and untreated cells. However, there are cases in which a priori grouping of samples will be less clear. For example, you may be examining airway brushings from a group of COPD patients that you hypothesize could be divided into multiple subgroups based on airway expression profiles. This type of exploratory analysis is well suited to clustering methods, which are able to systematically identify groups of samples based on similarity in patterns of expression alone, without needing any additional prior information. In addition to allowing one to discover heterogeneity in one's dataset, unsupervised clustering can also provide an important check on one's prior assumptions. Do treatment groups in a study really correspond to groupings in the dataset that are inferred when the analysis is blind to these imposed groups? In this section, we will cover clustering methodologies that can help one to identify groups of samples with similar expression profiles and also techniques for the visualizing the expression of genes that define such groups.

Gene Selection Prior to Clustering

The selection of genes on which to perform cluster analysis is critical to the quality and meaning of the clusters you will identify and will depend on the questions you are trying to answer. Let's return to the common scenario mentioned above, where you hypothesize that subgroups exist within a defined group of samples (e.g., subgroups of patients with different molecular forms of disease) but where the identity and number of any such subgroups are unknown. The best way to agnostically identify the genes that define these hypothesized groups is to select the most variable genes in the expression dataset (as described in Subheading 3.2.2). As a guide, we have found that clustering the top 50–200 variant genes in our airway epithelial cell datasets is usually adequate for successfully identifying subgroups, although the optimal number will depend greatly on the dataset analyzed and should be explored by the researcher.

An alternative situation is one in which a research question relates to clusters defined by a specific gene set of interest. For example, one may possess a set of known differentially expressed genes and would like to determine whether the level of differential expression of these genes is uniform across samples in the compared groups or whether subgroups exist, of which only a subset may be primarily driving the observed differential expression results. In this case, clustering with the differentially expressed gene set itself can identify group heterogeneity in expression. This type of clustering has proven important in the identification of molecular disease endotypes for asthma [60, 61].

Clustering and Visualization Methods: Overview

Three major types of algorithms that can be useful for clustering selected gene sets are (1) centroid-based clustering methods, (2) hierarchical clustering methods, and (3) dimensionality reduction methods. The last of these is not strictly a "clustering" methodology, but nonetheless can be used for cluster detection, and thus we include it here. We discuss each of these types of algorithms below. Before beginning, don't forget that you should always normalize your data using something like VST normalization prior to clustering.

Clustering and Visualization Methods: Centroid-Based Clustering

Among the most basic questions to ask about a dataset are, how many distinct groups are there, and, which samples belong to which group? Centroid-based clustering is most useful for answering the second of these questions, although there are ways to probe the first question as well. These methods start out by assuming that a given number of clusters exist in the dataset and then proceed to assign samples to those clusters that are most representative, based on some estimate of the cluster centroid.

One of the most commonly used centroid-based methods is k *means* clustering. This algorithm begins by randomly assigning samples to one of k specified clusters. The method then iteratively alternates through (1) estimation of the mean value of each cluster, (2) reassignment of samples to clusters with means that most closely match their expression profiles, and (3) estimation of cluster variance. The algorithm proceeds until variance within clusters can no longer be decreased. This method can easily be implemented using the *kmeans* function in R.

The biggest disadvantage of using this type of approach is that the "true" number of k clusters must be specified up front and that specifying the wrong number can produce misleading results. Methods exist for attempting to infer the optimal number of clusters if this information is not otherwise available. Most of these methods involve some sort of measurement of the additional information value (e.g., variance explained) achieved by successively invoking new clusters and then selecting a k based on some prespecified threshold of diminishing returns. One such approach is to use the "elbow plot" method in which one creates a scree plot with the sum of squares error (SSE) on the y-axis and the number of clusters, k, on the x-axis, and then chooses the k where the decrease in error (inversely proportional to the total variance explained) levels off such that there is an elbow in the trend. This will not necessarily provide an objective answer to the number of clusters in a dataset, but it can nonetheless be a useful guide (*see* **Note 19**).

One final disadvantage of the centroid-based approach is that it lacks a way to visualize the relative distance among clusters or the relative scatter within clusters inferred. However, cluster assignments inferred using a method such as k *means* can nevertheless be overlaid upon visualization plots such as dendrograms, heat maps, or PCA plots, which we discuss below.

Clustering and Visualization Methods: Hierarchical Clustering

One solution to the problem of not knowing the optimal number of clusters within a dataset is to cluster hierarchically. Hierarchical clustering sidesteps the need to specify the number of clusters a priori by simply clustering across all possible levels of data similarity/dissimilarity. The R function *hclust* implements a common hierarchical clustering methodology that works "bottom up." This algorithm begins with each sample in a separate cluster and then proceeds to merge the least dissimilar clusters in a stepwise fashion until all samples are in a single cluster (*see* **Note 20**).

The results of hierarchical clustering are then typically displayed in the form of a dendrogram, which allows one to visualize the nested patterns of expression dissimilarity among all samples in the dataset. In the dendrogram, the leaves at the tips of the tree are samples, and the length of an edge (i.e., branch) separating any two samples is proportional to the relative dissimilarity among them. Hierarchical clustering dendrograms are often displayed in conjunction with heat maps, which are matrices that show relative expression across genes (rows) and samples (columns) using colors. R packages such as heatmap3 [34] automatically generate clustering dendrograms for both samples and genes and then use the "leaves" of these dendrograms as the horizontal and vertical axes of an expression heat map.

With a dendrogram/heat map in hand, you can then proceed to look for groups separated by particularly long branches or groups that correspond to patterns of gene expression displayed in the accompanying heat map to draw cluster boundaries. You can also use the R function *cutree* to divide your dataset into such groups, either based on the desired number of groups or on a specified tree height at which to cut the dendrogram.

Clustering and Visualization Methods: Dimensionality Reduction

Unlike centroid-based and hierarchical clustering, which use the raw expression data or distances based directly on those data to cluster samples, dimensionality reduction approaches allow one to visualize clusters based on projections of the original data that summarize the major independent forms of variation in a dataset. Principle component analysis (PCA) and classical multidimensional scaling (MDS) methods, the latter of which we discussed earlier, are commonly used to extract such summaries. Natural clusters in an RNA-seq dataset can often be visualized by plotting the top summaries (i.e., dimensions or components) that explain the most variation. PCA can be run using the *prcomp* function in R (*see* **Note 21**), after which eigenvectors for the top principle components can be plotted against one another (*see* **Note 22**). As mentioned above, multidimensional scaling can be carried out using the *cmdscale* function.

Given the complexity and high dimensionality of RNA-seq data, it can be difficult to observe all clusters that may exist in a dataset by simply looking at two or three dimensions at a time. t-distributed stochastic neighbor embedding (t-SNE) is a nonlinear dimensionality reduction method that allows one to visualize high-dimensional data (such as represented by multiple principle components from a PCA) in a single two- or three-dimensional scatter plot. A machine learning approach, t-SNE attempts to combine information about sample similarities from across different dimensions and scales in the data [62]. In R, t-SNE plots can be created using the Rtsne package [35]. You can either use the normalized count data as input or a list of the top principle components derived from the count data (*see* **Note 23**).

3.2.5 Characterizing Heterogeneity Among Genes: Functional Enrichment, Pathway Analysis, and Co-expression Network Analysis

The high dimensionality of RNA-seq data makes it difficult not only to identify discrete groups of samples in a dataset but also to pinpoint the major shifts in the molecular functions and pathways that underlie differential expression among these groups. Given the complex web of interrelated genes and functions inherent in any biological system, one can literally spend hours inspecting the individual genes that contribute to the transcriptomic differences among groups and yet never manage to assemble the accumulated information into a holistic understanding of the cellular processes affected by a treatment. Thus, one necessary goal in analysis of RNA-seq data is to reduce the expression of thousands of genes into a more manageable and fundamental handful of gene clusters that can be associated with specific biological mechanisms.

Functional Enrichment Analysis

One way to accomplish this is by investigating how one's dataset maps onto existing knowledge. Over the past decades, researchers have amassed extensive amounts of information relating to the genes involved in various cellular and pathophysiological processes, and this knowledge has been assembled into vast searchable databases. Thus, one can identify biologically relevant gene clusters by assessing the degree of overlap between a set of genes in one's dataset and process-associated gene sets obtained from these established databases. For example, if genes known to play a role in type I interferon signaling comprise 0.01% of all genes expressed in the human airway but comprise 10% of the genes that are differentially expressed between two treatments, one can infer that interferon signaling in response to a pathogen has likely been provoked in one of the treatments. This method can thus tease out biologically relevant groups of genes from a dataset and allow one to make inferences about underlying processes and relationships that characterize a system.

Testing for overrepresentation ("enrichment") of annotated gene lists within a target dataset (from your RNA-seq data) is typically carried out using a Fisher's exact test or a hypergeometric test.

Both of these methods take as input a contingency table describing the number of genes in the target list and the annotated list, as well as in a "background" list (*see* **Note 24**). The background list of genes represents a broader pool of genes that might reasonably be expected to contain all the genes in the annotated list. This gene list allows one to calculate the null expected overlap between the target and annotated lists, against which statistical enrichment can be tested (*see* **Note 25**). If a Fisher's exact test (implementable using *fisher.test* in R) is constrained to be one-sided, then this test should produce the same results as a hypergeometric test (implementable using *phyper* in R) (*see* **Note 26**).

There is a very large number of publicly accessible online databases containing thousands of annotated gene lists, as well as web-based or downloadable tools for accessing them and carrying out enrichment tests. Some of the most useful functional annotation databases for studying functional pathways in the human airway are Gene Ontology (GO), Kyoto Encyclopedia of Genes and Genomes (KEGG), Reactome, and Interpro (which focuses on protein domains). Terms and mechanistic pathways annotated in these databases can relate to organismal associations, cellular functions, location of gene action in the body or cell, associations with specific diseases or chemical substances/drugs, etc. These annotations can range from general (e.g., chemical homeostasis) to specific (e.g., positive regulation of peptidyl-lysine acetylation) and, as is the case with GO terms, are often hierarchically structured. Some of the most useful programs for running enrichment tests using these databases, all of which are open source, are DAVID (the Database for Annotation, Visualization and Integrated Discovery; *see* **Note 27**), Enrichr (*see* **Note 28**), and GOstats (*see* **Note 29**).

One caveat with enrichment analyses is that functional and pathway annotations in all the available databases are incomplete and will often be based on a limited number of studies and/or studies that are specific to a particular tissue, cell type, organism, or disease. Thus, one must take care not to overinterpret or misinterpret one's data based on the rank order of term or pathway enrichment results, which will to some extent reflect the idiosyncrasies of the studies upon which these databases are based.

In-Depth Pathway Analysis and Transcription Factor Discovery

Aside from investigating possible functional or pathway associations, one may also want to carry out a more in-depth investigation about where a particular set of genes fits or acts within a broader mechanistic pathway. This involves making predictions concerning other genes or proteins—or networks of genes and proteins—that may be associated with a target gene list, as well as the direction of their effects (e.g., is transcription activated or inhibited by the associated genes or proteins; is an associated gene or protein an upstream regulator or downstream target?).

One of the best tools for gene association prediction is Ingenuity Pathway Analysis (IPA; [39]). Ingenuity, the commercial entity that developed IPA, hosts a large proprietary knowledge base against which target gene lists can be statistically compared. One can input not only a set of target genes but also other data, such as log fold changes from a differential expression analysis and associated p-values, which can be used to more precisely situate the target genes within one or more of Ingenuity's canonical pathways. These canonical pathways consist of causal networks of genes and proteins that describe data-supported molecular interactions and cascades. One extremely valuable capability of IPA is its potential to identify novel transcription regulators of a set of target genes. One disadvantage of IPA is that it levies a substantial licensing fee.

An alternative to predicting transcription factors using literature-based associations is to mine transcription factor sequence motifs. Because genes that are directly regulated by a transcription factor will share a binding site with that transcription factor, transcription factor sequence motifs that are enriched within a target group of co-expressing genes can be considered a strong candidate upstream regulator of those genes. Motif enrichment methods can be a powerful way to discover novel transcription factors for a set of genes. One easy-to-use tool for testing for transcription factor motif enrichment is iRegulon [40], which is available as a plugin for Cytoscape [41], an open source molecular network visualization software package (*see* **Note 30**). iRegulon relies on a very large library of motifs from across a variety of species. Some of the functional enrichment tools mentioned in the previous section can also test for transcription factor associations using transcription factor binding site libraries (e.g., TRANSFAC and JASPER available in Enrichr or UCSC_TFBS available in DAVID).

Co-expression Network Analysis

A final approach for rendering your RNA-seq dataset less granular is to cluster genes based on the correlated expression among them. This allows one to leverage information concerning discrete groups of genes that may be acting in concert directly from the data themselves, prior to doing any gene clustering based on the querying of preestablished databases. One of the most popular approaches to clustering genes based on gene co-expression is called weighted gene co-expression network analysis (WGCNA). This approach infers modularity in expression data from gene networks that are constructed based on pairwise correlations among genes.

These inferred "modules" of genes may correspond to discrete, biologically meaningful functions or pathways that are activated or inhibited within a group of samples.

The WGCNA method works, first, by measuring correlations in expression among pairs of genes. We note that as discussed above in regard to clustering and visualization, it is important to always normalize counts with something like VST prior to measuring expression correlations with WGCNA. Once a correlation matrix has been constructed, it is used to produce a gene network, where nodes are genes and connections among them (edges) correspond to correlations that surpass a specified threshold (*see* **Note 31**). To divide the genes in this network into discrete modules, a "topological overlap" metric is calculated among all pairs of nodes in the network, which is a similarity score defined by the number and strength of shared connections. Hierarchical clustering is then performed on these topological overlap scores, after which the resulting dendrogram is divided into modules based on specified tree cutting parameters (*see* **Note 32**). The method is fairly complex and using it requires surmounting a moderate learning curve. The founding paper [63] and available tutorials involving sample data (*see* **Note 33**) are very helpful for becoming acquainted with the various parameters and options, as well as the theory behind the method. The method is implemented using the WGCNA R package [64].

With a set of inferred gene modules in hand, one can proceed to ask how these modules relate to relevant external factors. To facilitate this, modules can be summarized using eigenvectors (called "eigengenes" in WGCNA lingo) representing the first principle component of the module genes (calculated using the function *moduleEigengene*). By identifying the modules with eigengenes that are most highly correlated with a set of variables (e.g., treatments applied to the data, clinical traits, etc.), one can isolate and study those sets of genes that are involved in a particular pathway of interest. WGCNA contains statistical and plotting functionality for internally carrying out these types of analyses. Module gene sets can also be used as input for functional enrichment, pathway analysis, and transcription factor inference methods described above.

Finally, in addition to using gene module assignments as a way to identify distinct biologically relevant functions and pathways, one can also harness information from the gene networks upon which module assignments are based. In particular, the most highly connected genes within a module network (called "hub genes") may play a central or driving role within a module. Genes that share a direct connection with a specific gene of interest may also represent good candidate regulators or targets of that gene and thus be worthy of further study. One flexible tool for visualizing and manipulating gene module networks is the program Cytoscape [41], and WGCNA contains a function (*exportNetworkToCytoscape*) for formatting gene networks for easy import into the program.

4 Notes

1. The first line of each sequence in a FASTQ file is a unique identifier and begins with an "@" symbol, followed by the instrument ID, run number, lane number, tile number, X coordinate of the cluster, Y coordinate of the cluster, read number (1 or 2 for paired sequencing), and index sequence.

2. Phred quality scores (Q) in Illumina output range from 0 to 93, where the probability of a base call being incorrect, P, can be calculated from Q using the following formula: $P = 1/10^{(Q/10)}$. For example, a Phred score of 30 corresponds to a 0.001 error rate, or 1 error in 1000 bases. In the FASTQ file, ASCII characters are used rather than raw Phred scores. Thus, to obtain the Phred score for a particular nucleotide in a FASTQ file, you need to convert the UTF-8 ASCII character assigned to the particular base to its corresponding integer and then subtract 33. This character conversion can be done using the "utf8ToInt" function in R. For example, if the quality score is "G," then Q can be calculated in R: 'utf8ToInt("G") − 33' = 38. This is equivalent to an error probability of 0.00016. Note that quality control software such as FastQC (discussed in the text) does all this conversion for you.

3. Although we only discuss quality checking here, we recommend carrying out quality checking after each step in the data processing pipeline (e.g., by calculating summary statistics for the number of reads per sample). This can alert you early on if some error has occurred.

4. A flow cell is a glass slide containing fluidic channels called lanes. Each lane has billions of "spots" with DNA primers for individual DNA templates to bind to. The most common causes of location-based sequence errors include bubbles that percolate through the flow cell, smudges that exist on the flow cell, or debris that has fallen into a flow cell lane.

5. One can download the most recent genome assembly and GTF files derived from UCSC from Illumina's iGenomes collection (see https://support.illumina.com/sequencing/sequencing_software/igenome.html). The assembly and GTF files from Ensembl are available here: https://www.ensembl.org/info/data/ftp/index.html. Please note, it is important to perform the alignment using a GTF file that is annotated according to the reference genome file used.

6. For an extensive description of how to read the data inside SAM files, see the document "Sequence Alignment/Map Format Specification" prepared by The SAM/BAM Format

Specification Working Group (currently available here: https://samtools.github.io/hts-specs/SAMv1.pdf).

7. kallisto is available for free download (Mac or Linux only) at https://pachterlab.github.io/kallisto.

8. Quantification can be in the form of either raw estimated counts or in units of transcripts per million (TPM), which normalizes abundances against differences in library size across samples and differences in effective length among transcripts.

9. Although [49] observed similar mapping and differential expression results across read lengths, splice junction detection was improved with longer reads.

10. The specified overlap mode can significantly alter the estimated abundances of genes in a dataset. "Union" (the default mode) is a fairly conservative approach and simply assigns any reads that overlap exons belonging to two or more genes as "ambiguous." "Intersection-strict" allows for overlap with a second, non-counted gene but requires that a read entirely overlaps the gene to which it is assigned. Reads that only partially overlap genes are not assigned at all ("no_feature"). "Intersection_nonempty" is the most liberal approach, as it allows for a read to be assigned to a gene that it only partially overlaps, even if it also partially overlaps a second gene. See the HTSeq website for a more thorough discussion of these counting rules (http://htseq.readthedocs.io/en/master/count.html).

11. Also, note that if your data are single end, you should specify "stranded = yes".

12. VST transformation can be carried out by setting the argument *blind* to either "TRUE" (the default) or "FALSE," where "blindness" is in respect to the experimental design. Thus, keeping this as "TRUE" will render normalized values that are irrespective of the treatments and groups in your dataset. For normalized expression values for use in most downstream analyses, the DESeq authors recommend setting *blind* to "FALSE" if one expects consistent differences in expression among treatment groups; otherwise, the method will overestimate dispersion. For more on this topic, see the DESeq vignette (load the DESeq library in R and then type 'vignette("DESeq2")').

13. For a good introduction for the differences among these normalization methods and how they are calculated, see http://www.RNA-seqblog.com/rpkm-fpkm-and-tpm-clearly-explained.

14. DESeq2 uses Cook's distance to identify outliers, and the threshold used (default equals the 0.99 quantile) can be

changed using the "cooksCutoff" argument in the *results* function, which is the DESeq2 function for extracting results from a DE analysis. Note that limma does not have this functionality, and thus outlier genes need to be identified and removed manually prior to analysis.

15. Note that DESeq2 uses the last variable listed in the formula as the one around which to summarize results (such as log fold changes). Thus, the main treatment of interest should typically be listed last in the formula.

16. The treatments in the design are treated as factors (an object class in R), and the ordering of those factors matters. The first factor will be the reference treatment (meaning, the treatment for which log fold changes in expression will be reported as in response to). One can use the R function *levels* to inspect the current ordering (e.g., 'levels(DESeqObject$treatment)') and *relevel* to change the ordering (e.g., 'relevel(DESeqObject$treatment, "control")').

17. https://www.bioconductor.org/help/workflows/rnaseqGene

18. https://www.bioconductor.org/help/workflows/RNAseq123

19. For more explanation on how to implement the elbow method, see http://www.mattpeeples.net/kmeans.html.

20. Hierarchical clustering is performed on a distance matrix computed from the raw data. Some methods (such as heatmap3) automatically create a distance matrix for you (the default is Euclidean distance). Using *hclust* directly requires that you do the conversion yourself, which can easily be done using the *dist* function.

21. The normalized count table should be transposed (using the *t* function in R) prior to running *prcomp*.

22. The eigenvectors for each dimension are stored in the *rotation* element in the *prcomp* output. In *rotation*, each row is a sample and each column is a dimension.

23. t-SNE plots can be easy to misinterpret given that distances and scatter among points are not linearly represented. Certain parameters of the method, particularly perplexity and the number of iterations, can greatly influence the characteristics of the plot. See http://distill.pub/2016/misread-tsne for a good explanation for how to understand and set these parameters.

24. Setting up the contingency table can be a bit tricky; the following blog post provides a good explanation for how to do this: http://mengnote.blogspot.qa/2012/12/calculate-correct-hypergeometric-p.html.

25. Selecting the appropriate background is not always obvious and will depend on the breadth of the question one is addressing. For genomic RNA-seq data from the human airway, the appropriate background will often simply be all genes in the human genome or all the genes in the human airway transcriptome. However, for more circumscribed questions, such as ones relating to the enrichment of a small subset of target genes relative to that of a broader set of genes in one's dataset, the background will be smaller (e.g., all differentially expressed genes from one's dataset). It is important to note that significance of enrichment tests will increase with the size of the background, so it is important to ensure that it is no larger than is reasonably necessary.

26. If using *phyper*, *lower.tail* should be set to "FALSE" if testing for enrichment or to "TRUE" if testing for depletion. Alternatively, when using *fisher.test*, *alternative* should be set to "greater" to test for enrichment or to "less" to test for depletion.

27. DAVID (the Database for Annotation, Visualization and Integrated Discovery) can be used either via a web server (https://david.ncifcrf.gov) or an R package [36], downloadable from Bioconductor (https://bioconductor.org/packages/release/bioc/html/RDAVIDWebService.html).

    The web server is good for quick enrichment of one or a few target gene lists. However, if you have a large number of gene lists to test, automation via the R package will make things much easier. The default background gene list is the entire genome of the species you specify; however, another background can be given. The DAVID method has other functionalities aside from functional enrichments, such as gene ID conversion and clustering of redundant annotation terms.

28. Enrichr [37] is a web-based tool (https://amp.pharm.mssm.edu/Enrichr) with annotations and background gene lists for mice and humans. The big advantage of Enrichr is its intuitive and interactive result layout and ease of use. However, background gene lists are preset by Enrichr and cannot be changed. Enrichr has an application program interface (API) that allows its function calls to be automated (for an example python tool for this, see https://github.com/russell-stewart/enrichrAPI).

29. GOstats [38] is an R package (https://bioconductor.org/packages/release/bioc/html/GOstats.html) that focuses mostly on GO term association. This is probably your best choice if your main interest is in automating GO term enrichment.

30. Cytoscape is a standalone Java-based software package available for download from http://www.cytoscape.org. The iRegulon plugin is available at http://iregulon.aertslab.org.

31. With WGCNA, this is a "soft" threshold that is typically assigned based on a power function. The power parameter, β, one specifies can have a large impact on the number of connections in the resulting network. Ideally, you choose a power parameter that maximizes the number of connections in the network while still adhering to the underlying assumptions the method makes about the shape of the network (i.e., a scale-free topology model). *See* [63], as well as their tutorial entitled "Step-by-step network construction and module detection" (https://labs.genetics.ucla.edu/horvath/CoexpressionNetwork/Rpackages/WGCNA/Tutorials/FemaleLiver-02-networkConstr-man.pdf) for how to specify the power parameter.

32. The selection of the optimal tree cutting parameters can be tricky and requires experimenting with the various options (e.g., *deepSplit*, *cutHeight*, *maxCoreScatter*, *minGap*, and *minSplitHeight*). A thorough discussion of these parameters can be found in the protocol "Dynamic Tree Cut: in-depth description, tests, and applications" written by the developers of WGCNA (see https://labs.genetics.ucla.edu/horvath/CoexpressionNetwork/BranchCutting/Supplement-published.pdf). The ultimate goal is to set these parameters such that you are cutting the tree at locations that capture "natural breaks" in the dendrogram. You don't want to over split what are really cohesive groups of genes into separate modules, nor do you want to under split modules, such that multiple co-expressed gene groups of biological relevance are lumped together. That being said, in our experience, it's often a good idea to start with fewer modules and then later break them apart if necessary.

33. https://labs.genetics.ucla.edu/horvath/CoexpressionNetwork/Rpackages/WGCNA/Tutorials/index.html.

References

1. Holtzman MJ, Byers DE, Alexander-Brett J, Wang XY (2014) The role of airway epithelial cells and innate immune cells in chronic respiratory disease. Nat Rev Immunol 14(10):686–698

2. Heijink IH, de Bruin HG, van den Berge M, Bennink LJC, Brandenburg SM, Gosens R, van Oosterhout AJ, Postma DS (2013) Role of aberrant WNT signalling in the airway epithelial response to cigarette smoke in chronic obstructive pulmonary disease. Thorax 68(8):709–716. https://doi.org/10.1136/thoraxjnl-2012-201667

3. Pilette C, Godding V, Kiss R, Delos M, Verbeken E, Decaestecker C, De Paepe K, Vaerman JP, Decramer M, Sibille Y (2001) Reduced epithelial expression of secretory component in small airways correlates with airflow obstruction in chronic obstructive pulmonary disease. Am J Respir Crit Care Med 163(1):185–194

4. Mall M, Grubb BR, Harkema JR, O'Neal WK, Boucher RC (2004) Increased airway epithelial Na+ absorption produces cystic fibrosis-like lung disease in mice. Nat Med 10(5):487–493. https://doi.org/10.1038/nm1028

5. Oglesby IK, Vencken SF, Agrawal R, Gaughan K, Molloy K, Higgins G, McNally P, McElvaney NG, Mall MA, Greene CM (2015) miR-17 overexpression in cystic fibrosis airway epithe-

lial cells decreases interleukin-8 production. Eur Respir J 46(5):1350–1360. https://doi.org/10.1183/09031936.00163414

6. Kuperman DA, Huang XZ, Koth LL, Chang GH, Dolganov GM, Zhu Z, Elias JA, Sheppard D, Erle DJ (2002) Direct effects of interleukin-13 on epithelial cells cause airway hyperreactivity and mucus overproduction in asthma. Nat Med 8(8):885–889. https://doi.org/10.1038/nm734

7. Hackett TL, Warner SM, Stefanowicz D, Shaheen F, Pechkovsky DV, Murray LA, Argentieri R, Kicic A, Stick SM, Bai TR, Knight DA (2009) Induction of epithelial-mesenchymal transition in primary airway epithelial cells from patients with asthma by transforming growth factor-beta 1. Am J Respir Crit Care Med 180(2):122–133. https://doi.org/10.1164/rccm.200811-1730OC

8. Craig VJ, Polverino F, Laucho-Contreras ME, Shi YY, Liu YS, Osorio JC, Tesfaigzi Y, Pinto-Plata V, Gochuico BR, Rosas IO, Owen CA (2014) Mononuclear phagocytes and airway epithelial cells: novel sources of matrix metalloproteinase-8 (MMP-8) in patients with idiopathic pulmonary fibrosis. PLoS One 9(5). https://doi.org/10.1371/journal.pone.0097485

9. Xu Y, Mizuno T, Sridharan A, Du YN, Guo MZ, Tang J, Wikenheiser-Brokamp KA, Perl AKT, Funari VA, Gokey JJ, Stripp BR, Whitsett JA (2016) Single-cell RNA sequencing identifies diverse roles of epithelial cells in idiopathic pulmonary fibrosis. JCI Insight 1(20):1–18. https://doi.org/10.1172/jci.insight.90558

10. Costa V, Aprile M, Esposito R, Ciccodicola A (2013) RNA-Seq and human complex diseases: recent accomplishments and future perspectives. Eur J Hum Genet 21(2):134–142. https://doi.org/10.1038/ejhg.2012.129

11. Wang Z, Gerstein M, Snyder M (2009) RNA-Seq: a revolutionary tool for transcriptomics. Nat Rev Genet 10(1):57–63. https://doi.org/10.1038/nrg2484

12. Hackett NR, Butler MW, Shaykhiev R, Salit J, Omberg L, Rodriguez-Flores JL, Mezey JG, Strulovici-Barel Y, Wang G, Didon L, Crystal RG (2012) RNA-Seq quantification of the human small airway epithelium transcriptome. BMC Genomics 13:82. https://doi.org/10.1186/1471-2164-13-82

13. Poole A, Urbanek C, Eng C, Schageman J, Jacobson S, O'Connor BP, Galanter JM, Gignoux CR, Roth LA, Kumar R, Lutz S, Liu AH, Fingerlin TE, Setterquist RA, Burchard EG, Rodriguez-Santana J, Seibold MA (2014) Dissecting childhood asthma with nasal transcriptomics distinguishes subphenotypes of disease. J Allergy Clin Immunol 133(3):670–678. https://doi.org/10.1016/j.jaci.2013.11.025

14. Tian B, Li XL, Kalita M, Widen SG, Yang J, Bhavnani SK, Dang B, Kudlicki A, Sinha M, Kong FP, Wood TG, Luxon BA, Brasier AR (2015) Analysis of the TGF beta-induced program in primary airway epithelial cells shows essential role of NF-kappa B/RelA signaling network in type II epithelial mesenchymal transition. BMC Genomics 16. https://doi.org/10.1186/s12864-015-1707-x

15. Nance T, Smith KS, Anaya V, Richardson R, Ho L, Pala M, Mostafavi S, Battle A, Feghali-Bostwick C, Rosen G, Montgomery SB (2014) Transcriptome analysis reveals differential splicing events in IPF lung tissue. PLoS One 9(3). https://doi.org/10.1371/journal.pone.0092111

16. Wesolowska-Andersen A, Everman JL, Davidson R, Rios C, Herrin R, Eng C, Janssen WJ, Liu AH, Oh SS, Kumar R, Fingerlin TE, Rodriguez-Santana J, Burchard EG, Seibold MA (2017) Dual RNA-seq reveals viral infections in asthmatic children without respiratory illness which are associated with changes in the airway transcriptome. Genome Biol 18(12):1–17. https://doi.org/10.1186/s13059-016-1140-8

17. Andrews S (2017) FastQC: a quality control tool for high throughput sequence data. Available online at http://www.bioinformatics.babraham.ac.uk/projects/fastqc

18. Jiang HS, Lei R, Ding SW, Zhu SF (2014) Skewer: a fast and accurate adapter trimmer for next-generation sequencing paired-end reads. BMC Bioinformatics 15:182. https://doi.org/10.1186/1471-2105-15-182

19. Bolger AM, Lohse M, Usadel B (2014) Trimmomatic: a flexible trimmer for Illumina sequence data. Bioinformatics 30(15):2114–2120. https://doi.org/10.1093/bioinformatics/btu170

20. Martin M (2011) Cutadapt removes adapter sequences from high-throughput sequencing reads. EMBnetjournal 17(1):10

21. Wu TD, Nacu S (2010) Fast and SNP-tolerant detection of complex variants and splicing in short reads. Bioinformatics 26(7):873–881. https://doi.org/10.1093/bioinformatics/btq057

22. Trapnell C, Pachter L, Salzberg SL (2009) TopHat: discovering splice junctions with RNA-Seq. Bioinformatics 25(9):1105–1111. https://doi.org/10.1093/bioinformatics/btp120

23. Wang K, Singh D, Zeng Z, Coleman SJ, Huang Y, Savich GL, He XP, Mieczkowski P, Grimm SA, Perou CM, MacLeod JN, Chiang DY,

Prins JF, Liu JZ (2010) MapSplice: accurate mapping of RNA-seq reads for splice junction discovery. Nucleic Acids Res 38(18):e178. https://doi.org/10.1093/nar/gkq622

24. Dobin A, Davis CA, Schlesinger F, Drenkow J, Zaleski C, Jha S, Batut P, Chaisson M, Gingeras TR (2013) STAR: ultrafast universal RNA-seq aligner. Bioinformatics 29(1):15–21. https://doi.org/10.1093/bioinformatics/bts635

25. Kim D, Landmead B, Salzberg SL (2015) HISAT: a fast spliced aligner with low memory requirements. Nat Methods 12(4):357–U121. https://doi.org/10.1038/nmeth.3317

26. Bray NL, Pimentel H, Melsted P, Pachter L (2016) Near-optimal probabilistic RNA-seq quantification. Nat Biotechnol 34(5):525–527. https://doi.org/10.1038/nbt.3519

27. Li H, Handsaker B, Wysoker A, Fennell T, Ruan J, Homer N, Marth G, Abecasis G, Durbin R, Genome Project Data P (2009) The sequence alignment/map format and SAMtools. Bioinformatics 25(16):2078–2079. https://doi.org/10.1093/bioinformatics/btp352

28. Anders S, Pyl PT, Huber W (2015) HTSeq-a Python framework to work with high-throughput sequencing data. Bioinformatics 31(2):166–169. https://doi.org/10.1093/bioinformatics/btu638

29. Trapnell C, Williams BA, Pertea G, Mortazavi A, Kwan G, van Baren MJ, Salzberg SL, Wold BJ, Pachter L (2010) Transcript assembly and quantification by RNA-Seq reveals unannotated transcripts and isoform switching during cell differentiation. Nat Biotechnol 28(5):511–515. https://doi.org/10.1038/nbt.1621

30. Liao Y, Smyth GK, Shi W (2014) feature-Counts: an efficient general purpose program for assigning sequence reads to genomic features. Bioinformatics 30(7):923–930. https://doi.org/10.1093/bioinformatics/btt656

31. Liao Y, Smyth GK, Shi W (2012) The Subread aligner: fast, accurate and scalable read mapping by seed-and-vote. Nucleic Acids Res 41:e108

32. Love MI, Huber W, Anders S (2014) Moderated estimation of fold change and dispersion for RNA-seq data with DESeq2. Genome Biol 15(12). https://doi.org/10.1186/s13059-014-0550-8

33. Ritchie ME, Phipson B, Wu D, Hu YF, Law CW, Shi W, Smyth GK (2015) Limma powers differential expression analyses for RNA-sequencing and microarray studies. Nucleic Acids Res 43(7):1–13. https://doi.org/10.1093/nar/gkv007

34. Zhao SL, Guo Y, Sheng QH, Shyr Y (2014) Advanced heat map and clustering analysis using Heatmap3. Biomed Res Int 2014:6. https://doi.org/10.1155/2014/986048

35. Krijthe JH (2015) Rtsne: T-distributed stochastic neighbor embedding using a Barnes-Hut implementation. https://github.com/jkrijthe/Rtsne

36. Fresno C, Fernández EA (2013) RDAVIDWebService: a versatile R interface to DAVID. Bioinformatics 29(21):2810–2811

37. Chen EY, Tan CM, Lou Y, Duan Q, Wang Z, Meirelles GV, Clark NR, Ma A (2013) Enrichr: interactive and collaborative HTML5 gene list enrichment analysis tool. BMC Bioinformatics 14:128. https://doi.org/10.1186/1471-2105-14-128

38. Falcon S, Gentleman R (2007) Using GOstats to test gene lists for GO term association. Bioinformatics 23(2):257–258

39. Krämer A, Green J, Pollard J, Tugendreich S (2014) Causal analysis approaches in Ingenuity Pathway Analysis. Bioinformatics 30(4):523–530

40. Janky R, Verfaillie A, Imrichova H, Van de Sande B, Standaert L, Christiaens V, Hulselmans G, Herten K, Sanchez MN, Potier D, Svetlichnyy D, Atak ZK, Fiers M, Marine JC, Aerts S (2014) iRegulon: from a gene list to a gene regulatory network using large motif and track collections. PLoS Comput Biol 10(7):e1003731. https://doi.org/10.1371/journal.pcbi.1003731

41. Shannon P, Markiel A, Ozier O, Baliga NS, Wang JT, Ramage D, Amin N, Schwikowski B, Ideker T (2003) Cytoscape: a software environment for integrated models of biomolecular interaction networks. Genome Res 13(11):2498–2504. https://doi.org/10.1101/gr.1239303

42. Robles JA, Qureshi SE, Stephen SJ, Wilson SR, Burden CJ, Taylor JM (2012) Efficient experimental design and analysis strategies for the detection of differential expression using RNA-sequencing. BMC Genomics 13:484. https://doi.org/10.1186/1471-2164-13-484

43. Williams AG, Thomas S, Wyman SK, Holloway AK (2014) RNA-seq data: challenges in and recommendations for experimental design and analysis. Curr Protoc Human Genet 83:11.13.1–11.13.20

44. Wu Z, Wu H (2016) Experimental design and power calculation for RNA-seq experiments. In: Mathé E, Davis S (eds) Statistical genomics. Methods in molecular biology, vol 1418. Humana Press, New York, NY

45. Peixoto L, Risso D, Poplawski SG, Wimmer ME, Speed TP, Wood MA, Abel T (2015) How data analysis affects power, reproducibility and biological insight of RNA-seq studies in complex datasets. Nucleic Acids Res

43(16):7664–7674. https://doi.org/10.1093/nar/gkv736

46. Tarazona S, Garcia-Alcalde F, Dopazo J, Ferrer A, Conesa A (2011) Differential expression in RNA-seq: a matter of depth. Genome Res 21(12):2213–2223. https://doi.org/10.1101/gr.124321.111

47. Sims D, Sudbery I, Ilott NE, Heger A, Ponting CP (2014) Sequencing depth and coverage: key considerations in genomic analyses. Nat Rev Genet 15(2):121–132. https://doi.org/10.1038/nrg3642

48. Engström PG, Steijger T, Sipos B, Grant GR, Kahles A, Ratsch G, Goldman N, Hubbard TJ, Harrow J, Guigo R, Bertone P, The RGASP Consortium (2013) Systematic evaluation of spliced alignment programs for RNA-seq data. Nat Methods 10(12):1185–1191. https://doi.org/10.1038/nmeth.2722

49. Chhangawala S, Rudy G, Mason CE, Rosenfeld JA (2015) The impact of read length on quantification of differentially expressed genes and splice junction detection. Genome Biol 16(131). https://doi.org/10.1186/s13059-015-0697-y

50. Robinson MD, Smyth GK (2007) Moderated statistical tests for assessing differences in tag abundance. Bioinformatics 23(21):2881–2887. https://doi.org/10.1093/bioinformatics/btm453

51. Law CW, Chen YS, Shi W, Smyth GK (2014) voom: precision weights unlock linear model analysis tools for RNA-seq read counts. Genome Biol 15(2):R29. https://doi.org/10.1186/gb-2014-15-2-r29

52. Oshlack A, Wakefield MJ (2009) Transcript length bias in RNA-seq data confounds systems biology. Biol Direct 4:14. https://doi.org/10.1186/1745-6150-4-14

53. Bullard JH, Purdom E, Hansen KD, Dudoit S (2010) Evaluation of statistical methods for normalization and differential expression in mRNA-Seq experiments. BMC Bioinformatics 11:94. https://doi.org/10.1186/1471-2105-11-94

54. McIntyre LM, Lopiano KK, Morse AM, Amin V, Oberg AL, Young LJ, Nuzhdin SV (2011) RNA-seq: technical variability and sampling. BMC Genomics 12:293. https://doi.org/10.1186/1471-2164-12-293

55. Soneson C, Delorenzi M (2013) A comparison of methods for differential expression analysis of RNA-seq data. BMC Bioinformatics 14:91. https://doi.org/10.1186/1471-2105-14-91

56. Roberts A, Trapnell C, Donaghey J, Rinn JL, Pachter L (2011) Improving RNA-Seq expression estimates by correcting for fragment bias. Genome Biol 12:R22

57. Soneson C, Love MI, Robinson MD (2016) Differential analyses for RNA-seq: transcript-level estimates improve gene-level inferences. F1000Research 4:1521

58. Rapaport F, Khanin R, Liang YP, Pirun M, Krek A, Zumbo P, Mason CE, Socci ND, Betel D (2013) Comprehensive evaluation of differential gene expression analysis methods for RNA-seq data. Genome Biol 14(9). https://doi.org/10.1186/gb-2013-14-9-r95

59. Seyednasrollah F, Laiho A, Elo LL (2015) Comparison of software packages for detecting differential expression in RNA-seq studies. Brief Bioinform 16(1):59–70. https://doi.org/10.1093/bib/bbt086

60. Wesolowska-Andersen A, Seibold MA (2015) Airway molecular endotypes of asthma: dissecting the heterogeneity. Curr Opin Allergy Clin Immunol 15(2):163–168. https://doi.org/10.1097/aci.0000000000000148

61. Woodruff PG, Modrek B, Choy DF, Jia GQ, Abbas AR, Ellwanger A, Arron JR, Koth LL, Fahy JV (2009) T-helper type 2-driven inflammation defines major subphenotypes of asthma. Am J Respir Crit Care Med 180(5):388–395. https://doi.org/10.1164/rccm.200903-0392OC

62. van der Maaten L, Hinton G (2008) Visualizing data using t-SNE. J Mach Learn Res 9:2579–2605

63. Zhang B, Horvath S (2005) A general framework for weighted gene co-expression network analysis. Stat Appl Genet Mol Biol 4:17. The Berkeley Electronic Press

64. Langfelder P, Horvath S (2008) WGCNA: an R package for weighted correlation network analysis. BMC Bioinformatics 9:559. https://doi.org/10.1186/1471-2105-9-559

Chapter 16

Application of Proteomics in Lung Research

Nichole A. Reisdorph, Cole Michel, Kristofer Fritz, and Richard Reisdorph

Abstract

Proteomics has enabled researchers to evaluate global protein changes in a relatively rapid and comprehensive manner. Applications of these technologies in lung research include biomarker and drug discovery, elucidating disease mechanisms, and quantitative clinical assays. Two common workflows exist for quantitative proteomics studies that are aimed at determining differences in protein levels: label-free and labeling methods. Here we describe specific techniques involved in both quantitative workflows; these include extensive sample preparation methods for several lung-specific sample types. Methods are also included for mass spectrometry-based sample analysis and data analysis. While the focus is on quantitative, clinical proteomics, these strategies are appropriate for a wide array of sample types and applications.

Key words Quantitative proteomics, Mass spectrometry, Proteomics, Labeling, Plasma proteomics, Label-free, Lung tissue, Bronchoalveolar lavage

1 Introduction

The proteome consists of all proteins expressed by a biological system; these include both posttranslationally modified (PTM) and unmodified proteins. Proteomics biomarker discovery has been applied to a variety of lung-related sample types that contain protein, including induced sputum, exhaled breath condensate (EBC), epithelial lining fluid (ELF), bronchoalveolar lavage (BAL) fluid, plasma, and serum [1, 2]. Overall, proteomic approaches have been widely used to identify the expression level and modification of proteins toward improved understanding of the pathophysiology of asthma, chronic obstructive pulmonary disease (COPD), and lung cancer. For example, S100 proteins, cytokeratins, actin, complement proteins, cytokines, and α2-macroglobulin are protein biomarkers associated with asthma that have been identified in multiple studies and/or multiple biofluids [3]. Overall, protein biomarker discovery faces several challenges, including depth of coverage, cost, and standardized sampling techniques [2]. However, given that proteins are the functional units of genes,

Scott Alper and William J. Janssen (eds.), *Lung Innate Immunity and Inflammation: Methods and Protocols*,
Methods in Molecular Biology, vol. 1809, https://doi.org/10.1007/978-1-4939-8570-8_16,
© Springer Science+Business Media, LLC, part of Springer Nature 2018

understanding the role of proteins in lung disease is an essential component to a comprehensive clinical treatment strategy.

Liquid chromatography-mass spectrometry (LCMS)-based methods have become powerful and popular means of profiling clinical samples for the purpose of biomarker discovery. Biomarkers can be used in several areas of clinical research and patient care; these include disease detection and diagnosis, drug discovery, and assessing drug efficacy. A wide range of biomolecules can be profiled using LCMS, including proteins, peptides, and modified peptides. Appropriately, there are a vast number of methods and workflows that can be used to compare relative quantities of molecules within a system [4]. The current work will focus on general mass spectrometry-based proteomics methods, which consist of sample preparation followed by data acquisition and data analysis. Because there are many options at each step of these workflows, alternatives are included throughout.

Quantitative LCMS-based proteomics studies generally focus on one of two basic workflows: (1) label-free quantitation and (2) metabolic and post-metabolic labeling strategies [4, 5]. In addition, proteomics workflows can be further divided based on the sample type (Fig. 1). For example, analysis of plasma generally requires removal of high-abundant proteins using affinity chromatography as a first step. Conversely, cellular proteins can be sub-fractionated based on organelle; for example, proteins can be enriched from nuclei or mitochondria using step-wise centrifugation, detergents, and/or sucrose gradients. Finally, proteins can be further fractionated based on chemistry including ion exchange chromatography. Initial first steps, including protein solubilization and fractionation, suitable for profiling of plasma, tissues, or cells generally include the following:

1. *Cell lysis and protein solubilization*—A variety of methods can be used to extract and solubilize proteins, depending on the sample type. Options include subcellular fractionation, precipitation, addition of various detergents, and centrifugation. In some cases, samples can be immediately denatured and digested (*see* **Note 1**).

2. *Fractionation*—Chromatography is often used to fractionate samples via chemistries such as strong cation exchange, in part because the step can be automated, although size exclusion or other methods can be used. Affinity chromatography, such as antibody-based immunoprecipitation or glycoprotein enrichment, is also used. Finally, immunodepletion to remove high-abundant proteins, such as albumin, can be used prior to plasma analysis.

3. *Protein digest*—Although quantitative LC/MS methods, such as molecular imaging and profiling [6], exist for intact proteins, more often proteins are further solubilized and digested with the protease trypsin prior to analysis by mass spectrometry.

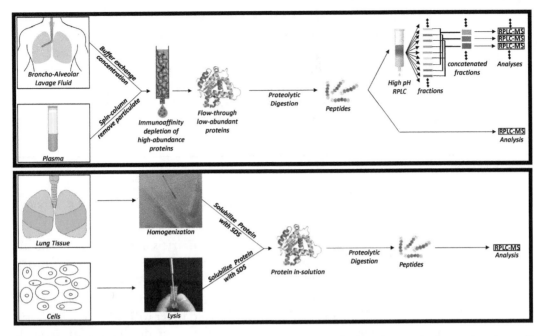

Fig. 1 Basic quantitative proteomics workflow for various sample types. Specific sample preparation steps will depend on the sample type being utilized. For example, bronchoalveolar lavage fluid and plasma are subjected to immunodepletion to remove high-abundant proteins such as albumin (top panel). Cells must be lysed, whereas lung tissue must first be homogenized (bottom panel). Following sample preparation, proteins are solubilized, digested, and analyzed using mass spectrometry. Due to a wide dynamic range in protein concentration, immunodepleted samples are often further separated into fractions (high pH reverse phase liquid chromatography). An optional labeling step occurs during protein digestion

Alternatives in this step include choice of proteolytic enzyme and buffer components (*see* **Note 2**).

4. *Mass spectrometry*—There are several classes of mass spectrometers that are used in proteomics such as quadrupole time of flight (QTOF), ion trap, ion cyclotron resonance/Orbitrap, and matrix-assisted laser desorption ionization time of flight (MALDI-TOF). The choice of instrument is based on the application and the features of the instrument itself.

5. *Data analysis*—Two general workflows exist for data analysis depending on the type of quantitation being performed: (a) label-free quantitation and comparison of the peptide intensities followed by a targeted mass spectrometry analysis to identify proteins or (b) quantitation of labels or peptide intensities from MS/MS data with concurrent protein identification. In both cases protein sequence databases are queried using a protein database search program.

Followings biomarker discovery, a targeted analysis can be developed using similar sample preparation and data acquisition parameters. Termed multiple reaction monitoring (MRM), this directed approach uses mass spectrometry and stable isotope-labeled internal standards. MRM is performed on both the labeled and the endogenous molecules and their amounts compared, providing quantitative information [7]. This method is more precise than antibody-based assays, which are often complicated by non-specific reactions. MRM assays can be a cost- and time-effective means of conducting large-scale validation studies for proteomics.

2   Materials

2.1   Solutions

Water always refers to HPLC-grade water (Burdick and Jackson) unless otherwise noted.

1. Protease for digestion—1 μg/μL trypsin solution: Add 100 μL ice-cold water to 1 × 100 μg vial of trypsin, mass spectrometry grade. Store immediately in 5 μL aliquots at −80 °C.

2. Protease for digestion—0.5 μg/μL lysyl endopeptidase (Lys-C) solution: Add 40 μL ice-cold water to 1 × 20 μg vial of lysyl endopeptidase, mass spectrometry grade (0.5 μg/μL Lys-C). Store at −80 °C (*see* **Notes 3** and **4**).

3. C18 clean up tips (Agilent) or columns (Pierce) for sample clean up. Equilibration, wash, and elution solutions require the following: trifluoroacetic acid (TFA), acetonitrile (ACN), and water. Solutions can be prepared in advance according to the manufacturer's instructions and placed in wide-mouth glass bottles. Solutions can be aliquoted into 1.5 mL microfuge tubes just prior to use for easy access and to limit cross-contamination between experiments.

4. BCA Protein Assay Kit (Pierce) for sample quantitation. All necessary solutions (2.0 mg/mL bovine serum albumin standard, Pierce BCA Protein Assay Reagent A, 500 mL × 2, and Reagent B) are provided in the kit except for water. Standard solutions should be made in 1.5 mL microfuge tubes following the protocol provided in the kit. Standards can be stored at 4 °C for up to 2 months. Prepare the BCA assay working reagent just prior to adding it to standards and samples. Based on estimated protein amounts, follow either the Standard Test Tube Protocol (working range = 20–2000 μg/mL) or the Enhanced Test Tube Protocol (working range = 5–250 μg/mL). Check the compatible substance concentration table provided in the instructions manual to ensure there will be no interfering reagents in your sample solution before you start.

5. 1 M Tris–HCl pH 8.5 stock solution: Dissolve 60.57 g of Tris Base, molecular biology grade (Tris) in ~490 mL of water. Adjust the pH of the solution to 8.5 using concentrated HCl

and a pH meter. Transfer the solution to a 500 mL volumetric flask and dilute to 500 mL total volume with water.

6. 100 mM Tris (100 mM Tris–HCl pH 8.5): For 500 mL solution, mix 50 mL of 1 M Tris–HCl pH 8.5 stock solution with 450 mL of water in a separate clean container. Use a graduated cylinder to measure and transfer each volume.

7. 1 M TCEP: Dissolve 0.2867 g of TCEP HCl (TCEP) in 1 mL of water. Make 20 µL aliquots and store at −20 °C for up to 2 years.

8. 8 M Urea (8 M urea in 100 mM Tris): Weigh 4.8048 g of urea and transfer to a 15 mL conical tube. Fill the 15 mL conical tube containing the urea to about 9 mL with 100 mM Tris. Dissolving urea in an aqueous solution is an endothermic reaction, and the temperature of solution will plummet. Roll the conical tube in your hands to warm the solution, and continue shaking until all urea pellets dissolve. Briefly centrifuge the tube and dilute to the 10 mL mark with 100 mM Tris. 8 M urea = 4.8048 g/10 mL 100 mM Tris. Make fresh before use.

9. 2 M Urea (2M urea in 100 mM Tris): Add one part 8 M urea to three parts 100 mM Tris (e.g., add 2 mL of 8 M urea to 6 mL of 100 mM Tris).

10. 500 mM IAA (500 mM iodoacetamide in 100 mM Tris): Dissolve 0.0925 g of iodoacetamide (IAA) in 1 mL of 100 mM Tris. Make fresh before use.

11. Protein Digestion Buffer (5% 2,2,2-trifluoroethanol (TFE) in 50 mM ammonium bicarbonate (ABC)): Dissolve 0.0415 g ABC in 10 mL of water in a 15 mL conical tube for a 52.5 mM ammonium bicarbonate solution (52.5 mM ABC). Transfer 9.5 mL of 52.5 mM ABC to a new conical tube, and add 500 µL of TFE. Make fresh before use.

12. 4% Sodium Dodecyl Sulfate (SDS) in 100 mM Tris: Dissolve 0.0400 g of SDS, electrophoresis grade, in 1 mL of 100 mM Tris (4% SDS = 0.04 g/mL). Make fresh before use.

13. 10% Sodium Dodecyl Sulfate (SDS) in 100 mM Tris: Dissolve 0.1000 g of SDS in 1 mL of 100 mM Tris (10% SDS = 0.1 g/mL). Make fresh before use.

14. Extraction buffer (1× HALT protease inhibitor single-use cocktail, 100 mM Tris): To 990 µL of 100 mM Tris, add 10 µL of 100× HALT.

15. Column Storage Solution (50% isopropanol): To 500 mL of water, add 500 mL of 2-propanol, LiChrosolv hypergrade for LCMS (IPA) in a 1 L wide-mouth bottle.

16. Plasma Standard: Add 5.0 mL of water to detergent-free glass vial containing 5 mL of lyophilized plasma (Sigma). Vortex

gently until plasma constituents dissolve. Store immediately in 200 μL aliquots in 1.5 mL microfuge tubes at −20 °C.

17. Resuspension buffer (3% aqueous acetonitrile (ACN) in 0.1% formic acid): To 969 mL of water, add 30 mL of ACN and a 1 mL ampule of formic acid, 99 + %, in a 1 L wide-mouth bottle. Mix well and store at 4 °C for up to 6 months (*see* **Note 5**).

18. Stock Ammonium Formate (AF) (stock 100 mM ammonium formate, pH adjusted to 10 with ammonium hydroxide): Weigh 1.5765 g of AF and quantitatively transfer to a 500 mL wide-mouth bottle with approximately 230 mL of water. Adjust the pH of the solution to 10 with ammonium hydroxide, extra pure, 25% solution in water using a pH meter. Rotate the bottle in a circular motion or use a clean stir bar as you add the ammonium hydroxide to get an accurate pH reading. When the solution is adjusted to pH 10 (3–10 mL ammonium hydroxide), transfer to a 250 mL volumetric flask and dilute to volume using water (*see* **Note 5**).

19. Buffer C (10 mM ammonium formate, pH adjusted to 10 with ammonium hydroxide): To 900 mL of water, add 100 mL of stock AF in a 1 L wide-mouth bottle (*see* **Note 5**).

20. Buffer D (10 mM ammonium formate in 90% acetonitrile (ACN), pH adjusted to 10 with ammonium hydroxide): To 900 mL of ACN, add 100 mL of Stock AF in a 1 L wide-mouth bottle (*see* **Note 5**).

21. Pump Seal Wash (10% isopropanol): To 900 mL of water, add 100 mL of 2-Propanol, LiChrosolv hypergrade for LCMS (IPA) in a 1 L wide-mouth bottle (*see* **Note 5**).

22. Buffer E (aqueous buffer, 0.1% formic acid): Add four 1 mL ampules of formic acid, 99+% to a 4 L bottle of water.

23. Buffer F (organic buffer, 90% acetonitrile (ACN), 0.1% formic acid): Combine 900 mL ACN, 100 mL HPLC-grade water, and 1 mL ampule of formic acid, 99+%.

2.2 Sample Preparation for Lung Tissue

2.2.1 Homogenization

1. Biopulverizer or homogenizer or glass tissue grinder.

2. HALT Protease Inhibitor Single-Use Cocktail (100×) (100× HALT, Thermo Fisher Scientific).

3. Extraction buffer (*see* above).

4. 25 mg lung tissue previously frozen at −80 °C. Lung tissue should have been rinsed with saline following excision and flash frozen in liquid nitrogen.

5. Dry ice.

6. Prechilled microfuge tubes.

7. Heat Block (water should be in the wells to ensure appropriate heat transfer).

2.2.2 Protein Assay	1. 10% SDS (*see* above).
	2. BCA Protein Assay Kit (*see* above).
	3. Heat Block (water should be in the wells to ensure appropriate heat transfer).
	4. Probe Sonicator.
	5. Microfuge 18 Centrifuge.
	6. NanoDrop 1000 Spectrophotometer (Thermo Fisher Scientific) or other appropriate spectrophotometer.
2.2.3 Protein Digestion	1. *See* reagents in Subheading 2.1.
	2. Urea (SigmaUltra).
	3. Amicon Ultra 0.5 mL 30 kDa centrifugal filters (YM30 filter, Millipore).
	4. Savant SPD1010 Integrated SpeedVac System (SpeedVac) or other appropriate SpeedVac.
2.3 Sample Preparation for Plasma *2.3.1 Immunodepletion*	1. High-performance liquid chromatography system (Agilent 1200 with binary pump, quaternary pump, or capillary pump, standard autosampler with thermostat equipped with 900 μL injection upgrade kit, diode array detector, and fraction collector with thermostat) (*see* **Note 6**).
	2. Multi Affinity Removal System (MARS) Column, Hu-14, 4.6 × 50 mm (Agilent).
	3. MARS Buffer A (Agilent): a proprietary buffer for Multiple Affinity Removal LC Columns.
	4. MARS Buffer B (Agilent): a proprietary buffer for Multiple Affinity Removal LC Columns.
	5. Pump Seal Wash (*see* above).
	6. LC storage solution (*see* above).
	7. Plasma standard (*see* above).
	8. Spin Filters, 0.22 μm cellulose acetate (0.22 μm spin filter, Agilent).
	9. 250 μL polypropylene snap top autosampler vials (ASV, Agilent).
	10. Snap Caps for ASV (Agilent).
	11. 2.0 mL microfuge tubes for fraction collection.
2.3.2 Plasma Protein Digestion	Supplies are the same as for lung tissue digestion.
2.3.3 High pH Peptide Fractionation (for Deep Proteome Coverage)	1. High-performance liquid chromatography system (Agilent 1200 with capillary pump, micro well-plate sampler with thermostat equipped with 40 μL loop capillary, column thermostat,

diode array detector, and fraction collector with thermostat G1330B) (*see* **Note 6**).

2. Stainless steel or peek capillaries. Preferably 0.075 mm internal diameter peek tubing. Silica-based capillaries will leach and break down over time in high pH solutions.

3. Waters XBridge Peptide BEH C18 130A, 3.5 μm, 1.0 × 150 mm Column (high pH peptide column, Waters).

4. Polypropylene autosampler vials (ASV) and caps.

2.4 Sample Preparation for Bronchoalveolar Lavage (BAL) Fluid

1. Amicon Ultra 0.5 mL 3 kD centrifugal filters (YM3 filter, Millipore).

2. MARS Buffer A (Agilent): a proprietary buffer for Multiple Affinity Removal LC Columns.

2.4.1 Concentration and Buffer Exchange

3. Spin filters, 0.22 μm cellulose acetate (0.22 μm spin filter, Agilent).

4. Polypropylene autosampler vials (ASV) and caps.

2.4.2 Immunodepletion and Digestion of BAL

All materials are the same for BAL and Plasma.

2.5 Sample Preparation for Cells

1. Phosphate-buffered saline, 1× without calcium or magnesium (PBS).

2.5.1 Collection

2. Centrifuge.

2.5.2 Digestion

Supplies are the same as those for lung tissue digestion.

2.5.3 High-Performance Liquid Chromatography (HPLC)

1. HPLC (e.g., an Agilent 1290 Infinity II) should include high-speed pump equipped with 1260 HIP degasser, capillary pump with degasser, multisampler, multiple column thermostat, UHPLC switching valve, and 1290 nanoadapter or other nano LC system configuration (e.g., Waters nanoACQUITY UPLC system or Thermo Ultimate 3000 nano system).

2. Buffer E and Buffer F (*see* above).

3. Acclaim Pepmap 100 75 μm × 2 cm, C18, 3 μm, 100A peptide enrichment column (trapping column, Thermo Fisher Scientific).

4. Acclaim Pepmap RSLC 75 μm × 25 cm, C18, 2 μm, 100A peptide analytical column (analytical column, Thermo Fisher Scientific).

2.6 Mass Spectrometry

LC/MS/MS instrument such as quadrupole time of flight (QTOF – Agilent 6550 or Bruker Impact) or Orbitrap (Thermo Fisher Scientific) equipped with nanoflow electrospray (ESI) source (Agilent G1988-6000).

2.7 ***Data Analysis***	1. Database searching programs utilizing MASCOT (a search engine that uses MS data to identify proteins from peptide sequences) include SpectrumMill (Agilent Technologies), Proteome Discoverer (Thermo Fisher Scientific), ProteinScape (Bruker), ProteinPilot (SCIEX), and MaxQuant (Max Planck Institute). Programs often are packaged with instruments and free versions can be found online. Resources can be found at www.expasy.org.

2.8 ***Materials for Labeling Methods***

2.8.1 Protein Extraction

Protein extraction for labeling methods follows the same general protocol as those for label-free methods. However, specific labeling reagents may require some minor modifications, including selection of detergents, proteases, and buffers. We strongly recommend reviewing the manufacturer's instructions before proceeding with protein extraction.

2.8.2 Labeling: Tandem Mass Tags (TMT)

1. TMT zero reagents (Thermo Fisher Scientific).
2. TMT duplex labeling kit (Thermo Fisher Scientific).
3. TMT 6-plex labeling kit (Thermo Fisher Scientific).
4. TMT 10-plex labeling kit (Thermo Fisher Scientific).

2.8.3 Labeling: Isobaric Tags for Absolute and Relative Quantitation (iTRAQ)

1. iTRAQ 4-plex labeling kit (SCIEX).
2. iTRAQ 8-plex labeling kit (SCIEX).

2.8.4 Labeling: Stable-Isotope Labeling by Amino Acids in Cell Culture (SILAC)

1. SILAC protein quantitation kit (Thermo Fisher Scientific).

3 Methods

We first discuss *label-free methods*. Please *see* Fig. 1 for an illustration of the steps involved in sample preparation for tissues, BAL, cells, and plasma.

3.1 ***Sample Preparation for Lung Tissue (Fig. 1, Bottom Panel)***

3.1.1 Homogenization

1. Maintain extraction buffer at 4 °C using ice during the entire homogenization procedure.
2. Label glass tissue grinders or label TissueLyser tubes (Qiagen) containing beads.
3. Freeze grinders or tubes on dry ice for at least 15 min prior to homogenization.
4. Weigh lung samples, place in prechilled glass tissue grinder or prechilled TissueLyser tubes, and then return tubes to dry ice.

5. Add 100 µL of ice-cold extraction buffer.

6. If using a tissue grinder, follow **step 7** and skip **step 8**. If using a TissueLyser, skip **step 7** and follow **step 8** instead.

7. If using a tissue grinder, use glass pestle to slowly homogenize lung tissue in upward and downward circular motions on ice until homogenization is complete. Homogenization is complete when there are no visible pieces of lung in the homogenate. Transfer to a 1.5 mL microfuge tube.

8. If using a TissueLyser, place samples in the TissueLyser. Screw on the lid; set TissueLyser oscillation to 50 Hz for 10 min and press start.

3.1.2 Protein Assay

1. Solubilize proteins by adding 2/3 the sample volume of 10% SDS (e.g., for 100 µL of sample, add 66.67 µL of 10% SDS).

2. Heat at 95 °C for 5 min using a heat block.

3. Allow samples to cool to room temperature and briefly centrifuge to remove liquid from caps.

4. Probe sonicate three times at 5 W in 5 s bursts (keep probe-sonicating tip close to the bottom of the microfuge tube to prevent foaming of the sample).

5. Briefly vortex samples and centrifuge at $18,000 \times g$ for 5 min at room temperature to clear lysate.

6. Follow **steps 1–8** in BCA Protein Assay Subheading 3.5.3.

7. Based on the concentration results from the protein assay, aliquot equal amounts of protein from each sample into their own 1.5 mL microfuge tube, and begin the digestion procedure. Make sure the lysate is cleared from the supernatant before transferring the appropriate volume to the new microfuge tube. Repeat **step 5** if necessary before transferring the appropriate volume of supernatant.

3.1.3 Protein Digestion

1. Note which sample had the largest volume of supernatant transferred; bring all the other samples up to this volume by adding the appropriate amount of 4% SDS.

2. All samples should now have equal amounts of protein and an equal volume.

3. Add 1/99th the volume of sample solution of 1 M TCEP to the sample so that the final concentration of TCEP is 10 mM. If the volume of 1 M TCEP to be transferred is too low, then dilute the 1 M TCEP 1:9 with 100 mM Tris and add 1/9th the volume of sample solution (*see* **Note 7**).

4. Heat samples at 95 °C for 10 min using a heat block (heat is necessary to help TCEP denature the proteins).

5. Allow samples to cool to room temperature and briefly centrifuge to remove liquid from caps.

6. Add 7x the sample volume of 8 M urea to the sample so that the concentration of SDS is approximately 0.5%

7. Follow protein digestion **steps 1–13** (Subheading 3.5.1).

3.2 Sample Preparation for Plasma (Fig. 1, Top Panel)

3.2.1 Immunodepletion Sample and Standard Preparation

1. Gently thaw plasma on ice. Plasma should have been previously frozen at −80 °C.

2. Mix plasma well by vortexing and aliquot 20 μL into 60 μL of MARS Buffer A.

3. Mix and transfer to a 0.22 μm spin filter.

4. Centrifuge at $18,000 \times g$ or max speed for 6 min at 4 °C.

5. Transfer to autosampler vial (ASV) and place in the autosampler (ALS) that has reached 4 °C.

3.2.2 Immunodepletion-HPLC Method Parameters

1. Pump gradient: Mobile phase A (MP A) is MARS Buffer A. Mobile phase B (MP B) is MARS Buffer B. The maximum pressure of the Hu-14 MARS column is 120 bar. The total method run time is 23.7 min. The gradient below is always used with this column.

Time (min)	B (%)	Flow (mL/min)
0.00	0.0	0.25
9.00	0.0	0.25
9.01	100.0	1.00
15.00	100.0	1.00
15.01	0.0	1.00
23.50	0.0	1.00
23.51	0.0	0.25

2. Autosampler: Thermostat set to 4 °C. The maximum amount of human plasma that can be injected on the Hu-14 MARS column is 20 μL (*see* **Note 8**). The maximum amount of prepped plasma that can be injected is 80 μL. Expect approximately 110 μg of low-abundant proteins from a maximum capacity injection.

3. The column is not temperature controlled.

4. Diode Array Detector: Recorded wavelength at 280 nm (requires UV lamp). Use a standard flow cell.

5. Fraction Collector: Thermostat set at 4 °C. The exact times at which the immunodepleted low-abundant proteins and the immuno-affinity high-abundant proteins elute will vary based on the dead volume of your HPLC system. Use the plasma

standard to determine the best times to collect each fraction. 1 mL is a sufficient volume to collect both fractions. A third "blank" fraction should be collected in the last minute of the gradient to ensure no protein is carried over in subsequent injections.

3.2.3 Immunodepletion-HPLC Procedure

1. Open purge valve and purge all mobile phase lines that will be in use with HPLC-grade water. Ensure at least 20 mL of freshwater is purged in each line (most HPLC systems will not have dead volumes greater than 20 mL from solvent bottle to pump heads).

2. Close purge valve and pump water at 2.0 mL/min for 10 min to ensure the entire flow path of the LC system has no solvents present other than water (MARS Buffer A and B have high concentrations of salts. If organic solvents are in the system, then the salts could precipitate damaging your LC system).

3. Switch Mobile phase A and B to MARS Buffer A and B, respectively. Purge these lines in the same manner as **steps 1 and 2**.

4. Pump MARS Buffer B and then MARS Buffer A through the system for 10 min each at 1.0 mL/min.

5. Stop flow and attach column in the proper flow orientation.

6. Equilibrate column for approximately 15 min with MP A at 0.25 mL/min.

7. Inject 50 μL of MARS Buffer A to serve as a blank. You do not need to use the fraction collector for blank injections. If the chromatogram looks clean proceed to the next step. A clean chromatogram should have only two notable peaks. The first peak should be due to MP B arriving in the flow cell in the middle of the run, and the second peak that has a slight shoulder should be due to MP A arriving in the flow cell toward the end of the run to regenerate the column.

8. Insert 2.0 mL microfuge tubes into the correct locations in the fraction collector.

9. Inject the plasma standard sample and compare the chromatogram with the manufacturer's specification provided with the column. If the two chromatograms are comparable, proceed with sample injections.

10. Always finish with a blank injection to store the column appropriately.

11. An enhanced BCA Protein Assay can be performed at this time to determine the concentration of each low-abundant fraction. Follow the applicable steps in BCA Protein Assay Subheading 3.5.3.

12. Samples can be stored at 4 °C overnight and digested the next day or they can be stored at −80 °C. If samples are stored at −80 °C, the salt in solution will precipitate, and possibly some

protein will too. The protein digestion method will solubilize any protein that may have precipitated. Best practice is to digest samples immediately or store at 4 °C overnight and then digest whenever possible.

13. To maintain the integrity and extend the lifetime of the HPLC system, do not leave MARS Buffer A and MARS Buffer B in the system for extended periods of time. Repeat **steps 1** and **2** with desalting buffer to desalt the LC system. Then repeat **steps 1** and **2** with LC storage buffer to store the system for long periods of time.

3.2.4 Protein Digestion

1. Add 63.8 μL of 10% SDS to 1 mL of low-abundant protein. Adjust all volumes and weights based on the total volume of low-abundant fraction (i.e., add 31.9 μL of 10% SDS to 0.5 mL of low-abundant protein).

2. Add 10.76 μL of 1 M TCEP per 1 mL of initial fraction volume.

3. Briefly vortex and heat at 95 °C for 10 min using a heat block. Use vials with lock caps to ensure they stay closed.

4. Allow samples to cool to room temperature and briefly centrifuge to remove liquid from caps.

5. Probe sonicate three times at 5 W in 5 s bursts (keep probe-sonicating tip close to the bottom of the microfuge tube to prevent foaming of the sample).

6. Briefly vortex samples and centrifuge to remove liquid from caps.

7. Add 0.4437 g ± 0.0030 g of urea per 1 mL of initial fraction volume. Dissolve by gently vortexing.

8. Measure the total volume of one sample using an appropriate micropipette. Use this value to represent all samples.

9. Follow protein digestion (*see* Subheading 3.5.1, **steps 1–13**).

3.2.5 High pH Peptide Fractionation (for Deep Proteome Coverage)

1. Sample and Standard Preparation: Resuspend digested sample or digested plasma standard in 40 μL of resuspension buffer by vortexing, let the sample sit for 5 min, and then vortex again. Briefly centrifuge and transfer to ASV and place in the autosampler set at 4 °C.

2. HPLC Method Parameters:

   (a) Cap pump gradient: Mobile phase A (MP A) is Buffer C. Mobile phase B (MP B) is Buffer D. The maximum pressure of the system is 400 bar. The HPLC gradient is dependent on the sample type, complexity, and goals of the experiment. The gradient listed below is appropriate for deep coverage of the plasma proteome.

Time (min)	B (%)	Flow (µL/min)
0.0	3	80
4.0	3	80
6.0	5	80
66.0	45	80
70.0	70	80
80.0	70	80
81.0	3	80
91.00	3	80

(b) Micro well-plate sampler: Thermostat set to 4 °C. Optimal injection range is 100–250 µg of sample on the high pH peptide column.

(c) Column temperature: 35 °C.

(d) Diode array detector: Recorded wavelength at 214, 254, and 280 nm (requires UV lamp). Use a micro flow cell.

(e) Fraction collector: Thermostat set at 4 °C. The fraction collection scheme is dependent on the goals of the experiment. Collecting 45 fractions over the time where the bulk of the peptides elute would be appropriate for deep coverage of the plasma proteome.

3. HPLC procedure (this is an appropriate procedure for deep proteome coverage of human plasma).

4. Purge MP A and MP B. Ensure at least 20 mL of fresh mobile phase is purged in each line (most HPLC systems will not have dead volumes greater than 20 mL from solvent bottle to pump heads).

5. Close purge valve and pump 50% MP B at 80 µL/min for 10 min to condition the system.

6. Stop flow and attach column in the proper flow orientation.

7. Pump 50% MP B through the column at 80 µL/min before switching the buffer composition to 3% MP B, and allow the column to equilibrate for approximately 10–15 column volumes worth of mobile phase.

8. Inject 40 µL of resuspension buffer to serve as a blank.

9. If the blank looks reasonably clean at 214 nm, inject approximately 100–250 µg of a digested plasma standard.

10. Ensure signal is observed at 214 nm and that you are content with the peak resolution. If you are not content with peak resolution, optimize the method further.

11. Insert 45 × 1.5 mL microfuge tubes into the correct locations in the fraction collector.

12. Inject 100–250 μg of sample.

13. Always finish with a blank injection to clean the column, and then store the column in 50% ACN with no modifiers.

14. Dry collected fractions in a SpeedVac at room temperature or 45 °C.

15. Freeze fractions at −80 °C.

16. To maintain the integrity and extend the lifetime of the HPLC system, do not leave high pH buffers in the system for extended periods of time. Repeat **steps 4** and **5** with desalting buffer to desalt the LC system. Then repeat **steps 4** and **5** with LC storage buffer to store the system for long periods of time.

3.3 Sample Preparation for Bronchoalveolar Lavage (BAL) Fluid (Fig. 1, Top Panel)

3.3.1 Concentration

1. Spin BAL fluid at $800 \times g$ for 5 min at 4 °C to remove cells.

2. Transfer supernatant to a new 1.5 mL microfuge tube, and add 1/99th the volume of supernatant of HALT 100× to the sample (can freeze samples at −80 °C after this step).

3. Filter the sample through a 0.22 μm spin filter by centrifuging at $18,000 \times g$ or max speed for 2 min at 4 °C to remove debris and possible mucus.

4. Perform a BCA assay on the sample at this point. Follow **steps 1–8** in Subheading 3.5.3.

5. Prewash the YM3 spin filter with 400 μL of water by vortexing gently and then centrifuge at $14,000 \times g$ for 10 min at 4 °C. Discard the flow-through, flip the filter over, and spin out the residual water at $1000 \times g$ for 30 s. Do not let the filter membrane dry out. Prewash the filters just before you are ready to concentrate the sample.

6. Concentrate BAL fluid sample to approximately 100 μL by spinning sample at $14,000 \times g$ for 15 min at 4 °C through the prewashed YM3 spin filter.

7. Perform buffer exchange with the same spin filter (to remove PBS used to collect BALF) by adding 400 μL of MARS Buffer A and centrifuging at $14,000 \times g$ for 15 min at 4 °C or until there is approximately 100 μL or less volume remaining.

8. Discard the flow-through, turn the filter over, and centrifuge the sample at $1000 \times g$ for 30 s to recover proteins.

9. Add three parts MARS Buffer A to one part sample volume, mix, and transfer to an ASV (400 μL is the maximum volume the ASV can hold).

3.3.2 Immunodepletion

The procedure is the same as described for plasma, Subheadings 3.2.1–3.2.3 (the concentrated BALF sample can immediately be placed in the autosampler).

3.3.3 Protein Digestion	The procedure is the same as described for plasma, Subheading 3.2.4.

3.4   Sample Preparation for Cells (Fig. 1, Bottom Panel)

3.4.1   Collection

1. At least one million cells are required for quantitative proteomics, which should generate at least 100 µg of extracted protein for most lung cell types.

2. Cells should be washed 3× on the plate with ice-cold PBS.

3. Harvest cells by scraping with 500 µL of PBS and transfer to a microfuge tube

4. Centrifuge at 18,000 × g for 5 min to pellet cells and discard the supernatant.

5. Store cell pellet at −80 °C.

3.4.2   Protein Assay

1. Gently thaw cell pellet on ice. The cell pellet should have been previously frozen at −80 °C.

2. Add the smallest reasonable volume of 4% SDS to solubilize cell pellet (usually 50–100 µL).

3. Heat at 95 °C for 5 min using a heat block

4. Allow samples to cool to room temperature and briefly centrifuge to remove liquid from caps.

5. Probe sonicate three times at 5 W in 5 s bursts (keep probesonicating tip close to the bottom of the microfuge tube to prevent foaming of the sample).

6. Briefly vortex samples and then centrifuge at 18,000 × g for 5 min at room temperature to clear lysate.

7. Follow **steps 1–8** in BCA Protein Assay Subheading 3.5.3.

8. Based on the concentration results from the protein assay, aliquot equal amounts of protein from each sample into their own 1.5 mL microfuge tube, and begin the digestion procedure. Make sure the lysate is cleared from the supernatant before transferring the appropriate volume to the new 1.5 mL microfuge tube. Repeat **step 5** if necessary before transferring the appropriate volume of supernatant.

3.4.3   Protein Digestion

The procedure is the same as described for lung tissue, Subheading 3.2.4.

3.5   Methods Common to All Procedures

3.5.1   Protein Digestion (Procedure Adapted from the Mann Lab [8])

1. Alkylate and reduce sulfhydryls by adding 1/19th the volume of sample solution of 500 mM IAA to the sample so that the final concentration of IAA is 25 mM. Incubate in the dark (wrapped in aluminum foil) for 30 min at room temperature, shaking gently on a vortexer or shaker.

2. Prewash the YM30 spin filter with 300 µL of 8 M urea by vortexing gently, and then centrifuge at 14,000 × g for 5 min at

room temperature. Discard the flow-through, turn the filter over, and remove the residual 8 M urea by centrifuging at 1000 × g for 30 s. Do not let the filter membrane dry out. Prewash the filters during the last 10 min of **step 1**.

3. Add denatured/reduced/alkylated sample to prewashed YM30 spin filter, and centrifuge at 14,000 × g for 6 min at room temperature and discard the flow-through. Sample can be loaded with multiple spins. Reducing the centrifugation temperature below ambient temperature will precipitate urea.

4. Once the sample is concentrated, add 300 μL of 8 M urea, and centrifuge at 14,000 × g for 6 min at room temperature and discard the flow-through. Repeat two more times (2×).

5. Add 300 μL of 2 M urea and centrifuge at 14,000 × g for 6 min at room temperature and discard the flow-through. Repeat two more times (2×).

6. Add 300 μL of digest buffer and centrifuge at 14,000 × g for 6 min at room temperature and discard the flow-through. Repeat two more times (2×).

7. Add 100 μL of digest buffer and transfer the filter to a fresh collection tube.

8. If you already determined the amount of protein in your sample, skip this step and move on to **step 9**. If you have not performed a protein quantitation assay on your sample, perform a BCA assay now. You can estimate the volume of sample by using the markings on the filter. Use this estimation when calculating your total protein after the assay.

9. Add 1:50 enzyme to protein of Lys-C to the sample. Gently vortex the sample and allow it to shake gently for 4 h at room temperature.

10. Add 1:40 enzyme to protein of trypsin to the sample. Gently vortex the sample for 1 min and incubate at 37 °C overnight.

11. Elute digested peptides by centrifuging at 14,000 × g for 10 min at room temperature.

12. Add 80 μL of digest buffer, vortex gently, and centrifuge at 14,000 × g for ~10 min at room temperature. Repeat this step one more time. Collect all flow-through into the same tube.

13. Transfer flow-through (eluted peptides) to a 1.5 mL microfuge tube and dry using a SpeedVac. Store in -80 °C freezer prior to LCMS analysis.

3.5.2 Cleanup

This step is optional and is used when heavy salts or urea is present in the sample, For example, when additional urea is used for the purpose of solubilization. Prepare samples, bind to C18 resin, wash, and elute according to the manufacturer's protocol. *See* **item 3** of Subheading 2.1 for a description of the supplies to use for this optional step.

3.5.3   BCA Protein Assay

1. Aliquot 10 µL or an appropriate aliquot with dilution based on the predicted concentration of each sample, and transfer to a 0.6 mL microfuge tube.

2. Follow the procedure provided with the Pierce BCA assay protein kit briefly.

3. Aliquot 10 µL of each BSA standard previously prepared using the manufactured protocol into a 0.6 mL microfuge tube. Based on the expected sample concentrations, use either the standard or enhanced assay standards. Ensure that you mix the standards and briefly centrifuge them before transferring them to the microfuge tube.

4. Aliquot 200 µL of BCA working reagent into both the standards and samples.

5. If using the standard assay, heat the samples at 37 °C for 30 min using a heat block.

6. If using the enhanced assay, heat the samples at 60 °C for 30 min using a heat block.

7. Allow samples and standards to cool to room temperature, mix by vortexing, and briefly centrifuge to remove liquid from the cap.

8. Measure samples and standards using an appropriate spectrophotometer.

3.6   Liquid Chromatography-Mass Spectrometry

3.6.1   High-Performance Liquid Chromatography

1. Most discovery-based relative quantitative label-free proteomic experiments will use C18-based columns for peptide separations. Various types of loading buffers and sample resuspension buffers can be used to achieve the necessary chromatographic results necessary for your experiment. A sample resuspension buffer of 3% ACN in 0.1% formic acid is appropriate for a nanoflow ESI coupled with a QTOF mass spectrometer. The optimal sample concentration will vary based on the sensitivity of the QTOF mass spectrometer in use. For a complex sample on an Agilent 6550 QTOF mass spectrometer equipped with a nanoflow ESI source, a target concentration of 0.2 µg/µL is a good starting point.

2. The HPLC gradient is dependent on the sample type and complexity. The method will also depend on if nano or capillary flow is used and the type of mass spectrometer being used. The gradient listed below is appropriate for a nanoflow ESI coupled with a QTOF mass spectrometer. Use a loading pump (e.g. capillary pump) to deliver the sample to the trapping column at 3.2 µL/min. Ensure that the sample will reach the trapping column based on the dead volume of your system from the autosampler needle seat to the inlet of the trapping column.

Time (Min)	B (%)	Flow (μL/min)
0.0	3	0.3
4.0	8	0.3
52.5	26	0.3
60.0	35	0.3
62.0	70	0.3
67.0	70	0.3
70.0	3	0.3
92.0	3	0.3

3.6.2 Mass Spectrometry

1. Mass spectrometry parameters and method specifics will vary somewhat depending on the instrument used. The method below is appropriate for an Agilent 6550 QTOF equipped with a nanoflow source.

2. Set instrument parameters as follows: Capillary voltage = 1200–1400 (this is empirically determined and is dependent on the number of hours the Picotip has been in use); drying gas temperature = 175 °C; flow rate = 11 liters/minute; Fragmentor = 360 V; acquisition mode = MS; TOF spectra range = 290–1700 m/z; acquisition rate = 1.5 spectra/s, 666.7 ms/spectra; ion polarity = positive. The acquisition rate can be increased to sample more spectra per cycle but with a trade-off of reduced response.

3. Empirically determine appropriate injection volume for analysis by running 0.5–1.0 μL of sample and inspecting the chromatogram. The most abundant peaks should not exceed an intensity of 107 (note that intensity is an arbitrary unit and set by the manufacturer; the value of 107 is appropriate for an Agilent QTOF). Inspect representative extracted ion chromatograms to determine if peaks are saturated. Overloading can result in chromatography artifacts such as retention time shifts and reduced mass accuracy. If necessary reduce the amount of analyte by lowering the injection volume or diluting samples.

4. To test the reproducibility of your system, analyze replicate injections of appropriate volumes (e.g., 0.5–3.0 μL). Analyze three replicates as a minimum; five to six replicates are recommended. Use appropriate metrics such as percent coefficient of variation (%CV), retention time drift, etc. of extractable compounds to determine if the method is robust enough for your quantitative purposes.

5. In general, it is recommended that you make a pooled quality control sample of equal parts of all samples being analyzed and bracket samples (e.g., after every five sample injections) at your own discretion to ensure the system does not change drastically over time. If the signal shifts by more than 20%, then consider rerunning those samples in that bracket.

3.6.3 Mass Spectrometry MS/MS (for Peptide Sequencing and Protein Identification)

1. Set up an MS/MS that is identical to the MS method listed in Subheading 3.6.2 with the following modifications: MS/MS parameters: MS = 290–1700; MSMS = 50–1700; MS acquisition rate = 10 spectra/s, 100 ms/spectra; MSMS acquisition rate = 3 spectra/s, 333 spectra/s; isolation width = narrow; collision energy = use formula; charge state 2 = slope 3.1 offset 1; charge state 3 and >3 = slope 3.6 offset −4.8; max precursors per cycle = 20 (this will depend on the complexity of the sample); active exclusion = exclude after 1 spectra and release after 0.4 min (this depends on the chromatographic peak shape and complexity of sample); isotope model = peptides; Only select precursors with a charge state of 2, 3, >3; sort precursors by abundance only; use the abundance dependent accumulation, target = 40,000 counts/spectrum (the lower the value, the lower the ion count needed in the MS/MS spectra, which could result in lower-quality spectra, but gain quicker scan rates and therefore more spectra collected per run), use MS/MS accumulation time limit; purity stringency = 100%; purity cutoff = 30%; or optimize acquisition parameters for your QTOF system based on the manufacturer's recommendations or users previous experience.

2. MS/MS (for targeted MS/MS analysis of candidate peptide masses). This step is performed following data analysis whereby select candidate peptides are not identified with the library.

3. Set up a method for targeted analysis of molecular features in the exported inclusion list. Except for acquisition parameters, this method should be identical to that used to collect MS level data. It is important that the same LC buffer preparations and LC columns are used for both MS and targeted MSMS analyses. Successful targeted analysis is dependent on consistent feature elution.

4. Data acquisition parameters for the instrument are the same as MS/MS parameters except: Import the inclusion list in the targeted table and leave the collision energy field blank.

5. For the targeted analysis, inject five to ten times more sample than was analyzed during MS acquisition. Adjust the amount as needed based on initial results.

6. Verify successful acquisition of targeted features by examining spectra. If targeted features were not properly selected within

designated RT windows, inspect extracted ion chromatograms to determine if RT drift occurred.

3.7 Data Analysis

Please *see* Fig. 2 for a schematic of database searching, including an example of search results.

3.7.1 Protein Database Searching

1. The following parameters are used for Spectrum Mill data extraction and database searching; however, the general concepts are valid for MASCOT or other search engines. All search engines will provide recommended search parameters which can be adjusted for specific experiments using the parameters listed here.

2. Extract spectra from the targeted MS/MS raw data files using the following parameters: signal to noise = 25:1, maximum charge state allowed = 4, precursor charge assignment = find.

3. Search parameters: SwissProt species-specific database (e.g., Human), carbamidomethylation as a fixed modification (if IAA used during digest), digest = trypsin, maximum of one missed cleavage, instrument = Agilent ESI Q-TOF, precursor mass tolerance = 20 ppm, product mass tolerance = 50 ppm, maximum ambiguous precursor charge = 3.

4. Validate protein identifications using a minimum of two peptides per protein, protein score > 20, individual peptide scores of at least 10, and scored percent intensity (SPI) of at least 70%. The SPI provides an indication of the percent of the total ion intensity that matches the peptide's MS/MS spectrum.

5. Perform manual inspection of spectra to validate spectrum match to predicted peptide fragmentation pattern as necessary to increase confidence in protein identifications.

6. Further validation should be performed by checking identified peptides against molecular features and retention times in the inclusion list (*see* **Note 9**).

3.7.2 MS-Level Quantitation

The accompanying chapter on Metabolomics (Chapter 17) includes a more detailed description of quantitative software.

1. Raw data quality control. Prior to performing detailed analyses, inspect total ion chromatograms (TIC) for abundance and retention time (RT) reproducibility. Abundance variance should be within 10–15%. Retention time variance should be less than 0.5 min. Exclude runs with poor reproducibility from further analysis, and repeat injections if necessary. An example of the data analysis workflow is shown in Fig. 3.

2. Molecular feature alignment, filtering, and statistical analysis. Prior to performing subsequent analysis, molecular features must be aligned for RT and mass. A detailed description of data import and analysis in MPP is beyond the scope of this

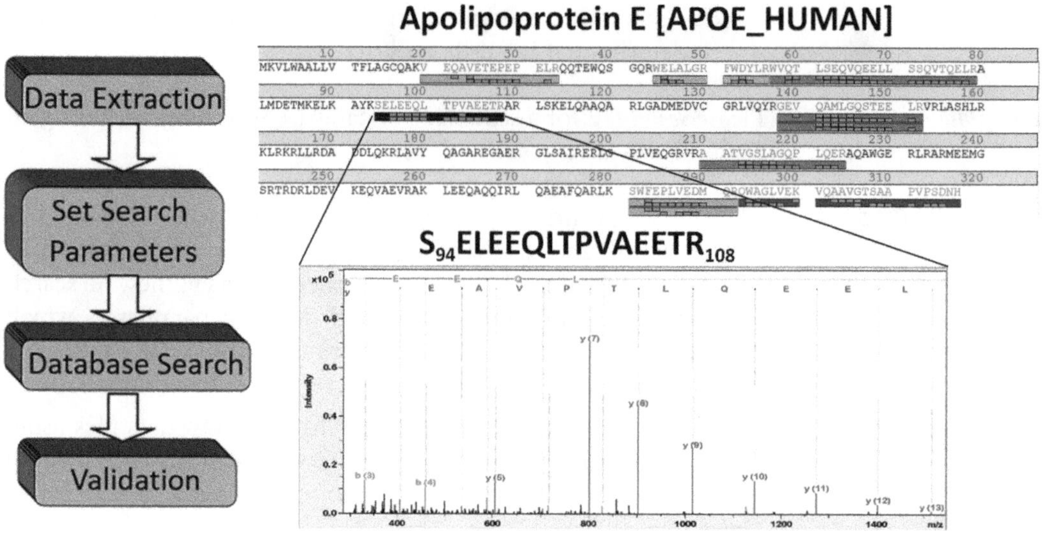

Apolipoprotein E [APOE_HUMAN]

Fig. 2 Database search workflow and results. Following data acquisition in tandem (MS/MS) mode, data are extracted to obtain a list of MS (peptides) and MS/MS (peptide fragments) peaks with the following information: neutral mass, isotopes, retention time, charge state, and peak intensity. Peak information is searched against a protein database using user-specified parameters. Outputs include extracted mass spectra that have been matched to peptide sequences. In the example shown, peptides from the human serum albumin (HSA) protein that have been matched to mass spectra are illustrated as checkered boxes; box series show peptide coverage obtained from multiple mass spectrometry experiments (top right illustration). Mass peaks from an individual mass spectrum (bottom right illustration) correspond to b- and y-series ions predicted from peptide fragments during database searching

chapter; however, you may refer to Chapter 17 on Metabolomics for basic parameters and considerations. The following differences must be included during peptide quantitation. Pre-alignment filters: mass defect = peptide-like; number of ions >2, charge state = multiple charge required, minimum abundance = 1000.

3. Normalization: Use the Standard MS scenario, select Use Recommended Order.

4. Export inclusion lists for targeted MSMS analyses: Select a Fold Change mass list. From the Tools menu, select Export Inclusion List. Set the RT tolerance appropriately for your data set. This value should be determined by inspection of raw chromatograms. A tolerance value of 0.3 min will result in MSMS acquisition 0.15 min on either side of the RT value assigned to the molecular feature. Ion selection criteria: start with All Z states and alter if necessary. For example if, on average, the number of features co-eluting within a given RT window exceeds the acquisition rate in the instrument MSMS parameters, you will need to simplify the list. Choosing Most Abundant will reduce the number of features. Alternatively you may need to edit the inclusion list manually to create multiple lists.

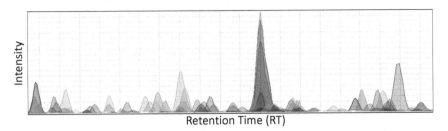

Step 1) Extract features → MH+, RT, intensity, z for each molecule

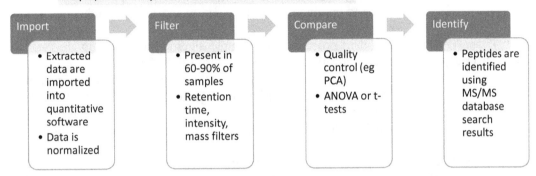

Step 2) Data Analysis in a Quantitative Software Program

Import	Filter	Compare	Identify
• Extracted data are imported into quantitative software • Data is normalized	• Present in 60-90% of samples • Retention time, intensity, mass filters	• Quality control (eg PCA) • ANOVA or t-tests	• Peptides are identified using MS/MS database search results

Fig. 3 Data analysis for quantitative proteomics. Following data acquisition in MS-only mode, data are extracted from raw files to obtain the following information: neutral mass, retention time, intensity, and charge state (Step 1). The resulting "Extracted Ion Chromatogram" can be visualized as a series of peptide peaks. This information is imported into a quantitative software program. Data are normalized using one of several options including normalization to total signal or quartile methods. Normalized data are filtered to reduce noise, and quality control visualizations such as principle component analysis (PCA) are used. Statistical tests, such as *t*-tests or ANOVA, are used to determine molecules that are differentially regulated. Differentially regulated molecules are identified using a separate tandem MS (MS/MS) analysis and database search

3.8 Protein Labeling Methods

3.8.1 Protein Extraction

Protein extraction procedures should follow those outlined above in Subheadings 3.1–3.4 for lung tissue, BAL fluid, plasma, and cells. Each protein extraction procedure should be verified and optimized for the labeling approach utilized. Manufacturer kit inserts and troubleshooting documentation should be consulted in order to optimize protein extraction buffers. Each buffer component, such as detergent (i.e., Triton X-100), chelator (i.e., EGTA), or buffer (i.e., Tris) should be documented as appropriate for the labeling assay.

3.8.2 Labeling: TMT and ITRAQ

These isobaric tagging methods utilize the power of multiplexed labels to examine a range of 2–10 total experimental variations. Chemical moieties are employed to label proteins with tags that are identical in mass, where all derivatized peptides are isobaric, resulting in identical elution times and spectral MS peaks for peptides originating from control and experimental samples. Relative quantitation among the samples is then revealed upon MS/MS fragmentation, where reporter ions with different masses are correlated back to each unique sample. Here, many samples are combined into one HPLC-MS/MS experiment, which reduces chromatography-related variability among multiple samples in a given study. The advantages and disadvantages of isobaric tagging are detailed elsewhere [9].

3.8.3 Labeling: SILAC

This metabolic labeling method involves the incorporation of isotopic labels, such as ^{13}C- or ^{15}N-labeled arginine or lysine, into protein when the sample is metabolically active (i.e., cell culture) [9, 10]. Please refer to "A practical recipe for SILAC" [11] for a thorough protocol detailing the adaptation and experimental phases of this technique in order to optimize sample preparation and analysis conditions.

3.8.4 Protein Digestion and Sample Cleanup

Standard procedures for sample digestion and cleanup are detailed above in Subheadings 3.5.1 and 3.5.2. These label-free methods should be adapted to specific labeling experimental strategies, where the manufacturer's kit insert will provide critical insight on buffers, temperature, detergents, and other chemicals that can interfere with labeling efficiency and the stability of isobaric tagging reactions.

3.8.5 High-Performance Liquid Chromatography

HPLC specifications for labeling methods are adapted from the label-free methods described above. Please *see* Subheading 3.6.1 for standard procedures relating to LC analysis. Including a method of LC fractionation prior to HPLC-MS/MS analysis is recommended in order to optimize detection of quantifiable peptides.

3.8.6 Mass Spectrometry

MS analysis should be adapted from Subheading 3.6.2 to meet the requirements of each labeled experiment. Each labeling strategy requires proper testing to achieve optimal MS sensitivity, MS peak resolution, and MS/MS fragmentation efficiency. Manufacturers often provide test kits that aid in workflow optimization.

3.8.7 Data Analysis

As detailed in Subheadings 2.7 and 3.7.1, there are many software platforms that integrate Mascot database searches into their quantitative analysis. Proteome Discoverer (Thermo Fisher Scientific) and Scaffold Q+ (Proteome Software) are just two examples of software platforms that may be used for data analysis. Each reagent and instrument manufacturer provides detailed recommendations for data analysis. These programs are capable of integrating Mascot search results with protein and peptide quantitation, with specific options for labeled results.

4 Notes

1. Sample collection and storage: Proteins and small molecules can degrade and/or become modified starting immediately upon collection. Freeze/thaw cycles can also have an effect on

certain molecules. Samples should therefore be processed, aliquoted in small volumes, and frozen at −80 ° C as quickly and as reproducibly as possible.

2. Protein sample processing: Keratin is a major contaminant of proteomics experiments and can not only affect your results, but can prevent identification of low-abundant proteins in a mixture. Several online resources exist for tips on preventing keratin contamination.

3. Low retention microfuge tubes are recommended, and some laboratories wash tubes in 50:50 MeOH:H_2O or ACN prior to use.

4. Protein digest: Trypsin will undergo rapid autolysis, especially at room temperature. When aliquoting, keep pre-labeled tubes on ice, work quickly, and freeze immediately at −80 °C. When ready to digest, thaw on ice and use immediately. Also, dithiothreitol (DTT) can be used in place of TCEP.

5. Mass spectrometry: In all cases, use bottles or consumables that have been designated for mass spec or HPLC use only. In-line filters are recommended prior to the column to prevent column clogging.

6. One HPLC system could be used for both immunodepletion and high pH peptide fractionation protocols. The system would consist of a capillary pump, degasser, standard autosampler with thermostat, TCC, DAD, and fraction collector with thermostat. The capillary pump can be converted to a binary pump by bypassing the 100 μL/min flow sensor and using the normal flow function. Capillaries would have to be changed for each method to accommodate the flow rate of each method. The dead volume of the HPLC system will affect peptide separation in the high pH peptide fractionation method, however.

7. In order to calculate the appropriate amount of 1000 mM TCEP or 100 mM TCEP to add to each sample to achieve a final concentration of 10 mM TCEP, use the following equation: $V_{TCEP} = 10\,mM \times V_I/(C_{TCEP} - 10\,mM)$, where V_{TCEP} = volume of TCEP to be added in μL, V_I = initial volume of solution in μL, and C_{TCEP} = concentration of TCEP solution to be added in mM.

8. If a larger quantity of plasma needs to be depleted, there are larger Hu-14 MARS columns that can be utilized. According to Mrozinski et al. "The average column capacity is about 20 μL of plasma or serum for the 4.6 mm id × 50 mm column, 40 μL for a 4.6 mm id × 100 mm column, 90 μL for a 7.5 mm id × 75 mm column, and 250 μL for a 10 mm id × 100 mm column" [8].

9. Database searching for protein identification: Repeat searches using different parameters; for example, include variable modifica-

tions such as phosphorylation, methylation, or missed cleavages. Make note of the database version used and all search parameters. These are generally required for manuscript submissions.

References

1. Paone G, Leone V, Conti V, De Marchis L, Ialleni E, Graziani C, Salducci M, Ramaccia M, Munafo G (2016) Blood and sputum biomarkers in COPD and asthma: a review. Eur Rev Med Pharmacol Sci 20(4):698–708

2. Priyadharshini VS, Teran LM (2016) Personalized medicine in respiratory disease: role of proteomics. Adv Protein Chem Struct Biol 102:115–146. https://doi.org/10.1016/bs.apcsb.2015.11.008

3. Terracciano R, Pelaia G, Preiano M, Savino R (2015) Asthma and COPD proteomics: current approaches and future directions. Proteomics Clin Appl 9(1–2):203–220. https://doi.org/10.1002/prca.201400099

4. Aebersold R, Mann M (2003) Mass spectrometry-based proteomics. Nature 422(6928):198–207. https://doi.org/10.1038/nature01511

5. Tao WA, Aebersold R (2003) Advances in quantitative proteomics via stable isotope tagging and mass spectrometry. Curr Opin Biotechnol 14(1):110–118

6. Spraggins JM, Rizzo DG, Moore JL, Noto MJ, Skaar EP, Caprioli RM (2016) Next-generation technologies for spatial proteomics: integrating ultra-high speed MALDI-TOF and high mass resolution MALDI FTICR imaging mass spectrometry for protein analysis. Proteomics 16(11–12):1678–1689. https://doi.org/10.1002/pmic.201600003

7. Quon BS, Dai DL, Hollander Z, Ng RT, Tebbutt SJ, Man SF, Wilcox PG, Sin DD (2016) Discovery of novel plasma protein biomarkers to predict imminent cystic fibrosis pulmonary exacerbations using multiple reaction monitoring mass spectrometry. Thorax 71(3):216–222. https://doi.org/10.1136/thoraxjnl-2014-206710

8. Mrozinski P, Zolotarjova N, Chen H (2008) Human serum and plasma protein depletion—novel high-capacity affinity column for the removal of the "Top 14"abundant proteins. Agilent Technologies Application Note, Santa Clara, CA

9. Chahrour O, Cobice D, Malone J (2015) Stable isotope labelling methods in mass spectrometry-based quantitative proteomics. J Pharm Biomed Anal 113:2–20. https://doi.org/10.1016/j.jpba.2015.04.013

10. Chen X, Wei S, Ji Y, Guo X, Yang F (2015) Quantitative proteomics using SILAC: principles, applications, and developments. Proteomics 15(18):3175–3192. https://doi.org/10.1002/pmic.201500108

11. Ong SE, Mann M (2006) A practical recipe for stable isotope labeling by amino acids in cell culture (SILAC). Nat Protoc 1(6):2650–2660. https://doi.org/10.1038/nprot.2006.427

Chapter 17

Application of Metabolomics in Lung Research

Nichole A. Reisdorph, Charmion Cruickshank-Quinn, Yasmeen Nkrumah-Elie, and Richard Reisdorph

Abstract

Advancements in omics technologies have increased our potential to evaluate molecular changes in a rapid and comprehensive manner. This is especially true in mass spectrometry-based metabolomics where improvements, including ease of use, in high-performance liquid chromatography (HPLC), column chemistries, instruments, software, and molecular databases, have advanced the field considerably. Applications of this relatively new omics technology in clinical research include discovering disease biomarkers, finding new drug targets, and elucidating disease mechanisms. Here we describe a typical clinical metabolomics workflow, which includes the following steps: (1) extraction of metabolites from the lung, plasma, bronchoalveolar lavage, or cells; (2) sample analysis via liquid chromatography-mass spectrometry; and (3) data analysis using commercial and freely available software packages. Overall, the methods delineated here can help investigators use metabolomics to discovery novel biomarkers and to understand lung diseases.

Key words Untargeted metabolomics, Mass spectrometry, Metabolite profiling, Targeted metabolomics, Lung tissue, BALF, Plasma, EBC

1 Introduction

Although its value is still being understood with respect to respiratory diseases [1], several groups have used metabolomics to explore various aspects of lung disease including phenotyping [2–6] and pathogenesis [7–9] and to reveal altered metabolism [7, 10]. For example, it has been found that the inflammatory marker leukotriene E4 (LTE4) can be used in some populations to predict response to asthma treatment, distinguish between severe and non-severe asthma, and to predict exacerbations [11, 12]. Others have discovered that pro-resolving lipid mediators are decreased in severe, uncontrolled asthma [13]. For metabolomics studies such as these, liquid chromatography-mass spectrometry (LC-MS)-based methods have become powerful and popular means of profiling clinical samples for the purpose of biomarker discovery. Biomarkers can be used in several areas of clinical research and patient care including

Scott Alper and William J. Janssen (eds.), *Lung Innate Immunity and Inflammation: Methods and Protocols*,
Methods in Molecular Biology, vol. 1809, https://doi.org/10.1007/978-1-4939-8570-8_17,
© Springer Science+Business Media, LLC, part of Springer Nature 2018

diagnosis, predicting response to medication, and monitoring drug efficacy. A wide range of biomolecules can be profiled using LC-MS, including proteins, peptides, and various classes of endogenous or exogenous small molecules and metabolites. Appropriately, there are a variety of methods and workflows that can be used to compare relative quantities of molecules within a system. The current work will focus on commonly used LC-MS-based profiling metabolomics methods; a general workflow consists of sample preparation → data acquisition → data analysis. Because there are many options at each step of these workflows, alternatives are included throughout.

Metabolomics is the global analysis of metabolites in a system and can be influenced by environmental, dietary, and genetic factors. While nuclear magnetic resonance (NMR) metabolomics studies are also routinely performed (9), LC-MS-based methods are now being used at an increasing rate; this is due to major advances in chromatography, instrumentation, and software. The basic LC-MS metabolomics workflow consists of the following basic components:

1. *Sample preparation*—protein removal and extraction of small molecules from tissue, cell, or biofluid matrix.

2. *Mass spectrometry*—detection of mass-to-charge (m/z) values of small molecules within samples. For high-resolution mass spectrometry data, this includes m/z values of isotope peaks and relative peak heights of all observed isotopic peaks for each molecule. Column retention time can also be included and utilized to improve annotation confidence.

3. *Data analysis*—comparison of compound abundances to determine differences. The result is a list of masses that are used to query a database.

4. *Database search*—a mass list is searched against a database of known and often well-characterized small molecules. A list of potential matches, i.e., candidate biomarkers, is the result.

5. *Purification/enrichment and identification*—small molecule candidate biomarkers are purified from the starting material, and structural studies are performed. If available, standards are purchased and their chemistry compared to that of the purified compound.

6. *Validation*—following identification of a candidate biomarker, large-scale clinical validation studies must be performed.

In metabolomics studies, sample preparation is an essential first step in obtaining robust and reproducible data sets toward meaningful data analysis. Because of this, **steps 1** and **2** are becoming standardized, and several individual laboratories offer extensive quality control procedures for these steps. Challenges in metabolite profiling include poor confidence during the identification step and large, inherent subject-to-subject variability when human and animal studies are performed. Procedures involved in the first four steps of a typical metabolomics workflow are included herein. Because metabolomics is highly sensitive to contamination from plastics, solvents, and reagents, extensive suggestions to minimize artifacts and some detail for recommended products are provided.

Following biomarker discovery, a targeted analysis can be developed using similar sample preparation and data acquisition parameters. An example of this technique is newborn screening in which specific amino acids, lipid mediators, and steroid hormones are assessed [11, 14]. Termed multiple reaction monitoring (MRM), this directed approach uses mass spectrometry and stable isotope-labeled internal standards. MRM is performed on both the labeled and the endogenous molecules and their amounts compared, providing quantitative information. MRM assays can be a cost- and time-effective means of conducting large-scale validation studies for metabolomics. Targeted analysis of specific small molecules entails **steps 1–3** and does not suffer from challenges in identification; examples of these assays include targeted methods for vitamins, amino acids, organic acids, steroid hormones, and lipid mediators. Due to space constraints, methods for these are not included.

2   Materials

2.1   Metabolite Extraction from Lung Tissue

2.1.1   Quality Control and Internal Standards

1. Internal standards are used to assess reproducibility during sample preparation and LC-MS. Internal standards (ISTD) are chosen based on sample type and experimental design. Isotopically labeled ISTDs that are similar in nature to those found biologically in bronchoalveolar lavage fluid (BALF), lung tissue, plasma, or urine would be ideal. These would include but are not limited to a selection of amino acids, hormones, or lipids. A few suggestions that are available from Sigma-Aldrich, Avanti Polar Lipids, and Cayman Chemical are as follows:

   (a) Hydrophilic standards: creatinine-D_3, lysine-D_4, valine-D_8, D-glucose-13C_6, alanine-D_3, methylmalonic acid-D_3.

   (b) Hydrophobic standards: heptadecanoic acid, *cis*-10-nonadecenoic acid, testosterone-d_2, triglyceride (14:0/16:1/14:0)-d_5, ceramide (d18:1/17:0), phosphatidylethanolamine (PE) (17:0/17:0), phosphatidylcholine (PC) (15:0/15:0) (*see* **Note 1**).

2. Solvents for hydrophilic stock and standard solutions include HPLC-grade or better water or methanol.

3. Solvent for hydrophobic stock and standard solutions include methanol or chloroform (*see* **Note 1**).

4. Create hydrophilic stock solutions at 2 mg/mL using a variety of isotopically labeled standards and/or other polar compounds which are exogenous to the sample being analyzed. From each stock solution, create a standard solution in 1:1 methanol/water with all standards to a final concentration of 25 µg/mL creatinine-D_3, 100 µg/mL lysine-D_4, and 200 µg/mL valine-D_8, for example.

5. Create hydrophobic stock solutions using a variety of isotopically labeled standards and/or other nonpolar compounds which are exogenous to the sample being analyzed; example stock concentrations include 17:0 fatty acid (4 mg/mL), 19:1 fatty acid (4 mg/mL), 17:0 ceramide (2 mg/mL), 17:0 PE (1.75 mg/mL), 15:0 PC (2 mg/mL), and testosterone-D_2 (1 mg/mL). From each stock, create a solution in 1:1 chloroform/methanol with all standards to a final concentration of 50 µg/mL 17:0 ceramide, 100 µg/mL 15:0 PC, 100 µg/mL testosterone-D_2, 200 µg/mL 17:0 fatty acid, 200 µg/mL 19:1 fatty acid, and 200 µg/mL 17:0 PE in the standard mix (*see* **Notes 1** and **2**).

6. If preparing urine samples on multiple days, it is recommended to prepare a batch prep QC each day using Liquichek Urinalysis Control Sample, Bio-Rad Laboratories, Irvine, California (UAC1 436).

2.1.2 Tissue Homogenization

1. Cold methanol stored at −20 °C.

2. 25 mg lung tissue previously frozen at −80 °C. Lung tissue should have been rinsed with saline following excision and then flash frozen in liquid nitrogen.

3. Biopulverizer or homogenizer (Qiagen TissueLyser LT) or glass tissue grinder (Fisher Kimble 3 mL), or suitable alternative.

4. Dry ice.

5. Pre-chilled microcentrifuge tubes (1.5 mL low retention).

2.1.3 Protein Precipitation

1. Protein precipitation organic solvent: Methanol stored at −20 °C.

2. Refrigerated centrifuge.

3. Vortexer.

4. Microcentrifuge tubes (1.5 mL low retention).

2.1.4 Liquid-Liquid Extraction

1. Organic solvent: Methyl *tert*-butyl ether (MTBE) (*see* **Note 2**).

2. Organic solvent: Methanol.

3. Sample reconstitution solvents include HPLC-grade water, methanol, acetonitrile, or formic acid.

4. HPLC-grade water.

5. Centrifuge.

6. Refrigerated centrifuge.

7. Glass culture tubes.

8. Microcentrifuge tubes (1.5 mL low retention).

9. Nitrogen dryer.

10. Pasteur pipet (9″).

11. Vacuum centrifugal concentrator.

12. Autosampler vials and caps for mass spectrometry analysis.

13. Autosampler vial glass inserts (250 μL).

2.2 Metabolite Extraction from Bronchoalveolar Lavage (BAL) Fluid

1. Materials required for the *precipitation of BAL fluid proteins* are the same as for lung tissue (*see* Subheading 2.1.3).

2. Materials required for the *extraction of small molecules* from BAL fluid following protein precipitation are the same as for lung tissue (*see* Subheading 2.1.4).

2.3 Metabolite Extraction from Other Samples

1. Materials required for the precipitation of proteins and extraction of small molecules from *plasma* are the same as for lung tissue (*see* Subheadings 2.1.3 and 2.1.4).

2. In most cases, *urine* can be analyzed neat (i.e., without sample preparation). Therefore, no additional materials are required.

3. In the event that high protein content is suspected, a methanol precipitation can be used as described for other biofluids, or brief vortexing, followed by centrifugation as described in Subheading 3.3.2.

2.4 Liquid Chromatography- Mass Spectrometry- Based Untargeted Metabolomics of Lung Tissue and BAL Fluid

2.4.1 High- or Ultra Performance Liquid Chromatography

High-performance liquid chromatography system [Agilent 1290 Infinity with binary pump (G4220A) with degasser (G1379B), autosampler (G4226A) with thermostat (G1330B), and column compartment (G1316C)]. For reference mass infusion, a separate isocratic pump (G1310A) with splitter (G1607-6000) is recommended but not required. Alternatives include Acquity UPLC, Waters Corp., Milford, MA; Shimadzu Nexera, Shimadzu, Columbia, MD; Dionex UltiMate 3000, Thermo Fisher Scientific, Waltham, MA.

2.4.2 Separation of Aqueous Fraction from any Extraction Method (Hydrophilic Normal Phase)

1. Column: Phenomenex Kinetex 2.6 μM HILIC 100 Å (50 × 2.1 mm) LC column (00B-4461-AN) and Agilent Zorbax Eclipse Plus-C18 5 μm (2.1 × 12.5 mm) Narrow Bore Guard Column (PN 821125-937). Alternative columns are available that are very similar in length and diameter; however, they may differ slightly in pore size and angstroms, or column

chemistry. Researchers are cautioned to find columns that best suit the need of their individual sample type, LC/MS system, and overall research goals, acknowledging that some method optimization may be required. Other columns that may be suitable for this analysis include Waters XBridge BEH HILIC XP Column, 130 Å, 2.5 μm, 2.1 × 50 mm, (186006077) with XBridge BEH HILIC XP VanGuard Cartridge, 130 Å, 2.5 μm, 2.1 × 5 mm guard column (186007790); Thermo Fisher Scientific Accucore HILIC 2.6 μM 80 Å (50 × 2.1 mm) LC Column (17526-052130).

2. 1 M ammonium acetate stock solution: Dissolve 3.84 g of ammonium acetate in 50 mL of LC-MS grade water. This stock solution lasts a few weeks depending on how often the HILIC MS method is run.

3. 50 mM ammonium acetate solution: Mix in a 100 mL solvent bottle by adding 5 mL of 1 M ammonium acetate solution to 95 mL LC-MS grade water. Adjust to pH 5.8 with ~17 μL of glacial acetic acid using the pH probe. This solution must be freshly prepared each day.

4. Buffer A: 50% acetonitrile in 10 mM pH 5.8 ammonium acetate. Buffer A (for 50–60 injections) is prepared *FRESH* using 32.5 mL of 50 mM ammonium acetate, 162.5 mL acetonitrile, and 130 mL LC-MS grade water.

5. Buffer B (For 50–60 injections): 90% acetonitrile with 10 mM pH 5.8 ammonium acetate is prepared *FRESH* using 65 mL of freshly prepared 50-mM ammonium acetate and 585 mL acetonitrile.

2.4.3 Separation of BAL Fluid (Hydrophilic Reverse Phase)

1. Column: Agilent Zorbax Narrow Bore RRHT SB-AQ (1.8 μm, 2.1 × 100 mm, 80 Å) analytical column (828700-914) and an Agilent Zorbax SB-AQ (5 μm, 2.1 × 12.5 mm) guard column (821125-933). Alternative columns are available that are very similar in length and diameter; however, they may differ slightly in pore size and angstroms, or column chemistry. Researchers are cautioned to find columns that best suit the need of their individual sample type, LC-MS system, and overall research goals; note that some method optimization may be required. Other columns that may be suitable for this analysis include Thermo Fisher Scientific Hypersil GoldTM aQ C18 Polar Endcapped 1.9 μm (2.1 × 100 mm) column (25302-102130) and Waters Acquity UPLC HSS T3 Column, 100 Å, 1.8 μm, 2.1 × 100 mm (186003539) with a Acquity UPLC HSS T3 VanGuard Pre-column, 100 Å, 1.8 μm, 2.1 × 5 mm (186003976) guard column.

2. Buffer A: Water with 0.1% formic acid. In a 1000 mL bottle, add 1000 mL LC-MS grade water. Add 1 mL of formic acid to

the water and invert a few times to mix. It is recommended to open and add the full-concentration (i.e., "neat") formic acid under a fume hood to reduce exposure.

3. Buffer B: 90:10 acetonitrile/water with 0.1% formic acid. In a 1000 mL bottle, add 900 mL acetonitrile, then add 100 mL of LC-MS grade water, and then add 1 mL of formic acid to the solution, and invert to mix.

2.4.4 For Separation of All Lipid Fractions (Hydrophobic Reverse Phase)

1. Column: Agilent Zorbax Rapid Resolution HD (RRHD) SB-C18, 1.8 μm (2.1 × 100 mm) analytical column (858700-902) and an Agilent Zorbax SB-C18, 1.8 μm (2.1 × 5 mm) guard column (821725-902). Alternative columns are available that are very similar in length and diameter; however, they may differ slightly in pore size and angstroms, or column chemistry. Researchers are cautioned to find columns that best suit the need of their individual sample type, LC-MS system, and overall research goals; note that some method optimization may be required. Other columns that may be suitable for this analysis include Waters Acquity UPLC HSS C18 SB Column, 100 Å, 1.8 μm, 2.1 × 100 mm [186004119] with Waters Acquity UPLC HSS C18 SB VanGuard Pre-column, 100 Å, 1.8 μm, 2.1 × 5 mm (186004136) and Sigma-Aldrich's Kromasil C18 HPLC Column, 1.8 μm particle size, pore size 100 Å, L × I.D. 100 × 2.1 mm (K08971295) with a Kromasil C18 Guard Cartridge.

2. Buffer A: LC-MS grade water with 0.1% formic acid. In a 1000 mL bottle, add 1000 mL LC-MS grade water. Then add 1 mL of formic acid to the water and invert a few times to mix.

3. Buffer B: 0.1% formic acid in isopropanol/acetonitrile/water (60:36:4). In a 1000 mL bottle, add the following: 600 mL LC-MS grade isopropyl alcohol, 360 mL acetonitrile, 40 mL LC-MS grade water, and then 1 mL formic acid. Invert several times to mix.

2.4.5 Separation of Urine Samples (Reverse Phase)

1. Column: Agilent Zorbax SB-AQ 2.1 × 100 mm, 1.8 μm column, 600 Bar (Part Number 828700-914), coupled to an Agilent Zorbax SB-AQ Narrow Bore Column 2.1 × 12.5 mm, 5 μm guard column (821125-953). Alternatives include UCT (Bristol, PA) Selectra Aqueous C18 1.8 μm, 100 × 2.1 mm UPLC column (SLAQ100ID21-18UM) with 5 μm, 10 × 2.1 mm Guard Cartridge (SLAQGDC20-5UM) and Waters Acquity UPLC HSS C18 SB Column, 100 Å, 1.8 μm, 2.1 × 100 mm [186004119] with Waters Acquity UPLC HSS C18 SB VanGuard Pre-column, 100 Å, 1.8 μm, 2.1 × 5 mm (186004136).

2. Buffer A: 0.1% formic acid in LC-MS grade water. In a 1000 mL bottle, add 1000 mL LC-MS grade water. Then add 1 mL of formic acid to the water and invert a few times to mix

3. Buffer B: 90% acetonitrile and 0.1% formic acid in LC-MS grade water. In a 1000 mL bottle, add 900 mL acetonitrile, then add 100 mL of water, then add 1 mL of formic acid to the solution, and invert a few times to mix.

2.5 Data Analysis

1. Feature extraction software. Open-source options: XCMS (https://xcmsonline.scripps.edu/), MZmine 2 (http://mzmine.github.io/). Commercial options: Mass Hunter Qualitative Analysis or Mass Hunter Profinder (Agilent Technologies), DataAnalysis (Bruker) or other commercial software.

2. Differential analysis software. Open-source options: MZmine 2 (http://mzmine.github.io/), MetaboAnalyst (http://www.metaboanalyst.ca), Metabolomics Workbench (http://www.metabolomicsworkbench.org), Mass Profiler Professional (MPP) (Agilent Technologies), Metaboscape (Bruker) or other commercial software.

3. Small molecule databases are available online and many can be downloaded at no cost. These include the Human Metabolome Database (HMDB) (http://www.hmdb.ca/) and Lipid Maps (http://www.lipidmaps.org/). Others are not available for download but can be searched online. These include Metlin (http://metlin.scripps.edu/), Chemspider (http://www.chemspider.com/) and PubChem (https://pubchem.ncbi.nlm.nih.gov/).

4. A partial listing of additional resources is available at the Metabolomics Workbench (http://www.metabolomicsworkbench.org/).

3   Methods

3.1 Metabolite Extraction in Lung Tissue

3.1.1 Quality Control and Internal Standards

1. Ensure that the chosen ISTD covers a wide range of the chromatogram; for example, if the acquisition time is 20 min, standards that elute every 5 min could be used.

2. Test the degree of ionization of each standard ahead of time using at least five different concentrations for each standard to determine the limit of detection and linearity of the instrument for these compounds. The concentration of the stock solution for each individual standard will vary depending on experimental design, and the concentration of each standard in the combined spiked standard may range from 20 μg/mL to 2 mg/mL depending on how well they ionize.

3. Create a spike mix of positive controls, and add them to the samples at 1×, 2×, or 4× concentration levels to quantitatively monitor the strength of the sample preparation and the accu-

racy of the instrumental data. Final concentration of positive controls can be 2 mg/mL D-glucose-13C$_6$, 100 μg/mL alanine-D$_3$, 200 μg/mL methylmalonic acid-D$_3$, 20 μg/mL triglyceride-D$_5$, and/or other hydrophobic and hydrophilic standards. Adjust the standard concentrations based on the sensitivity of the MS and HPLC instrumentation used for analysis (also *see* **Notes 1** and **2**).

3.1.2 Tissue Homogenization

1. Chill methanol to −20 °C.

2. Label glass tissue grinders or TissueLyser tubes containing beads.

3. Freeze grinders or tubes on dry ice at least 15 min prior to homogenization.

4. Weigh lung samples, and place in pre-chilled glass tissue grinder or pre-chilled TissueLyser Tubes, and return tubes to dry ice (*see* **Note 3**).

5. Add 0.5 mL cold methanol to each grinder or tube.

6. If using a tissue grinder, use a glass pestle to slowly homogenize lung tissue in upward and downward circular motions. Then transfer homogenized tissue to microcentrifuge tube (*see* **Note 4**).

7. If using a TissueLyser, place samples in the TissueLyser. Screw on the lid. Set TissueLyser oscillation to 50 Hz for 10 min and press start.

8. Centrifuge homogenized samples in chilled centrifuge at 14,000 × g and 4 °C for 10 min.

9. Aliquot supernatant into a clean microcentrifuge tube. Store at −80 °C, or proceed to the next step.

3.1.3 Protein Precipitation

1. Thaw samples (if previously frozen at −80 °C after the homogenization step).

2. Add 10 μL of hydrophilic standard, 10 μL of hydrophobic standard, 10 μL of either the 1×, 2×, or 4× hydrophilic positive control, and 10 μL of either the 1×, 2×, or 4× hydrophobic positive control to each sample.

3. Add 400 μL of ice-cold methanol (stored at −20 °C) to each sample.

4. Vortex for 10 s per tube.

5. Centrifuge at 0 °C for 15 min at 18,000 × g.

6. Transfer all the supernatant to a new glass culture tube, and then dry under nitrogen (N$_2$).

7. Discard the protein pellet or store at −80 °C for later use in proteomics.

3.1.4 Liquid-Liquid Extraction

1. Using a glass pipet, add exactly 3.1 mL MTBE to the dried methanol residue, cap after adding MTBE to each tube, and vortex for 30 s. Please *see* Fig. 1 for a visual representation of these steps (*see* **Notes 5** and **6**).

2. Add 750 µL of water and vortex for 10 s.

3. Spin at 250 × *g* for 10 min at room temperature to form bilayer.

4. Remove exactly 2.5 mL of top/MTBE layer (without touching bottom layer), and transfer to a new, clean glass culture tube.

5. Lay out caps first before adding MTBE below.

6. Add exactly 3 mL MTBE to remaining water part of sample, cap after adding MTBE to each tube, and vortex for 10 s.

7. Spin 800 × *g* for 10 min in centrifuge at room temperature to form bilayer.

8. Remove exactly 3 mL of MTBE (without getting water) and combine with previous MTBE tube. Cap and dry later (*see* **Notes 1** and **2**).

9. Dry residual aqueous solution under N_2 (~1.5 h). This will become the aqueous fraction.

10. Lay out caps first before adding water below.

11. Quickly add 100 µL of water—to minimize oxidation (with plastic tip).

12. Add 400 µL of ice-cold methanol (with plastic tip), vortex for 10 s, and perform a quick spin (<10 s) in the centrifuge to remove residue from sides of tube.

13. Transfer as much liquid as possible to a 1.5 mL low retention microcentrifuge tube (use Pasteur pipet).

14. Freeze at −80 °C for 25 min (to precipitate any remaining residual proteins).

15. Spin at 0 °C for 15 min at 14,000 × *g*.

16. Remove supernatant (with plastic tip) and transfer to a clean microcentrifuge tube.

17. Dry in vacuum centrifugal concentrator at 45 °C (~1.5–2 h).

18. Quickly resuspend dried supernatant in 5% ACN in H_2O.

19. Vortex for 30 s at medium speed and perform a quick spin (~10 s) to collect all liquid.

20. Transfer to a glass autosampler vial with glass insert with plastic tip and freeze at −80 °C (make two sets of autosampler vial aliquots). This is the hydrophilic fraction.

21. Dry the MTBE fraction (from **step 8** above) under nitrogen at 35 °C (after freezing water + methanol aqueous fraction above).

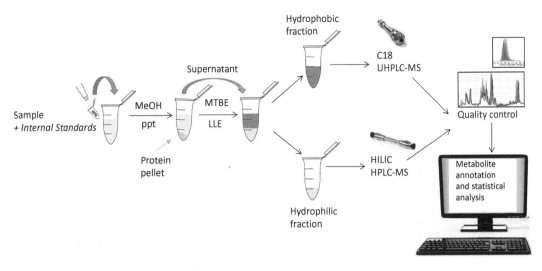

Fig. 1 Metabolomics workflow. Internal standards are added to the sample. Methanol is then added to precipitate protein; following centrifugation, the pellet is discarded or used in proteomics. The supernatant is removed, and the organic solvent methyl *tert*-butyl ether (MTBE) is added to separate the hydrophilic and hydrophobic metabolites. The MTBE layer contains dissolved nonpolar compounds, and the water layer contains dissolved polar compounds. Following centrifugation, the MTBE layer (top layer) is removed, concentrated under a nitrogen dryer and analyzed using C18 chromatography. The bottom, mainly hydrophilic layer, is also concentrated under a nitrogen dryer, and analyzed using hydrophilic interaction liquid chromatography (HILIC). Quality control is performed using the internal standards. Data analysis is performed as described in the text. For more detail please *see* citation [17]. *Ppt* precipitation, *LLE* liquid-liquid extraction

22. When completely dry, quickly resuspend in methanol.

23. Vortex for 5 s (one tube at a time to get good resuspension) and transfer to a glass autosampler vial with a glass insert using a Pasteur pipet (make two sets of autosampler vial aliquots). This is the hydrophobic fraction.

3.2 Metabolite Extraction in Bronchoalveolar Lavage (BAL) Fluid

3.2.1 Protein Precipitation

1. For BAL fluid, add 400 μL of ice-cold methanol to 200 μL BAL fluid and vortex for 30 s.

2. Incubate at −20 °C for 30 min with vortexing every 10 min.

3. Centrifuge at $14,000 \times g$ for 10 min at 4 °C.

4. Place supernatant in new tube and discard the pellet.

5. The sample can be aliquoted into autosampler vials and frozen at −80 °C until further use; alternatively, continue on to Subheading 3.2.2.

3.2.2 Metabolite Extraction

Dry down supernatant in speed vac or under nitrogen and resuspend in HPLC starting buffer: 100 μL of 5% acetonitrile in water for HILIC analysis, 100 μL of 100% methanol for C18 analysis, or 100 μL of water with 0.1% formic acid for SB-AQ analysis.

3.3 Metabolite Extraction in Other Samples

3.3.1 Plasma

1. For plasma samples, add 400 μL ice-cold methanol to 100 μL plasma and vortex for 30 s.

2. Incubate at -20 °C for 30 min with vortexing every 10 min.

3. Centrifuge at 14,000 × g for 10 min at 4 °C.

4. Place supernatant in new tube and discard the pellet.

5. The sample can be aliquoted and frozen at −80 °C until further use; alternatively, continue on to **step 6**.

6. Liquid-liquid extraction is conducted as described above in Subheading 3.1.4.

3.3.2 Urine

1. Immediately after sample collection urine is placed in a 15–45 mL conical tube and centrifuged at 3000 × g for 10 min at 4 °C (make certain the centrifuge is at 4 °C before spinning samples). If necessary, the sample can remain at 4 °C (or on ice) for up to 1 h prior to centrifugation. Following centrifugation, the supernatant is aliquoted and placed in new 15 mL tubes or into microfuge tubes (preferred) and immediately frozen at −80 °C. The sample can also be analyzed fresh.

2. Sample preparation for LC-MS analysis:

3. Thaw urine samples on ice.

4. Once thawed, vortex for 10–15 s.

5. Centrifuge the samples at 3000 × g at 4 °C for 10 min to pellet any urinary proteins.

6. Remove 20 μL of supernatant and add to an autosampler vial. Internal standards can be added at this time, though not required. If preparing urine samples on multiple days, prepare a batch prep QC each day using Liquichek Urinalysis Control Sample, Bio-Rad Laboratories, Irvine, California (UAC1 436), or suitable alternative.

7. Create a pooled urine sample for use as an instrument QC (on a single day of sample prep only). It is recommended that you prepare a single instrument QC in one autosampler vial, and then aliquot accordingly. You will need to calculate the required volume and number of aliquots for your analysis. Remember to prepare QCs for both positive and negative ionization modes, plus 10%.

8. Samples can be refrozen at −80 °C or maintained at 4 °C if analysis will be conducted immediately (*see* **Note 7**).

3.4 Liquid Chromatography-Mass Spectrometry-Based Untargeted Metabolomics of Lung Tissue and BAL Fluid

3.4.1 Hydrophilic Samples: High-Performance Liquid Chromatography

1. For hydrophilic samples, set HPLC gradient as follows:

Time (min)	B (%)	Flow (mL/min)
0	100	0.6
2.1	90	0.6
8.6	50	0.6
8.7	0	0.6
14.8	100	0.6
24.8	100	0.6

2. Mass spectrometry parameters and method specifics will vary somewhat depending on the instrument used. The method below is appropriate for an Agilent 6200 Series TOF or Agilent 6500 Series QTOF equipped with an electrospray source.

3. Set instrument parameters as follows:

    (a) Drying gas temperature: 300 °C

    (b) Drying gas flow rate: 12 L/min

    (c) Sheath gas temperature: 300 °C

    (d) Sheath gas flow: 12 L/min

    (e) Nebulizer: 35 psig

    (f) TOF spectra range: 50–1700

    (g) Scan rate (spectra/s): 2.01

    (h) Ion polarity: positive

    (i) Isocratic pump flow: 3.0 mL/min

    (j) Binary pump flow: 0.6 mL/min

    (k) Column compartment temperature: 20 °C

    (l) Injection volume: 1.0 μL

4. Analyze 1 μL of prepared sample with 10% of samples for each batch run in duplicate.

5. An example of the resulting total ion chromatogram (TIC) is shown in Fig. 2.

3.4.2 Hydrophobic Samples: High-Performance Liquid Chromatography

1. For hydrophobic samples, set the HPLC timetable as follows:

Time (min)	B (%)	Flow (mL/min)
0	30	0.7
0.5	70	0.7
7.42	100	0.7
10.4	100	0.7
10.5	30	0.7
15.1	70	0.7

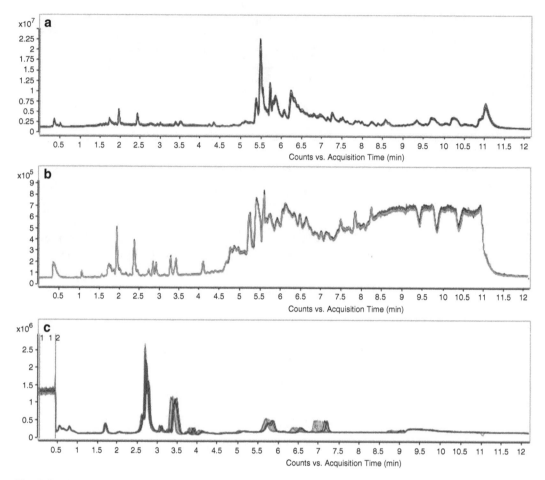

Fig. 2 Overlay of 12 total ion chromatograms showing intensity of mouse lung tissue metabolites. (**a**) Lipid fraction of extracted lung tissue from mice analyzed in positive ionization mode using C18 reversed phase chromatography. Small molecules were extracted from mouse lung using methods described herein, (**b**) lipid fraction analyzed in negative ionization mode using C18 reversed phase chromatography, (**c**) aqueous fraction analyzed in positive ionization mode using hydrophilic interaction liquid chromatography

2. Mass spectrometry parameters and method specifics will vary somewhat depending on the instrument used. The method below is appropriate for an Agilent 6200 Series TOF or Agilent 6500 Series QTOF equipped with an electrospray source.

3. Set instrument parameters as follows:

   (a) Drying gas temperature: 300 °C

   (b) Drying gas flow rate: 12 L/min

   (c) Sheath gas temperature: 275 °C

   (d) Sheath gas flow: 1 L/min

   (e) Nebulizer: 35 psig

   (f) TOF spectra range: 50–1700

(g) Scan rate (spectra/s): 2.0

(h) Ion polarity: positive (it is also recommended to repeat in negative to maximize coverage)

(i) Isocratic pump flow: 1.0 mL/min

(j) Binary pump flow: 0.7 mL/min

(k) Column compartment temperature: 60 °C

(l) Injection volume: 4.0 μL

4. Analyze 4 μL of prepared sample with 10% of samples for each batch run in duplicate.

5. An example of the resulting total ion chromatogram (TIC) is shown in Fig. 2.

3.4.3 Urine Samples: High-Performance Liquid Chromatography

1. For hydrophobic samples, set the HPLC timetable as follows:

Time (min)	B (%)	Flow (mL/min)
0	2	0.25
3.0	2	0.25
10.0	60	0.25
15.0	100	0.25
20.0	100	0.25
20.1	2	0.25
30.0	2	0.25

2. Mass spectrometry parameters and method specifics will vary somewhat depending on the instrument used. The method below is appropriate for an Agilent 6200 Series TOF or Agilent 6500 Series QTOF equipped with an electrospray source.

3. Set instrument parameters as follows:

(a) Drying gas temperature: 225 °C

(b) Drying gas flow rate: 18 L/min

(c) Sheath gas temperature: 325 °C

(d) Sheath gas flow: 12 L/min

(e) Nebulizer: 35 psig

(f) TOF spectra range: 75–1700

(g) Scan rate (spectra/s): 2.0

(h) Ion polarity: positive (it is also recommended to repeat in negative to maximize coverage)

(i) Binary pump flow: 0.25 mL/min

(j) Column compartment temperature: 30 °C

(k) Injection volume: 1.0 μL

4. Analyze 1 μL of prepared sample with 10% of samples for each batch run in duplicate.

3.5 Data Analysis

In this discussion the term "molecular features" is used to describe individual ions including all detected isotopic peaks and charge carriers. "Compounds" is used to describe metabolites or other small molecules. Compounds consist of one or more molecular features detected by the mass spectrometer.

Prior to performing differential analysis, raw data must be prepared by extraction of molecular features. This results in a list of entities (compounds, metabolites, etc.) that can be compared across experimental groups. Prior to data preparation, the quality of the raw data should be assessed to avoid expenditure of time and resources on poor quality data that are unlikely to produce useful results. The following section deals with examination and quality checks of raw data, molecular feature extraction, and differential analysis.

3.5.1 Raw Data Quality Control

It is critical to perform quality control checks and overall data assessment prior to performing detailed analyses. Inspect total ion chromatograms (TIC) or base peak chromatograms (BPC) for abundance and retention time (RT) reproducibility. Abundance variance should be within 10–20%. Retention time variance should be as minimal as possible. Optimize and validate liquid chromatography (LC) methods and column chemistries to maximize reproducibility. Mass and time alignment artifacts will increase as RT variance increases. Expected RT variance will vary by LC methodology. For example, with normal phase applications such as hydrophobic interaction chromatography (HILIC), RT variance will typically be higher than for reversed phase methods, (e.g., 0.2–0.5 min). When reversed phase methods are used, RT variance should be low (e.g., 0.05–0.15 min). Although there is no strict maximum, if RT variance exceeds these ranges, when possible, correct the problem, and reinject the samples. When RT variance is high and it is not feasible to reinject samples, RT correction (performed in addition to RT alignment) will likely be necessary. If overall reproducibility is good except for a small number of aberrant samples, it is reasonable to discard runs with poor reproducibility from further analysis without reanalyzing the entire sample set, provided the study will still retain adequate statistical power in the absence of these samples. Alternatively, the questionable samples can be reanalyzed.

3.5.2 Molecular Feature Extraction

A variety of software tools are available to perform molecular feature extraction. Some of these such as XCMS and MZmine 2 are freely available at no cost. There are a variety of commercial solutions as well. Whether you choose free or commercial solutions, it is important that you understand the various parameters and settings and how they will affect your results. Make sure that you understand how the extraction software that you choose deals with different charge carriers, charge states, noise thresholds, and peak height thresholds. It is helpful to start with a proof-of-principal data set, for example, one that compares neat matrix to matrix that is spiked with several standards. Experiments such as this help you to understand the limitations and capabilities of the chosen software (both for data preparation and data analysis). Regardless of the software used, any molecular feature extraction method should include the following critical steps:

1. Determine noise and peak thresholds. It is important to evaluate the approximate level of electronic and chemical noise. High-resolution data mining algorithms scan and evaluate spectra to determine which ions should be grouped into discrete ion clusters. If the applied thresholds are too low, this will result in a large number of artifacts in the extracted compound list. To evaluate peak thresholds, use the data analysis software specific to the mass spectrometer used to acquire the data, e.g., Mass Hunter Qualitative Analysis (Agilent), DataAnalysis (Bruker), or other. Extract spectra over three or more regions of the chromatogram. Examine spectra to determine the approximate spectral peak height value below which discrete ion clusters are no longer evident. As a general rule you can safely use a data extraction peak threshold that is two to three times that value. Use lower values if you want to include very low abundance compounds, but be aware that this will likely result in an increased number of feature extraction artifacts.

2. Determine relevant charge carriers and charge states. The charge carriers, and their respective compound ion species, will vary based on the sample source, sample preparation method, LC-MS mobile phases utilized in the acquisition method, and whether the data was acquired in positive or negative ionization mode. The most common charge carriers in positive ion mode data are H+, Na+, K+, and NH4+, with H+ and Na+ being the predominant species. Common negative mode charge carriers include Cl−, CHOO−, and CH3COO−. It is important to note that data extraction algorithms do not all deal with multiple charge carriers and ion species in the same way. For example, Agilent Mass Hunter software deconvolutes and groups spectra of different charge carriers assigned to the same compound. The total relative abundance of the compound is then reported as volume, which is the total combined area of all ion species assigned to that compound. XCMS does not group charge carriers. So compounds that have multiple ion species will be represented multiple times in the extracted

compound list. Additional software is required to group and annotate ion species. An example of such software is *Camera* which is available at Bioconductor.org. To determine likely charge carriers to include, allow all of those listed above and process a representative data file using the data extraction software. Evaluate the charge carriers assigned by the feature extraction software and the quality of the spectra to which they are assigned. Make sure they look reasonable. If any particular charge carrier appears to result in dubious looking spectra, disallow it from the extraction method.

3. Metabolites can also be detected as more than one charge state although most will be singly charged. Lipids however will often be detected as +2 charge species. To reduce the likelihood of charge state-related feature extraction artifacts, limit the charge states to those you expect to be included in your data set. Most metabolites in urine or in the aqueous phase (if your sample preparation includes phase separation) will be limited to +1 charge species. Lipids will often be detected as +1 and +2 species.

4. Evaluate overall feature extraction results. After thresholds and charge carriers have been determined, execute feature extraction on one or more representative data files. Examine the spectra for the low abundance compounds. If the majority of the compounds at the low end of the abundance range appear to be of poor spectral quality, or if the chromatograms appear to be extracted from noise, increase your extraction height threshold and reevaluate. If the software you are using allows you to sort the compound list by the number of ions they include, evaluate charge carriers and charge states in the same way.

3.5.3 Molecular Feature Alignment and Normalization

1. After feature extraction has been completed, prior to performing differential analysis, molecular features must be aligned for retention time and mass. Mass Profiler Professional (Agilent Technologies) and other programs typically perform this function during data import. As listed in Materials, a variety of commercial and open-source differential analysis programs are available. All programs will differ in the alignment algorithms employed, as well as the selection of user-definable parameters that are available. A detailed description of data import and analysis in any specific software program is beyond the scope of this chapter. Basic parameters and considerations that are relevant regardless of the software used are included here.

2. Mass and time alignment. All compounds must be aligned between samples for both mass and time. The mass alignment window is typically described in parts per million (ppm) mass error (some programs may describe mass error in millidaltons). The width of the ppm window should be based upon the mass accuracy capabilities of the mass spectrometer and whether ref-

erence (e.g., lock mass) mass correction was applied successfully. The window should be set slightly wider than the mass error expected based on typical instrument performance. For example, if the expected error for a time-of-flight instrument is 1–5 ppm, set the mass alignment window to 10–15 ppm. This is because matrix effects and mass extraction error can lead to additional mass error for extracted compounds. Avoid setting the ppm window too wide as alignment artifacts will occur.

3. The retention time (RT) window should be based on the retention time variation observed in the raw data (*see* discussion on raw data quality control above). As with the mass error window, the RT window should also be set slightly higher than the drift observed in the raw data. For example, if the observed RT drift is 0.05 min, use an alignment window of approximately 0.10 min. Again, using an alignment window that is too narrow will result in alignment artifact. When in doubt, increase the width of the alignment window; a window that is too narrow will result in far more alignment artifacts than one that is too wide.

4. Normalization. It is standard practice to normalize data prior to performing differential analysis. There are several concepts to keep in mind when choosing whether to normalize and which normalization algorithm to use. First, not all data need to be normalized. If the total ion chromatograms in the raw data overlay tightly with little variation, normalization may not have a significant effect, and the effects may not be beneficial. It is often useful to analyze data with and without normalization.

5. There is a variety of normalization algorithms used for metabolomics data sets. Determining which algorithm is most appropriate for a particular data set is not always straightforward. Some algorithms work best with large or small data sets. Others don't function well if there are a higher number of missing values due to feature extraction error. Consult a biostatistician for recommendations specific for your data. Two algorithms that work reasonably well for many data sets are percentile shift and total useful signal. The percentile shift algorithm is a "per sample" normalization tool. First the compounds within a sample are ranked by intensity. Next, the abundance of each compound is normalized against a percentile rank intensity. The percentile is typically chosen by the user, typically 75%. The percentile shift algorithm operates under the assumption that sum intensity of all samples is approximately equal. This algorithm should not be applied to data sets with large intensity differences between samples. The total useful signal approach may be better for data sets with a higher degree of variability between samples. In this approach, the total abundance of all compounds within each sample is used as a correction factor to normalize the abundance of each compound within the sample.

1. To assess the quality of the aligned data, covariance analysis and/or unsupervised clustering should be performed. It is typically important to apply some basic filtering tools prior to performing these analyses. For example, application of a minimum frequency filter can help remove spurious compounds that occur in only one or very few samples. Frequency filters can be applied such that only compounds that appear in a minimal number or percentage of samples are retained for analyses. Compounds that are removed in this way can complicate covariance and statistical analysis results and rarely contribute any useful information.

2. Various covariance analyses can be accomplished in most differential software programs. Alternatively, R scripts can be used to perform these and other differential analysis steps. A principal components analysis (PCA) can be very informative regarding the overall variance in the data set. Samples that are similar tend to cluster together, whereas those that are more dissimilar cluster farther apart. Clustering is determined by the loadings of the metabolites included in the analysis. When a group of metabolites within certain samples have similar abundance profiles, these samples tend to cluster together. It is common practice to use PCA to demonstrate that samples cluster in discreet groups based on a response to an experimental condition or because of a shared clinical phenotype. However it is equally, if not more important, to ask whether nonexperimental parameters have noticeable effects on PCA clustering. For example, plasma samples that are collected from multiple sites, differences in collection tubes, phlebotomy practices, or storage conditions can exert effects on the data. In many cases these effects are evident in PCA results. In such a case the "collection site" parameter would account for more variance within the data than experimental parameters such as clinical phenotype or therapy. As a result, clinically relevant information would likely be lost or at the very least heavily diluted, making the information difficult to extract. It is important to look for effects from any nonexperimental parameter that could potentially exert effects on the data. A few examples of factors that may affect clinical data results include gender, age, sample collection/storage methods, patient compliance, and diet. A variety of factors related to sample preparation (such as date of sample preparation or changes in reagents mid-study) and LC-MS acquisition (such as instrument breakdown and repair mid-study or a change in the LC-MS column) can also exert effects on complex data sets. PCAs are one tool that aid in identifying these effects.

3. PCA can be used effectively to discover important patterns in complex data sets that might be important for a clinical study.

For example, PCA clustering may reveal that there are important patient subgroups within a particular phenotype, perhaps in response to a specific therapy. However, there are important concepts to consider regarding PCA. First, there is no statistical significance implied in the results. Metabolites with high loadings are those that drive the observed clustering the most. The loadings may be indicative of the importance of the metabolite; some analysts use PCA loadings to find metabolites of potential importance to a study. However, not all metabolites with high loadings are relevant, and not all will pass a significance filter. Moreover, a lack of discernable clustering does not mean that there are no metabolites that exhibit discernable and/or important abundance differences between experimental groups. If most compounds in an analysis do not change, but a small number change significantly, these differences may not be enough to drive PCA clustering to a noticeable extent. Most often, statistical tests are required to identify these important compounds.

4. Unsupervised hierarchical clustering is another tool that can be used to identify patterns in a data set that may indicate either influences of nonexperimental parameters or important clinical factors. Although the PCA and unsupervised hierarchical clustering algorithms are very different, the results can be used in similar ways. An example of PCA and unsupervised hierarchical clustering results are shown in Fig. 3.

3.5.5 Differential Analysis: Statistics and Fold-Change Filters

1. ANOVA/t-test. Although there are more sophisticated statistical tools that may be required for some data sets, t-tests for pairwise comparisons and analysis of variance (ANOVA) for more complex comparisons are the most commonly utilized. Generally speaking, it is considered incorrect practice to use pairwise comparisons when there are more than two experimental conditions or parameters. In these cases, an ANOVA is more appropriate.

2. There are several types of t-tests and ANOVA algorithms, and it is important to select the type(s) that are most suited to your data. For example, parametric tests (such as the t-test) assume normal distribution and equal variance within the data set. Nonparametric tests do not assume normal distribution or equal variance. Using an incorrect test to analyze your data can result in obtaining a statistically "incorrect" result. There are tools that can be used to evaluate distribution and to determine whether variance is equal or unequal. There are other factors to consider including whether a paired or unpaired t-test is appropriate for your experimental design. If you are unsure about what type of test to use, consult a biostatistician.

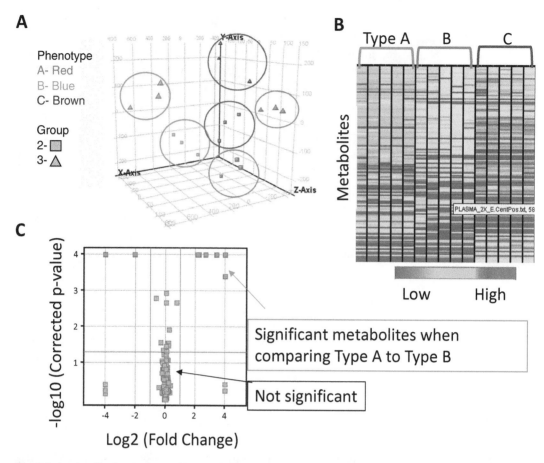

Fig. 3 Data visualization tools used in metabolomics. Examples of principle components analysis (PCA), hierarchical clustering, and volcano plots are shown. (**a**) PCA shows distribution of six sample types (Groups 2A, 2B, 2C, and 3A, 3B, 3C) for approximately 1000 urine metabolites. (**b**) Metabolites are clustered according to three sample types, A, B, or C. Metabolites shown in blue are upregulated (high) compared to those in red (low). (**c**) Resulting volcano plot following a *t*-test comparing Type A to Type B samples. Red boxes in upper quadrants indicate metabolites that reached significance ($p < 0.05$) and fold change >1.5 thresholds

3. Multiple testing correction (MTC) should be applied to tests of significance. The higher the number of tests performed, the higher the likelihood of false positives (metabolites that appear to be different but really are not). Each metabolite is considered a separate test so the number of tests can be quite high. MTC algorithms help account for this and reduce false positives. There are various MTC algorithms; some are more stringent than others. The Benjamini-Hochberg algorithm is commonly used and is not excessively stringent. It is a good option for clinical studies. The Bonferroni algorithm is quite stringent and will typically result in a very low number of metabolites that pass the significance test, particularly in clinical data sets where the variance can be quite high. In some cases you may need to disable MTC in order to get any metab-

olites to pass a statistical test. It is likely that many of these metabolites will not be reliable, so use discretion and verify that the metabolites actually have different mean abundances between experimental groups in the raw data.

4. To increase the likelihood that metabolites will pass a significance filter when MTC is applied, use metabolite lists that are as small as reasonably possible. MTC algorithms become more stringent on larger metabolite lists. Apply a minimum frequency filter to remove spurious compounds. Limit the statistical analysis to only the metabolites that are relevant to the question you are trying to answer.

5. Fold change. A *t*-test or ANOVA will provide a mass list of metabolites that are statistically significant. However, not all of these metabolites will have relevant fold changes between experimental groups. Very low fold changes such as 1.1- or 1.2-fold often are observed, but it is difficult to know if such subtle changes are relevant to your study. Therefore, it is recommended that you apply a fold-change filter after performing statistical tests. Filters of 2.0- or 1.5-fold are typically appropriate.

3.5.6 Metabolite Identification

1. Mass spectral peaks are extracted using vendor software and are collapsed into compounds using either instrument vendor tools or publicly available software such as XCMS [15] or CAMERA [16]. Molecular formulas are generated based on the exact mass and isotope ratios. In some software programs,

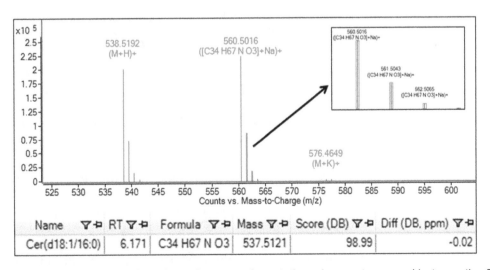

Fig. 4 Generation of molecular formulas and compound annotation using exact mass and isotope ratios. The spectral plot shows the proton, sodium, and potassium spectral peaks for an annotated compound. Inset represents the actual ion height and isotope ratios (black lines) compared to the theoretical distribution (red box) for the annotated compound based on the sodiated ion. The table underneath shows the compound name, chromatographic elution time of the compound (6.171 min), the database score (98.99 out of a possible 100), and the mass error which is very close to 0 ppm

predicted isotope distribution is compared to actual ion height, and a score is generated (Fig. 4).

2. Exact mass and/or molecular formulas are used to search metabolite databases. A Metlin search can be performed online at http://metlin.scripps.edu/. An HMDB search can be performed online at http://www.hmdb.ca/. Alternatively, in-house databases can be constructed and searched. These can comprise multiple free-ware databases including those listed above (*see* **Note 8**).

4   Notes

1. Always use glass for storing (glass culture tubes, glass autosampler vials) or transferring (glass pipettes) lipids and organic solvents.

2. Minimize exposure of all lipid samples and standards to air. Seal polytetrafluoroethylene (PFTE) caps tightly to avoid air exposure and evaporation. Immediately resuspend dried lipids in the next solvent, or keep them under a steady stream of nitrogen.

3. Turn on centrifuge and set to 0 ° C prior to start of sample preparation.

4. Keep samples on ice at all times unless otherwise indicated.

5. These techniques use many volatile solvents so keep all solvents capped during sample preparation.

6. These procedures require the use of hazardous, flammable, or volatile solvents such as MTBE. Perform all steps in a fume hood.

7. Variation in urine dilution factors requires normalization for reporting, thus additional urine analysis may be required. Urine normalization methods include quantitative urinary creatinine analysis, osmolality, specific gravity, or sum of the total area.

8. While an error rate has not been established for the process of database annotation, it is generally assumed to be quite high. The reasons for this are several-fold: (1) many compounds may have the same mass within the database search error tolerance, (2) many compounds have closely related isomers, (3) multiple compounds listed as possible database hits result in investigators using the data "as is" or applying their best judgment to assign the "best" annotation, (4) incorrect annotations can be eliminated based on knowledge of the experimental system. However, selection of the correct isomers is nevertheless subjective, and the likelihood of error is unfortunately high.

Therefore, to be confident in the annotations and the interpretation of the clinical relevance, the structural identity of annotated compounds must be confirmed. Confirmation can be performed using nuclear magnetic resonance spectrometry, or by tandem mass spectrometry (MS/MS), and the use of standards where available.

References

1. Snowden S, Dahlen SE, Wheelock CE (2012) Application of metabolomics approaches to the study of respiratory diseases. Bioanalysis 4(18):2265–2290. https://doi.org/10.4155/bio.12.218

2. Mattarucchi E, Baraldi E, Guillou C (2012) Metabolomics applied to urine samples in childhood asthma; differentiation between asthma phenotypes and identification of relevant metabolites. Biomed Chromatogr 26(1):89–94. https://doi.org/10.1002/bmc.1631

3. Reisdorph N, Wechsler ME (2013) Utilizing metabolomics to distinguish asthma phenotypes: strategies and clinical implications. Allergy 68(8):959–962. https://doi.org/10.1111/all.12238

4. Bowler RP, Bahr TM, Hughes G, Lutz S, Kim YI, Coldren CD, Reisdorph N, Kechris KJ (2013) Integrative omics approach identifies interleukin-16 as a biomarker of emphysema. OMICS 17(12):619–626. https://doi.org/10.1089/omi.2013.0038

5. Bowler RP, Jacobson S, Cruickshank C, Hughes GJ, Siska C, Ory DS, Petrache I, Schaffer JE, Reisdorph N, Kechris K (2015) Plasma sphingolipids associated with chronic obstructive pulmonary disease phenotypes. Am J Respir Crit Care Med 191(3):275–284. https://doi.org/10.1164/rccm.201410-1771OC

6. Cruickshank-Quinn CI, Mahaffey S, Justice MJ, Hughes G, Armstrong M, Bowler RP, Reisdorph R, Petrache I, Reisdorph N (2014) Transient and persistent metabolomic changes in plasma following chronic cigarette smoke exposure in a mouse model. PLoS One 9(7):e101855. https://doi.org/10.1371/journal.pone.0101855

7. Ho WE, Xu YJ, Xu F, Cheng C, Peh HY, Tannenbaum SR, Wong WS, Ong CN (2013) Metabolomics reveals altered metabolic pathways in experimental asthma. Am J Respir Cell Mol Biol 48(2):204–211. https://doi.org/10.1165/rcmb.2012-0246OC

8. Jung J, Kim SH, Lee HS, Choi GS, Jung YS, Ryu DH, Park HS, Hwang GS (2013) Serum metabolomics reveals pathways and biomarkers associated with asthma pathogenesis. Clin Exp Allergy 43(4):425–433. https://doi.org/10.1111/cea.12089

9. Quinn KD, Schedel M, Nkrumah-Elie Y, Joetham A, Armstrong M, Cruickshank-Quinn C, Reisdorph R, Gelfand EW, Reisdorph N (2017) Dysregulation of metabolic pathways in a mouse model of allergic asthma. Allergy. https://doi.org/10.1111/all.13144

10. Ried JS, Baurecht H, Stuckler F, Krumsiek J, Gieger C, Heinrich J, Kabesch M, Prehn C, Peters A, Rodriguez E, Schulz H, Strauch K, Suhre K, Wang-Sattler R, Wichmann HE, Theis FJ, Illig T, Adamski J, Weidinger S (2013) Integrative genetic and metabolite profiling analysis suggests altered phosphatidylcholine metabolism in asthma. Allergy 68(5):629–636. https://doi.org/10.1111/all.12110

11. Armstrong M, Liu AH, Harbeck R, Reisdorph R, Rabinovitch N, Reisdorph N (2009) Leukotriene-E4 in human urine: comparison of on-line purification and liquid chromatography-tandem mass spectrometry to affinity purification followed by enzyme immunoassay. J Chromatogr B Analyt Technol Biomed Life Sci 877(27):3169–3174. https://doi.org/10.1016/j.jchromb.2009.08.011

12. Rabinovitch N, Reisdorph N, Silveira L, Gelfand EW (2011) Urinary leukotriene E(4) levels identify children with tobacco smoke exposure at risk for asthma exacerbation. J Allergy Clin Immunol 128(2):323–327. https://doi.org/10.1016/j.jaci.2011.05.035

13. Levy BD (2012) Resolvin D1 and resolvin E1 promote the resolution of allergic airway inflammation via shared and distinct molecular counter-regulatory pathways. Front Immunol 3:390. https://doi.org/10.3389/fimmu.2012.00390

14. Armstrong M, Jonscher K, Reisdorph NA (2007) Analysis of 25 underivatized amino

acids in human plasma using ion-pairing reversed-phase liquid chromatography/time-of-flight mass spectrometry. Rapid Commun Mass Spectrom 21(16):2717–2726. https://doi.org/10.1002/rcm.3124

15. Smith CA, Want EJ, O'Maille G, Abagyan R, Siuzdak G (2006) XCMS: processing mass spectrometry data for metabolite profiling using nonlinear peak alignment, matching, and identification. Anal Chem 78(3):779–787. https://doi.org/10.1021/ac051437y

16. Kuhl C, Tautenhahn R, Bottcher C, Larson TR, Neumann S (2012) CAMERA: an inte-grated strategy for compound spectra extrac-tion and annotation of liquid chromatography/mass spectrometry data sets. Anal Chem 84(1):283–289. https://doi.org/10.1021/ac202450g

17. Cruickshank-Quinn C, Quinn KD, Powell R, Yang Y, Armstrong M, Mahaffey S, Reisdorph R, Reisdorph N (2014) Multi-step prepara-tion technique to recover multiple metabo-lite compound classes for in-depth and informative metabolomic analysis. J Vis Exp 89:e51670. https://doi.org/10.3791/51670

Chapter 18

Functional Genomics in Murine Macrophages

Frank Fang-Yao Lee and Scott Alper

Abstract

In this chapter, we describe methods for functional genomics studies in mouse macrophages. In particular, we describe complementary methods for gene inhibition using RNA interference (RNAi) and gene overexpression. These methods are readily amenable to medium- and high-throughput functional genomics investigations. These complementary loss-of-function and gain-of-function genomic approaches provide a rapid means of investigating the function of candidate genes prior to initiating more cumbersome studies in vivo.

Key words Macrophage, RAW264.7, RNAi, siRNA, Overexpression

1 Introduction

While a genetic lesion is the gold standard to investigate gene function on a case by case basis [1], the ability to manipulate gene activity in medium- or high-throughput fashion remains a mainstay of biological investigation [2]. In this chapter, we outline methods for functional genomics studies in mouse macrophages. In particular, we describe (1) methods using RNA interference (RNAi) to inhibit the function of candidate genes and (2) gene overexpression approaches to activate candidate genes. These complementary loss-of-function and gain-of-function assays offer a powerful approach to test the function of novel gene candidates prior to the initiation of in vivo investigations.

Macrophages can be a challenging cell type in which to perform such experiments. Macrophages are difficult to transfect efficiently without affecting cell viability [3, 4], and macrophages are somewhat resistant to RNAi compared to other cell types. These challenges are even more pronounced in primary macrophages. To overcome these issues, we have optimized transfection, RNAi using siRNAs for gene inhibition, and gene overexpression techniques in one of the most commonly used immortalized mouse macrophage cell lines, RAW264.7 [5–8]. While working with this cell line may not recapitulate all of the biology present in primary cells derived

Scott Alper and William J. Janssen (eds.), *Lung Innate Immunity and Inflammation: Methods and Protocols*,
Methods in Molecular Biology, vol. 1809, https://doi.org/10.1007/978-1-4939-8570-8_18,
© Springer Science+Business Media, LLC, part of Springer Nature 2018

directly from mice, the ability to manipulate these cells is a positive trade-off.

First, we describe a method to efficiently deliver siRNA to RAW264.7 macrophages using an Amaxa electroporation system (*see* **Note 1**). We have optimized transfection with this system to obtain close to 100% transfection efficiency, excellent cell viability post-transfection, and good siRNA-mediated gene knockdown. Macrophages perform many critical immune functions [9], and methods to monitor some of these functions are described in other chapters in this volume. As an example of one possible immune assay, we describe the ability to test the effect of siRNA-mediated inhibition of candidate genes on lipopolysaccharide-induced inflammatory cytokine production.

Second, we describe a method to deliver plasmids engineered to overexpress candidate genes. With this optimized technique, we routinely obtain approximately 50% transfection efficiency. Because of imperfect transfection in macrophages, follow-up studies require approaches to identify transfected cells. These approaches include using plasmids engineered to express either drug resistance genes or fluorescent proteins. In the former, toxic agents (such as neomycin) are added to the transfected cells. Only those that have been successfully transfected can survive. In the latter, successfully transfected cells express fluorescent proteins and can be purified using flow cytometry with cell sorting. To illustrate an approach that we have found is amenable to high-throughput study, we describe the use of cytokine-luciferase reporters to investigate the immune effects of overexpression of candidate genes.

2   Materials

2.1   RNAi

1. RAW264.7 cells (immortalized mouse macrophage cell line).

2. Amaxa Nucleofector 96-well shuttle and associated equipment (Nucleofector 2 device and attached computer) (Lonza).

3. SF Cell Line 96-well Nucleofector Transfection Kit (Lonza). This kit contains the 96-well electroporation plate, the transfection solution, and the transfection supplement. The supplement solution should be added to the transfection solution prior to use.

4. Centrifuge.

5. Macrophage culture medium: DMEM (containing glucose, L-glutamine, and sodium pyruvate) supplemented with 10% FBS.

6. 0.25% Trypsin/EDTA.

7. PBS.

8. RPMI-1640.

9. siRNA(s) targeting gene(s) of interest.

10. Negative and positive control siRNAs (*see* **Note 2**).

11. Fluorescently labeled siRNA (*see* **Note 2**).

12. Tissue culture treated dishes or flasks.

13. 96-well tissue culture plates.

14. 96-well round-bottom plates.

15. *E. coli* Lipopolysaccharide (LPS).

16. Mouse cytokine ELISA kit.

2.2 Overexpression

1. Macrophage culture medium: DMEM (containing glucose, L-glutamine, and sodium pyruvate) supplemented with 10% FBS.

2. 0.25% Trypsin/EDTA.

3. PBS.

4. FuGene HD (Roche) (*see* **Note 3**).

5. OptiMem (Thermo Fisher Scientific).

6. Tissue culture treated dishes or flasks.

7. 24-well tissue culture plates.

8. Plasmid(s) driving expression of candidate gene(s) (*see* **Note 4**).

9. Plasmid(s) driving expression of control gene(s) (*see* **Note 4**).

10. Plasmid driving expression of GFP (*see* **Note 4**).

11. Plasmid driving expression of firefly luciferase reporter (*see* **Note 5**).

12. Plasmid driving expression of Renilla luciferase internal control reporter (*see* **Note 5**).

13. Dual luciferase assay kit that can measure firefly and Renilla luciferase activity.

3 Methods

3.1 RNAi: Setting Up Amaxa System for Macrophage Transfection

1. Turn on the Amaxa 96-well shuttle, the Amaxa Nucleofector II device, and the attached laptop.

2. Start the Amaxa 96-well shuttle software. Choose *new parameter file* from the file menu.

3. Select the wells to be transfected on the 96-well schematic and apply program DS-136 to these wells (*see* **Note 6**).

4. Add the supplement solution to the Solution SF in the 96-well Nucleofector Solution SF kit prior to use. Allow this supplemented solution to equilibrate at room temperature.

3.2 RNAi: Preparation of Macrophages for Transfection

1. Maintain RAW264.7 cells in macrophage culture medium at 37 °C in 5% CO_2. Grow cells to approximately 70–80% confluence (*see* **Note 7**). Prepare enough cells to transfect the planned number of samples (200,000 cells per sample).

2. Prewarm macrophage culture medium, Trypsin/EDTA, and RPMI-1640 to 37 °C.

3. To collect macrophages for transfection, aspirate off culture medium from plates containing adherent RAW264.7 cells and wash with PBS. Aspirate the PBS wash, and then add prewarmed Trypsin-EDTA to the cells (1 mL Trypsin-EDTA is sufficient to cover the cells on a 10 cm cell culture dish). Incubate cells for 2–3 min with occasional plate agitation. Do not over-trypsinize as this can damage the macrophages and diminish transfection efficiency. Once the cells are detached, add macrophage culture medium (9 mL for a 10 cm culture dish) and pipette up and down to finish detaching macrophages from the plate.

4. Count the macrophages using a hemacytometer. Calculate how many cells are needed for the experiment (200,000 cells/transfection well). Be sure to include a few extra wells worth of cells to allow for pipetting errors.

5. Centrifuge the calculated number of macrophages at $150 \times g$ for 10 min at room temperature (*see* **Note 8**).

6. Aspirate off the supernatant and resuspend the pelleted macrophages in the appropriate volume (20 µL per 200,000 cells) of nucleofector SF transfection solution containing the transfection supplement. This solution is toxic to the cells, so once the cells are resuspended, additional transfection steps should be performed rapidly with the goal of returning the cells to macrophage culture medium as soon as feasible.

3.3 RNAi: Macrophage Transfection and Cell Recovery

1. Aliquot 2 µL of siRNA to each well of a 96-well round-bottom plate (*see* **Note 9** about siRNA dose). This step can be performed, while the RAW264.7 cells are being centrifuged (Subheading 3.2, **step 5**).

2. Add 20 µL of the resuspended cells to each well containing siRNA (this can be facilitated by using a multichannel or repeat pipettor). Mix gently by pipetting up and down. Transfer 20 µL of this mixture into the 96-well nucleofection plate. Avoid pipetting air bubbles to prevent electroporation errors.

3. Place covered nucleofection plate in the 96-well nucleofector shuttle, and press the *upload and start* button to initiate transfection. Note and censor any wells that have a transfection error (as reported by the software during transfection); if the solutions are prepared correctly and no bubbles are present, there should rarely if ever be errors.

4. Add 80 µL of prewarmed RPMI-1640 to each well (*see* **Note 10**); be sure to add the liquid gently as the cells are very fragile after electroporation. Allow the cells to recover at 37 °C in 5% CO_2 for 2 min.

5. Transfer the cells to a fresh 96-well tissue culture plate; add an additional 100 μL of macrophage culture medium. Once the macrophages have attached to the plastic several hours later, the medium can be removed and fresh macrophage culture medium added (although in our experience, this is not necessary). Incubate macrophages in 96-well tissue culture plate until time of analysis (24–36 h after transfection).

3.4 RNAi: Follow-up Studies

1. As outlined in other chapters in this volume, macrophages perform numerous immune functions that can be monitored following siRNA delivery. In other cell types, such studies are typically performed 48–72 h after siRNA delivery to allow time for the endogenous mRNA to be degraded. In our experience in mouse macrophages, we find that the optimal time follow-up study is much earlier, 24–36 h post-transfection. This timing can be further optimized on a gene by gene basis.

2. A variety of controls should be performed to validate these RNAi studies. siRNA-mediated gene knockdown should be quantified at either the mRNA level (by isolating RNA from the adherent cells and monitoring gene knockdown by qPCR) or even better at the protein level (by western blot or flow cytometry). Knockdown varies from siRNA to siRNA and is not as robust as in other cell lines, but 70–90% knockdown for most genes should be achievable. A nontargeting siRNA or other negative control siRNA should be used as a control, because the electroporation can affect the cells. Transfection efficiency should be monitored using a fluorescently labeled siRNA in conjunction with flow cytometry. We routinely obtain close to 100% transfection efficiency with this technique.

3. Cell viability should also be monitored for two reasons. First, improper transfection technique (in particular transfecting unhealthy cells or being too rough with the cells) can lead to significant cell death. Second, inhibition of essential target genes could alter cell viability. A variety of kits are available commercially to monitor cell viability.

4. Optimization considerations include modifying the dose of siRNA (we typically start with a high dose, and then titrate the dose down if we identify siRNAs that induce a phenotype), timing after transfection for assay, and health and passage number of cells. To avoid off-target siRNA effects [10], various validation steps should be performed to confirm any siRNA-induced phenotypes observed. These include using multiple siRNA duplexes targeting the same gene to test if multiple duplexes induce the same phenotype, titrating the siRNA to the lowest dose that induces a phenotype, monitoring knockdown efficiency, and ultimately, rescuing the induced phenotype by reintroducing a transgene that cannot

be targeted by the siRNA. Identifying a gene overexpression-induced phenotype (*see* Subheading 3.5) that is opposite to that induced by RNAi also strengthens the argument that the effect of the siRNA is specific.

5. To illustrate one application of siRNA-mediated gene inhibition, we knocked down genes in the TLR4 signaling pathway that mediate the response to lipopolysaccharide (LPS) from gram-negative bacteria. In wild-type macrophages, LPS induces significant inflammatory cytokine production; mutation or inhibition of genes in the TLR4 signaling pathway greatly weakens this response. In this sample experiment, 24 h after siRNA transfection the macrophage culture medium was replaced with macrophage culture medium containing 20 ng/mL LPS. Six hours later, the cell culture supernatant was collected for analysis of pro-inflammatory cytokine production by ELISA (Fig. 1).

3.5 Gene Overexpression: Macrophage Transfection

1. Maintain RAW264.7 cells in macrophage culture medium at 37 °C and 5% CO_2. Grow cells to approximately 70–80% confluence. Prepare enough cells to transfect the planned number of wells (200,000 cells per well).

2. Prewarm to 37 °C macrophage culture medium and Trypsin/EDTA.

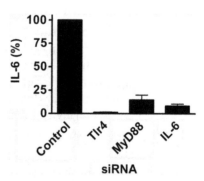

Fig. 1 siRNA-mediated inhibition of genes in the TLR4 signaling pathway inhibits LPS-induced pro-inflammatory cytokine production. Pools of 4 siRNA duplexes (Dharmacon) targeting the indicated genes or a pool of control siRNA duplexes that did not target any mouse gene (nontargeting control siRNA, Dharmacon) were transfected into RAW264.7 macrophages. After 24 h, 20 ng/mL LPS was added, and 6 h later, IL-6 release into the supernatant was monitored by ELISA. Inhibition of TLR4 (the LPS receptor), MyD88 (a downstream signaling adaptor), or IL-6 itself with siRNA all greatly weakened LPS-induced IL-6 production without altering cell viability (not shown)

3. To collect macrophages for transfection, aspirate off macrophage culture medium from plates containing adherent RAW264.7 cells and wash with PBS. Aspirate the PBS wash, and then add prewarmed Trypsin-EDTA to the cells (1 mL Trypsin-EDTA is sufficient to cover the cells on a 10 cm cell culture dish). Incubate cells for 2–3 min with occasional plate agitation. Do not over-trypsinize as this can damage the macrophages and diminish transfection efficiency. Once the cells are detached, add macrophage culture medium (9 mL for a 10 cm culture dish) and pipette up and down to finish detaching macrophages from the plate.

4. Count the macrophages using a hemacytometer. Aliquot 200,000 cells/well of a 24-well plate and bring total volume of macrophage culture medium in each well to 1 mL. Incubate macrophages in 24-well format overnight at 37 °C and 5% CO_2.

5. Prewarm cell OptiMem, macrophage culture medium, and FuGene HD to room temperature.

6. Mix DNA (300 ng–1 μg total), FuGene HD (3.75 μL), and OptiMem to a total volume of 100 μL. Incubate for 15 min at room temperature.

7. While incubating DNA and FuGene HD, wash the cells with 1 mL OptiMem and replenish with fresh macrophage culture media.

8. Add the entire mixture to one well of the 24-well plate containing RAW264.7 cells (*see* **Note 11**).

9. Incubate cells at 37 °C in 5% CO_2 for 24 h.

3.6 Gene Overexpression: Follow-up Assays

1. The design of follow-up studies should take into account the imperfect transfection efficiency of macrophages. If the overexpression plasmid contains a suitable selectable marker, then the appropriate antibiotic can be added 24–48 h after transfection to select stable lines. Alternatively, if the plasmid expresses a fluorescent protein, then flow cytometry can be used to sort macrophages carrying the transfected plasmid.

2. We have also co-transfected cytokine-luciferase reporters to monitor the effect of gene overexpression on an inflammatory readout. In these studies, we used a high ratio of overexpression plasmid to luciferase reporters to maximize the likelihood that the reporter will be present in cells also containing overexpression plasmid. We outline this procedure below.

3. We transfect 600 ng of the candidate gene overexpression plasmid, 300 ng of Elam firefly luciferase reporter (Elam-Luc), and 100 ng of an internal control gene Renilla luciferase reporter (SV40-rluc). Elam-Luc is an NFκB-dependent reporter; SV40-rluc was used to normalize/control for transfection efficiency between wells.

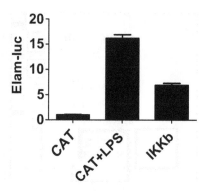

Fig. 2 IKKβ overexpression increases transcription of an NFκB-dependent luciferase reporter. Three plasmids expressing 300 ng Elam-Luc, 100 ng SV40-rluc, or 600 ng candidate gene overexpression were co-transfected into RAW264.7 macrophages in 24-well format. Elam-luciferase is an NFκB-dependent firefly luciferase reporter (gift of D. Golenbock). SV40-rluc is a Renilla luciferase reporter used for normalization to control for transfection efficiency (Promega). The candidate genes overexpressed include either the known NFκB activator HA-IKK2 (Addgene, [11]) or negative control Chloramphenicol Acetyltransferase (CAT) (Invitrogen). 24 h after transfection, cells were stimulated with 20 ng/mL LPS for 24 h or not as indicated. Luciferase activity was then monitored using the dual Luciferase assay kit (Promega). LPS treatment or IKKβ overexpression stimulated transcription of the NFκB-dependent reporter

4. Transfect and recover cells as described in Subheading 3.5 (*see* **Note 12**).

5. 24 h after transfection stimulate cells with 20 ng/mL LPS for 24 h and then monitor luciferase activity using a commercially available dual luciferase activity monitoring kit. We provide sample data for a gene that enhances NFκB reporter activity in Fig. 2.

4 Notes

1. A variety of approaches have been used to transfect macrophage cell lines. While lipid-mediated transfection can be used to deliver siRNA on a case-by-case basis, in our experience, electroporation is more consistently robust and more amenable to medium- and high-throughput approaches. The trade-off is that electroporation is more expensive than lipid-mediated transfection.

2. In addition to delivering siRNAs targeting candidate genes, several control siRNAs should be transfected in each experiment. These include fluorescently labeled siRNAs, which can be used to monitor transfection efficiency with flow cytometry or microscopy. Appropriate controls also include

siRNAs that do not target any particular gene or that target a gene known to not affect the phenotype of interest. This serves as a negative control. It is important to compare the results with candidate siRNAs to these negative control siRNAs instead of (or at least in addition to) untransfected control cells, because the mere act of transfecting the cells does alter them biologically. Additionally, positive control siRNAs known to affect the phenotype of interest should also be included if available.

3. There are numerous chemical transfection reagents available. In our experience, FuGene HD provides very efficient transfection of plasmids into macrophages (on the order of 50%). In contrast, we have not found FuGene HD to reliably transfect siRNAs into macrophages.

4. There are many commercial sources for plasmids driving cDNA expression for most mouse or human genes, all engineered with different promoters and selectable markers. We routinely use plasmids that drive gene expression with the strong CMV promoter, but other promoter-driven plasmids are available. Control plasmids should be used in these overexpression studies, including negative control and positive control plasmids that express genes that either don't or do affect the phenotype of interest, respectively. Transfection efficiency should also be monitored using plasmids expressing a fluorescent protein such as GFP in conjunction with flow cytometry or microscopy.

5. Transfection efficiencies above 50% or so are hard to obtain in macrophages. These luciferase reporters will be used to overcome this inefficiency.

6. We also have optimized siRNA-mediated transfection conditions for J774A.1, another commonly used mouse macrophage cell line. We treat this other macrophage cell line exactly the same as the RAW264.7 macrophages, except that program DS-130 should be used to transfect J774A.1 cells.

7. The key to obtaining good transfection is working with very healthy cells. Cells should not be allowed to overgrow, should be used at a relatively early passage number, should not be centrifuged too rapidly, and in general should be treated in a delicate fashion.

8. Don't centrifuge too fast, as this can damage the cells.

9. The exact siRNA dose should be optimized for each siRNA. As a good starting point, we use a relatively high siRNA dose in macrophages, 2 μL of a 20 mM siRNA solution or 40 pmol of siRNA per transfection.

10. The cells are very fragile after electroporation, so treat them gently. To avoid introducing a calcium spike that can damage

the cells at this stage, RPMI-1640 (which is low in calcium) rather than DMEM is added. Once the cells recover for a few minutes, DMEM can be added to the cells too.

11. The amount of DNA and FuGene HD used (and their relative ratio) may need to be titrated by the individual investigator.

12. LPS contamination of the plasmid preparations can confound these results. Be sure to use a Miniprep Kit for plasmid purification that removes LPS contamination, and use a control plasmid prepared with the same kit.

Acknowledgments

This work was funded by NIH grant R01ES025161 and the Wendy Siegel Fund for Leukemia and Cancer Research. Thanks to D. Golenbock for the Elam-Luc reporter construct.

References

1. Capecchi MR (2005) Gene targeting in mice: functional analysis of the mammalian genome for the twenty-first century. Nat Rev Genet 6(6):507–512

2. Taylor J, Woodcock S (2015) A perspective on the future of high-throughput RNAi screening: will CRISPR cut out the competition or can RNAi help guide the way? J Biomol Screen 20(8):1040–1051

3. Carralot JP, Kim TK, Lenseigne B, Boese AS, Sommer P, Genovesio A, Brodin P (2009) Automated high-throughput siRNA transfection in raw 264.7 macrophages: a case study for optimization procedure. J Biomol Screen 14(2):151–160

4. Lee G, Santat LA, Chang MS, Choi S (2009) RNAi methodologies for the functional study of signaling molecules. PLoS One 4(2):e4559

5. De Arras L, Alper S (2013) Limiting of the innate immune response by SF3A-dependent control of MyD88 alternative mRNA splicing. PLoS Genet 9(10):e1003855

6. De Arras L, Guthrie BS, Alper S (2014) Using RNA-interference to investigate the innate

immune response in mouse macrophages. J Vis Exp (93):e51306

7. De Arras L, Laws R, Leach SM, Pontis K, Freedman JH, Schwartz DA, Alper S (2014) Comparative genomics RNAi screen identifies Eftud2 as a novel regulator of innate immunity. Genetics 197(2):485–496

8. De Arras L, Seng A, Lackford B, Keikhaee MR, Bowerman B, Freedman JH, Schwartz DA, Alper S (2013) An evolutionarily conserved innate immunity protein interaction network. J Biol Chem 288(3):1967–1978

9. Burke B, Lewis CE (2002) The macrophage, 2nd edn. Oxford University Press, Oxford; New York

10. Editorial (2003) Whither RNAi? Nature cell biology. Nat Cell Biol 5(6):489–490

11. Nakano H, Shindo M, Sakon S, Nishinaka S, Mihara M, Yagita H, Okumura K (1998) Differential regulation of IkappaB kinase alpha and beta by two upstream kinases, NF-kappaB-inducing kinase and mitogen-activated protein kinase/ERK kinase kinase-1. Proc Natl Acad Sci U S A 95(7):3537–3542

Part III

Analysis of Lung Innate Immunity and Inflammation

Chapter 19

Assessment of Ozone-Induced Lung Injury in Mice

Vandy P. Stober and Stavros Garantziotis

Abstract

Ozone is a major pollutant in the air we breathe, and elevated levels lead to significant morbidity and mortality. As the climate warms, levels of ozone are predicted to increase. Accordingly, studies to assess the mechanisms of ozone-induced lung diseases are paramount. This chapter describes mouse models of ozone exposure and methods for assessing the effects of ozone in the lungs. These include bronchoalveolar lavage, necropsy, and measurement of lung function. Lavage allows for assessment of cell infiltration, cytokine production, tissue damage and capillary leakage in the airspaces. Necropsy provides tissue for gene expression, histology, and protein assessment in the whole lung. Lung physiology is used to assess airway hyperresponsiveness, tissue and total lung resistance, compliance, and elastance.

Key words Ozone, Bronchoalveolar lavage, flexiVent, Lung function measurements, Inflammation

1 Introduction

Many infectious and environmental exposures induce lung injury, and many methods have been developed to assess exposure-induced lung injury in mice. In this chapter, we describe two such methods: (1) collection of bronchoalveolar lavage fluid, which can be used to assess inflammatory cell infiltration and inflammatory cytokine production, and (2) airway physiology measurements using the flexiVent apparatus. To demonstrate the utility of these assays, we also describe methods to expose mice to a key environmental toxin, ozone.

1.1 Ozone as an Important Environmental Toxin

Since the Clean Air Act of 1970, ambient ozone (O_3) has been identified as a major pollutant. Lung injury by ozone leads to airway inflammation and airway hyperresponsiveness (AHR). As a result, ozone is an important trigger of lung disease exacerbations. Well-conducted epidemiology studies suggest that in the United States, ozone leads annually to 800 deaths, 4500 hospitalizations, 900,000 school absences, and over 1 million restricted activity days at a cost of $5 billion [1–5]. Each 10-ppb increase in 1-h daily

Scott Alper and William J. Janssen (eds.), *Lung Innate Immunity and Inflammation: Methods and Protocols*,
Methods in Molecular Biology, vol. 1809, https://doi.org/10.1007/978-1-4939-8570-8_19,
© Springer Science+Business Media, LLC, part of Springer Nature 2018

maximum ozone levels is associated with a mortality increase in heart-lung disease patients by 0.39–0.87% [2, 4, 6, 7]. In children, 27,000 hospital admissions and 19,000 emergency room visits for asthma were attributed to ozone in 2005 [8]. The US Environmental Protection Agency (EPA) states that due to climate change, ozone levels and effects will increase substantially [9]. It is thus imperative to understand the mechanism of ozone-induced lung disease.

Mouse models of ozone-induced lung inflammation have aided us substantially in understanding the pathobiology of lung after ozone exposure. In the sections below, we describe the methods for ozone exposure; invasive measurement of airway physiology parameters in intubated, sedated, and paralyzed mice; bronchoalveolar lavage; and necropsy.

1.2 Bronchoalveolar Lavage

Bronchoalveolar lavage is widely used to study cellular and humoral components of the epithelial lining fluid in both humans and animals. Advantages of this procedure include the ability to reliably assess indices of inflammation and epithelial damage after lung injury, ease of performance, and high reproducibility. The main disadvantage is that the procedure is terminal in small rodents. Additionally, it should be noted that changes in lavage fluid do not directly reflect interstitial cellularity or cytokine expression. Lavage techniques vary considerably, and there is a paucity of studies comparing different techniques. Our unpublished observations, as well as personal communications by other investigators, suggest that euthanasia agents and lavage methodology subtly influence results. This may be important, particularly in low-grade injury models such as ozone-induced injury, described below. While performing lavage, it is crucial to avoid over- and underinflation of the lung, as both will lead to unreliable samples. Underinflation of the lungs during lavage leads to nonrepresentative sampling of the airspace. Conversely, overinflation of the lungs can lead to injury, artificial increases in cytokine levels, and emphysema-like changes in histology [10]. We believe these risks are best avoided by allowing for passive inflation of the lungs at a standard pressure equivalent to 20 cm H_2O pressure. In the interest of best utilization of tissue for studies, we provide a protocol for lavage, with subsequent snap freezing in of the right lung liquid N_2 for genomic or proteomic analyses, and paraformaldehyde fixation of the left lung for histologic assessment.

1.3 Airway Physiology Measurements

Measurement of airway physiology parameters is crucial for the assessing how environmental insults affect lung function. Several methods have been described [11], ranging from whole-body plethysmography in unrestrained animals to invasive measurement of airway pressures in anesthetized, paralyzed, and intubated animals. There are advantages and disadvantages to all methods, and the ideal approach depends on the purpose of the measurement. We have found the invasive measurement of airway pressures to be a reasonably reproducible and reliable method of obtaining relevant information on airways obstruction, airways hyperresponsiveness (AHR), and lung tissue compliance. This chapter does not intend to endorse a specific product. However, at present all airway physiology assessment methods are specific to available apparatuses. In the following we describe the use of a widely accepted and used product, flexiVent. flexiVent measures respiratory mechanics by using volume and pressure measurements to assess lung function. The equipment has transducers on both the inspiratory and expiratory lines that record data as the ventilator cylinder delivers measured volume or pressure perturbations to the animal's lungs. In brief, the anesthetized animal is attached to the machine by a cannula inserted in the trachea. The animal is anesthetized, paralyzed, and ventilated, and when stable breathing is attained, the lungs are assessed using the chosen pressure or volume perturbations. Prior to a perturbation, an airway constrictive agent (like methacholine) can be delivered through a nebulizer attached to the inspiratory lines, in order to evaluate airway responsiveness.

2 Materials

2.1 Ozone Exposure

1. Cork with hole.
2. Erlenmeyer flask, vented.
3. Humidity/temperature meter and probe.
4. 55-L Hinners-style exposure chamber [12].
5. Luer connectors.
6. Manometer.
7. Picopure distilled water.
8. Oxygen bottle (100%) with regulator and connector.
9. Stainless steel cage to house mice during exposure, custom-made.
10. Tape.
11. PVC Tubing (3/8 in. and 1/4 in.).
12. Teledyne 400A ozone analyzer (Teledyne API, San Diego, CA).
13. Ozone generator, Yanco model (Matheson Gas, Montgomeryville, PA).
14. Flow meter.

2.2 Bronchoalveolar Lavage

1. PVC tubing (1.27 mm).
2. 70% Ethyl alcohol.
3. Butterfly needles (any gauge).
4. Needles (21, 23 gauge).

5. Tuberculin syringe (if blood sampling is desirable).

6. Forceps (3″ sharp-tip; 3″ serrated).

7. Scissors (large, dissecting).

8. Hemostat.

9. 10 mL syringe.

10. Three-way stopcock.

11. Braided silk suture material.

12. Liquid N_2 with appropriate container.

13. Ice.

14. 4% paraformaldehyde.

15. Phosphate-buffered saline (PBS).

2.3 Airway Physiology Measurements: Equipment

1. Computer and monitor (minimum requirements: 2.8 GHz quad core, 8 GB RAM, 1 TB hard drive, Windows 7).

2. flexiVent Legacy System (Scireq respiratory equipment, Legacy System no longer available from Scireq, but newer systems are) including flexiVent base unit, flexiVent EC controller unit, flexiVent XC accessories controller, flexiVent module M1, Scireq FV-AN-A1 aerosol base, PEEP trap, or flexiVent FX with Module 1 or 2.

3. Scireq Aeroneb plug-in.

4. Scireq Aeroneb plug-in adapter cord.

5. Aeroneb pro aerosol cup.

6. PVC tubing (3/8 in.).

7. Luer connector kit (Harvard apparatus).

8. FlexiWare 7.6 software.

9. Manometer (Scireq).

10. Heating pad (42 °C).

2.4 Airway Physiology Measurements: Other Materials

1. Ethanol, 70%.

2. Methacholine (acetyl B-methacholine chloride).

3. Pancuronium bromide.

4. Saline, 0.9%.

5. Urethane.

6. Cannulae 18–20 ga.

7. Compressed air.

8. Forceps (3″ sharp-tip; 3″ serrated).

9. Kim wipes.

10. P200 pipette with tips.

11. Scissors (spring and dissecting).

12. Suture thread 4.0.

13. Syringes 1 cc with 26 gauge needles.

2.5 Airway Physiology Measurements: Stock Solutions

1. Anesthesia: Urethane (125 mg/mL stock solution)—add 12.5 g urethane into 100 mL ddH$_2$O. Store protected from light at room temperature.

2. Paralytic: Pancuronium bromide (8 mg/mL stock solution)—add 50 mg pancuronium bromide to 6.25 mL 0.9% saline and store protected from light at −4 °C. Working solution is 0.08 mg/mL, a 1:100 dilution of the stock solution.

3. Methalcholine (100 mg/mL stock solution, prepared fresh each time): add 1 g of methacholine to 10 mL 0.9% saline. Working solutions are 100, 50, 25, and 12.5 mg/mL. Dose range can be changed to suit experimental setup.

3 Methods

3.1 Ozone Exposure

Ozone is a highly reactive gas, and exposure can be hazardous. Thus, it is important to avoid exposure of the operator. This is best achieved by placing the setup in a ventilated fume hood. Proper and undisturbed circulation of air in the exposure chamber is crucial in order to generate a uniform distribution of ozone throughout the chamber. We recommend placing the ozone monitoring tube in different positions within the chamber to ensure that ozone concentrations throughout the chamber are uniform.

1. Set up the ozone generator inside a fume hood using PVC tubing and Luer connectors as per Fig. 1.

2. Turn on the ozone generator and oxygen flow, which will direct 100% oxygen gas through an ultraviolet (UV) light ozone generator upstream from the exposure chamber.

3. Turn on air supply, and run it through a water source to maintain the relative humidity in the chamber (*see* **Note 1**).

4. Ensure the chamber is "pre-charged" to 2 ppm ozone.

5. Place mice individually in stainless steel cages and place cages in chamber (*see* **Note 2**).

6. Record the ozone reading, and adjust the air and ozone flow until the detector reads 2 ppm ozone (*see* **Note 3**).

7. Continue exposure for desired length of time (for AHR models: 3 h).

3.2 Bronchoalveolar Lavage

1. Euthanize mouse. This is best achieved pharmacologically through intraperitoneal pentobarbital injection (*see* **Note 4**).

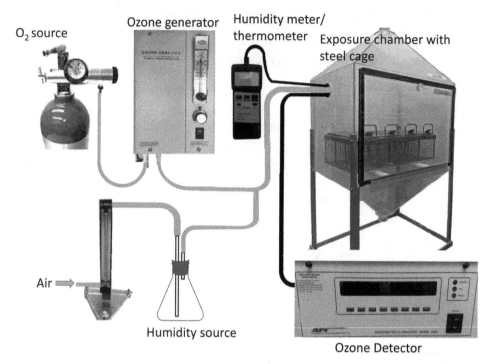

Fig. 1 Ozone exposure setup. 100% oxygen is conducted through PVC tubing into the ozone generator, which contains a UV bulb. Ozone is then routed through PVC tubing into the exposure chamber in a mixed stream of air that has been humidified. Humidification can be simply achieved by bubbling the air (not the ozone!) through purified water. Humidification to a reading of approximately 50–60% should be targeted. The ozone level is sampled in the chamber by the ozone detector, which has its exhaust port attached to a carbon filter and vented to the room

2. Soak fur with 70% ethanol to ensure that it is matted and does not interfere with necropsy.

3. Perform a cross-shaped incision in the abdominal area, from the suprapubic area to the subxiphoid area, with cross snips in both directions in the mid-abdomen area. Gently push bowel loops aside and identify inferior vena cava (IVC).

4. If blood collection is desired, this can be now performed using a tuberculin (1 mL) syringe and 23-gauge needle (*see* **Note 5**). If no blood collection is desired, snip the IVC to decompress the vasculature.

5. Use scissors to thoroughly sever the diaphragm, aorta, and esophagus. The lungs are now collapsed, thus allowing safer handling of the thorax.

6. Use scissors to perform median sternotomy from the xiphoid process through the tip of the sternum. Carefully expand section through the skin of the neck until the mandible is reached.

7. Open the two exposed thorax flaps, and cut through the ribs on either side of the sternum, approximately midway between

sternum and anterior axillary line to expose the mediastinum (heart, vessels, lungs).

8. Bluntly dissect the salivary glands and musculature anterior to the trachea using forceps.

9. Thread forceps under the trachea, bluntly expand the area under the trachea, and sever the esophagus using dissecting scissors (*see* **Note 6**). Visualize trachea from thyroid cartilage to carina.

10. Thread a suture (approx. 3″ long) under the trachea.

11. Make small incision between two tracheal rings in the upper third of the exposed tracheal segment (*see* **Note 7**).

12. Cut PVC tube (approx. 1″ long) to create a bevel at approximately a 45° angle to the axis of the tube. Connect this to butterfly tubing, from which you have previously removed the needle. Connect butterfly/PVC set-up to a plunger-free 10 mL syringe, which is secured vertically on a stand (Fig. 2). Flush tubing. A three-way stopcock ensures flow control (*see* **Note 8**).

13. Insert PVC tube bevel down into the trachea, such that the tip of the tube gently lifts the tracheal tissue and enables easy advancement into the trachea (*see* **Note 9**). Tie in place with a simple knot.

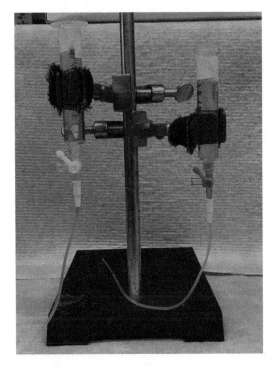

Fig. 2 Mouse lung lavage setup. 10 mL syringes are held upright and connected to the cannula via a three-way stopcock (shown in locked position)

14. Measure the height of the PBS and paraformaldehyde fluid columns. These should be 20 cm above the trachea. Adjust if necessary.

15. Move stopcock to the "open" position to start lavaging the lungs with PBS. All lobes should inflate (*see* **Note 10**).

16. When the lungs have stopped inflating, disconnect butterfly tubing from 10 mL syringe/stopcock setup, and flip tubing around so that it can drain into your sampling tube. When done, pinch tubing to prevent air from entering, and reconnect to syringe/stopcock. Repeat lavage process as needed (*see* **Note 11**).

17. BAL sample can be used to measure: (1) cell counts using a hemacytometer, cytokines by ELISA, and tissue damage/leakage via total protein content.

18. Flush the pulmonary vasculature to remove the blood. Grasp right ventricle with serrated forceps, and gently provide traction. Insert a 21-gauge needle into the right ventricle (bevel pointing downward) (*see* **Note 12**). Flush with approximately 3–6 mL saline. The lungs should turn from pink-red to tan-white.

19. Grasp right lung lobes with forceps, and use hemostat to pinch the lung off at the hilum (*see* **Note 13**).

20. With dissecting scissors, cut right lung lobes close to the hemostat, place in cryo-resistant tubes and flash-freeze in liquid nitrogen. The flash-frozen tissue can be used to analyze gene expression, protein expression, and extracellular matrix composition.

21. Remove saline-containing butterfly tubing from PVC, and attach paraformaldehyde-containing tubing.

22. Open stopcock and allow the left lung to passively expand.

23. When the lung is fully expanded, stop flow at stopcock, and remove PVC tubing from the trachea while holding the suture threads under tension (*see* **Note 14**).

24. If lung histology is to be performed, sever the trachea above the suture, lift the trachea by the suture thread, and carefully dissect the left lung and attached heart from the posterior mediastinum. Place into sample container with 4% paraformaldehyde (*see* Chapter 20 for lung histologic methods).

3.3 Airway Physiology Measurement

3.3.1 Study Definition and Planning

1. Scireq provides detailed manuals and technical notes about the flexiVent on their website; these manuals can be accessed by requesting an account on their website. Read both the flexiVent manual and the flexiWare manual. Scireq also has a Jove Video for instruction on using the FX flexiVent [13].

2. Scireq provides premade scripts for different pulmonary assessments. These can be altered, or new scripts can be written in the Script Editor program.

3. flexiVent monitors four main types of computer-controlled pressure or volume waveforms that are described in-depth in the flexiVent manual. In brief, the primary perturbations are TLC, Snapshot, Prime waves (Quick Prime), and pressure-volume (PV) loops. TLC is total lung capacity. Snapshot is a volume-based perturbation that assesses total lung resistance and compliance. Quick Prime pressure perturbation allows for differentiation of tissue and airway contributions to lung resistance and compliance. Pressure-volume loops are perturbations that assess total lung function. The standard perturbations are listed in Table 1.

4. For airway hyperresponsiveness monitoring, Quick Prime and Snapshot with TLC should be used between methacholine doses to fully inflate the lungs.

3.3.2 Experimental Methods

1. Ensure apparatus is fully assembled (*see* **Note 15**).

2. For legacy model: fill PEEP trap with deionized water. Set interior tube 2–3 cm below water.

3. Weigh the experimental animal.

4. Follow program prompts for calibration.

5. Anesthetize experimental animal by injecting 1–2 g/kg urethane intraperitoneally.

6. Use a toe pinch to ensure the mouse is fully anesthetized, and secure it to the heating pad using lab tape.

7. Clean and wet fur of throat area with 70% EtOH.

Table 1
Standard perturbations on flexiVent

Name	Purpose	Description
TLC	Open airspaces Standardize volume	Deep inflation of lungs to a pressure of 30 cm H_2O followed by a breath hold of a few seconds
Snapshot	Resistance and compliance data	Single frequency, sinusoidal forced oscillation waveform matched to respiratory rate
Quick Prime	Measure input impedance Distinguish between airways and tissues	Broadband (multifrequency) forced oscillation waveforms for fixed duration
PV loop	Assess nonlinearities in PV loops Measure quasi-static compliance	Slow stepwise or continuous inflation to total lung compliance (TLC) and deflation back to functional residual capacity (FRC), controlling either volume or pressure

8. With serrated forceps, lift the skin and make a vertical cut no lower than the clavicles up to larynx, exposing salivary glands.

9. To visualize the muscles around the trachea, bluntly separate salivary glands.

10. Using spring scissors and forceps, lift the muscle, and part it to expose the trachea (*see* **Note 16**).

11. Using sharp forceps, slide their tip around and under the trachea between the esophagus. Spread forceps, grasp a 2–4 in. piece of suture, and pull it under the trachea.

12. Make a horizontal incision between the second and third tracheal rings below the larynx, and insert the cannula, which is tied in place with suture thread (*see* **Note 17**).

13. Attach mouse to the flexiVent "Y" tubing via the cannula (*see* **Note 18**).

14. Turn on default ventilation.

15. Ensure that the mouse is at the same height and in a straight line from the "Y" tubing to the mouse's lungs to ensure good ventilation and no blockage of cannula or trachea. (Fig. 3).

16. Run a single TLC (deep inhale).

17. Inject paralytic, 0.08 mg/mL pancuronium bromide intraperitoneally. This will stop the mouse from breathing on its own; breathing will now be controlled by the ventilator.

18. Observe the trace of Pcyl until the trace is a smooth cyclic trace, thus assuring that the paralytic has taken effect (Fig. 4).

19. Activate desired script, and follow Software prompts for methacholine nebulization and measurements.

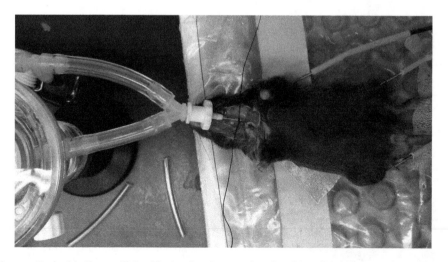

Fig. 3 Mouse attached to the ventilator. The trachea is cannulated and in proper alignment with "Y" tubing

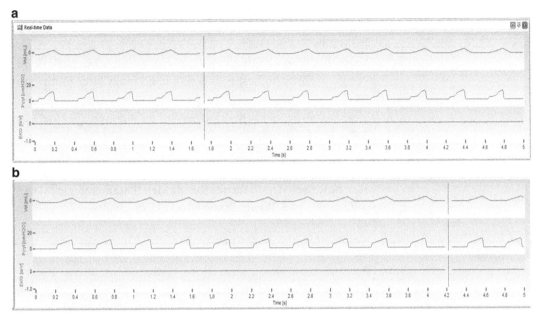

Fig. 4 Real-time data trace in flexiWare software. (**a**) The Pcyl trace shows pressure fluctuations due to the incompletely paralyzed mouse breathing against the ventilator (arrows). (**b**) Pcyl trace of a fully anesthetized and paralyzed mouse. Note the smooth, straight pressure curve rise

20. After the final dose, change the fluid in the nebulizer to PBS, and activate the nebulizer two times to clean the high-dose solution out of the tubing.

21. Stop the ventilation by pushing the stop ventilation button on the Task bar.

22. Remove the nebulizer from the stand and disconnect the Y tubing from the flexiVent.

23. Blow out all liquid in it with compressed air, dry, and reattach everything.

24. Assign next animal to the active site, and follow Software prompts to enter information about next experimental animal.

25. Clean out the cannula by blowing compressed air through it.

26. Reattach cannula to "Y" tubing, and repeat tube calibration.

27. Repeat process until all animals have been assayed.

3.3.3 Data Analysis

1. Review data by choosing "Review and Reporting" from welcome screen. Select an experiment to review. Data can be directly analyzed in this program; however, to enter numbers into other charting software or graphing programs, you will need to export the data, which puts it in excel format. To export data, click "Data" and then "export" from the toolbar. Export "Parameters" (the raw data) and "Dose Response" (a precalculated maximum dose response). Every time you export

data, the software will use the same file names. To avoid problems, rename the files with the experimental identifier as soon as they are exported.

2. Determine if you want to analyze data as a specific peak point, average of three highest points or a fixed point after methacholine dose.

3. Data can also be normalized as % baseline average for each parameter if desired.

4. Data quality can be assessed by examining the coefficient of determination (COD) for each parameter read. The COD indicates how well the data fit the model. Parameter reads with COD closer to 1.00 are considered better quality.

4   Notes

1. Chamber air should be exchanged at the rate of 20 changes/h, with ~50% relative humidity and a temperature of 20–25 °C.

2. This will cause a temporary drop in ozone concentrations in the chamber. Start timing the exposure when the levels have stabilized back at 2 ppm ozone.

3. Ozone level should stabilize with a fluctuation of ±0.05 ppm.

4. We find that pentobarbital causes much less vascular congestion and alveolar hemorrhage than euthanasia via CO_2 or cervical dislocation and is less likely to alter results due to the presence of the blood and exudative immune cells in the lung parenchyma.

5. Larger-bore needles may lead to IVC perforation; smaller-bore needles may lead to hemolysis. We prefer IVC sampling to cardiac sampling, if possible, as this approach does not include puncturing the heart tissue and activating tissue factor.

6. The natural aperture of the forceps keeps the trachea under mild tension and allows for more precise handling.

7. Scissors can be used for this activity, but we prefer using the tip of a 19 gauge needle, which acts as a scalpel and prevents severing the trachea.

8. Placing the stopcock lever at 45° to the outlets ensures that there is no flow and no air enters the tubing.

9. Do not advance beyond the carina, as this is likely to severe the airway.

10. If a lung lobe is not inflating, try repositioning the PVC tube (usually pulling back will resolve this issue). Note that PBS may interfere with some assays such as flow sorting of cells. Some investigators prefer to use other solutions such as Hanks'

Balanced Salt Solution or add 0.5% EDTA or 2% BSA or FBS to their PBS for this reason.

11. Usually three repetitions ensure a return of approximately 2–2.5 mL (for a 20–25 g mouse). If a more concentrated lavage fluid is desired, instead of draining to collection tube after the first pass, use tuberculin syringe to repeatedly draw and slowly re-instill fluid.

12. Flushing ensures that there are no artifacts from pooled blood elements in the pulmonary vasculature.

13. Some operators tie off the lung at the hilum to ensure there is no leakage of fixative, but this is not absolutely necessary. In our experience, the hemostat crushes the tissue such that the hilum remains collapsed and does not leak for inflation pressures of 20 cm H_2O pressure.

14. Thus, when PVC tube is removed, the suture knot ties the trachea off.

15. There are certain software-dependent steps required for data entry and storage. These are likely to change with Software upgrades and thus are not described here.

16. Do not cut too far down around the edges of the muscle around trachea because there is risk of cutting the carotid artery.

17. Ensure the cannula is placed so it only reaches the last tracheal ring.

18. Cannulae for flexiVent are available in several sizes ranging from 18 to 22 gauge. They also come ridged or smooth. We recommend obtaining endotracheal tube, 1.0 × 20 mm from Harvard apparatus since it has ridges for the suture to lock the cannula in place and reference lines to ensure proper placement.

Funding

This work was supported through funding from the Division of Intramural Research, National Institute of Environmental Health Sciences. Grant number 1ZIAES102605.

References

1. Dockery DW, Pope CA 3rd, Xu X, Spengler JD, Ware JH, Fay ME, Ferris BG Jr, Speizer FE (1993) An association between air pollution and mortality in six U.S. cities. N Engl J Med 329(24):1753–1759. https://doi.org/10.1056/NEJM199312093292401

2. Bell ML, McDermott A, Zeger SL, Samet JM, Dominici F (2004) Ozone and short-term mortality in 95 US urban communities, 1987-2000. JAMA 292(19):2372–2378. https://doi.org/10.1001/jama.292.19.2372

3. Gryparis A, Forsberg B, Katsouyanni K, Analitis A, Touloumi G, Schwartz J, Samoli E, Medina S, Anderson HR, Niciu EM, Wichmann HE, Kriz B, Kosnik M, Skorkovsky J, Vonk JM, Dortbudak Z (2004) Acute effects of ozone on

mortality from the "air pollution and health: a European approach" project. Am J Respir Crit Care Med 170(10):1080–1087. https://doi.org/10.1164/rccm.200403-333OC

4. Katsouyanni K, Zmirou D, Spix C, Sunyer J, Schouten JP, Ponka A, Anderson HR, Le Moullec Y, Wojtyniak B, Vigotti MA et al (1995) Short-term effects of air pollution on health: a European approach using epidemiological time-series data. The APHEA project: background, objectives, design. Eur Respir J 8(6):1030–1038

5. Hubbell BJ, Hallberg A, McCubbin DR, Post E (2005) Health-related benefits of attaining the 8-hr ozone standard. Environ Health Perspect 113(1):73–82

6. Ito K, De Leon SF, Lippmann M (2005) Associations between ozone and daily mortality: analysis and meta-analysis. Epidemiology 16(4):446–457

7. Levy JI, Chemerynski SM, Sarnat JA (2005) Ozone exposure and mortality: an empiric bayes metaregression analysis. Epidemiology 16(4):458–468

8. Fann N, Lamson AD, Anenberg SC, Wesson K, Risley D, Hubbell BJ (2012) Estimating the national public health burden associated with exposure to ambient PM2.5 and ozone. Risk Anal 32(1):81–95. https://doi.org/10.1111/j.1539-6924.2011.01630.x

9. Grambsch A; Hemming BL, CP. Weaver Gilliland A, Grano D, Hunt S, Johnson T, Loughlin D, Winner D (2009) Assessment of the Impacts of Global Change on Regional US Air Quality: a synthesis of climate change impacts on ground-level ozone. www.epa.gov. 2016

10. Cilley RE, Wang JY, Coran AG (1993) Lung injury produced by moderate lung overinflation in rats. J Pediatr Surg 28(3):488–493. discussion 494–485

11. Hoymann HG (2007) Invasive and noninvasive lung function measurements in rodents. J Pharmacol Toxicol Methods 55(1):16–26. https://doi.org/10.1016/j.vascn.2006.04.006

12. Hinners RG, Burkart JK, Punte CL (1968) Animal inhalation exposure chambers. Arch Environ Health 16(2):194–206

13. McGovern TK, Robichaud A, Fereydoonzad L, Schuessler TF, Martin JG (2013) Evaluation of respiratory system mechanics in mice using the forced oscillation technique. https://www.jove.com/video/50172/evaluation-respiratory-system-mechanics-mice-using-forced-oscillation. Accessed 24 July 2017

Chapter 20

Lung Histological Methods

Aneta Gandjeva and Rubin M. Tuder

Abstract

The lung is ideally suited to the application of histological methods to study its structure, cellular composition, and molecular characteristics of more than 30 types of cells. The key in these endeavors are proper tissue preservation/fixation, well-established protocols aimed at sectioning and staining, and understanding of lung morphology. Molecular studies can be performed in laser-captured cells and microscopic structures.

Key words H&E staining, Pentachrome staining, Formalin-fixed paraffin-embedded (FFPE), Laser capture microdissection (LCM), Immunohistochemistry (IHC), PCR array pathway-focused gene expression

1   Introduction

Lung histology forms the foundation of analysis of pulmonary structure. Given that the lung is organized in a fractal pattern of the parallel airway and circulatory branching structures, containing more than 30 different cell types, histological analyses provide key data in both the clinical and experimental setting.

In the context of the normal lung, seminal studies performed by Weibel and Gomez more than 50 years ago, shed light on how critical are the structural characteristics of the mammalian lung [1]. These studies involved the stereological analysis of the human lung using histological sections. Based on approaches developed to study the composition of materials in complex mineral structures, Weibel and Gomez derived key information related to the number, volume, surface area, and length of lung structures from histological sections. The highlights of their data include the determination that the adult lung has about 300×10^6 alveoli, with an average size of approximately 240 µm in diameter; alveoli represent 60% of the total lung volume. The gas exchange area approximates that of a singles tennis court (80 m²), packed in approximately 5 L of volume. The average length of alveolar capillaries ranged between

Scott Alper and William J. Janssen (eds.), *Lung Innate Immunity and Inflammation: Methods and Protocols*,
Methods in Molecular Biology, vol. 1809, https://doi.org/10.1007/978-1-4939-8570-8_20,
© Springer Science+Business Media, LLC, part of Springer Nature 2018

8.2 and 13 μm; there are approximately 277 billion capillaries with an average diameter of 8 μm. The capillary exchange area is 10% lower than the alveolar area, accommodating 140 mL of blood.

These data are nicely complemented by the data from Crapo and collaborators [2]. They determined that the adult lung contains 29×10^9 type I cells, 37×10^9 type II cells, 68×10^9 endothelial cells, 84×10^9 interstitial cells, and 23×10^9 alveolar macrophages. The surface area of type I cells is about 5100 μm^2, while type II cells have an average surface area of only 180 μm^2. In comparison, the surface area of capillary endothelial cells is 1350 μm^2. Notably, endothelial cells cover only 27% of the alveolar exchange area (lined largely by type I cells); however due to relative differences in their size, there are 3.6 times more endothelial cells than type I cells.

Several histological-based methods lend themselves to the experimental interrogation of the human lung. These involve structural and functional assessments, largely based on gene expression profiling or protein expression. We review below some key methods applied to the histology of the normal lung.

2 Materials

Prepare all solutions using ultrapure water (prepared by purifying deionized water, to attain a sensitivity of 18 MΩ cm at 25 °C) and analytical grade reagents. Prepare and store all reagents at room temperature, and store all flammable reagents (ethanol, xylene, eosin, etc.) in flame-resistant cabinets. Diligently follow all waste disposal regulations when disposing of materials.

2.1 Tissue Processing

1. 10% Buffered formalin.
2. Ethanol (100%, 95% and 74% in deionized water).
3. Xylene.
4. Paraffin.

2.2 Hematoxylin and Eosin (H&E) Staining

All reagents are commercially available.

1. Xylene.
2. Ethanol (100%, 95%, 80%, and 75% in deionized water).
3. Eosin Y- Alcoholic.
4. Hematoxylin 7211.
5. Clarifier histological clearing agent.
6. Bluing Reagent—histological pH-controlled solution for bluing hematoxylin.
7. Deionized water.

2.3 Movat's Pentachrome Staining

1. Bouin's Mordant.
2. Alcian blue, 1%.
3. Alkaline alcohol.
4. Weigert's reagent.
5. Ferric chloride, 2%.
6. Sodium thiosulfate.
7. Crocein scarlet/acid fuchsin.
8. Phosphotungstic acid, 5%.
9. Acetic acid, 1%.
10. Ethanol, 95%.
11. Alcoholic saffron.

2.4 Immunohistochemistry Using HRP Conjugated Secondary Antibody

1. Antigen retrieval solution (citrate based, pH 6.0 or Tris based, pH 9.0).
2. Steamer or rice cooker (*see* **Note 1**).
3. Pap-pen.
4. Peroxidase—0.03% hydrogen peroxide containing sodium azide.
5. Primary antibody.
6. Isotype antibody for negative control to primary antibody.
7. Secondary antibody, horseradish peroxidase (HRP) conjugated.
8. DAB chromogen kit, containing DAB (3,3′-diaminobenzidine) chromogen solution, and DAB substrate.
9. PBS solution (phosphate-buffered saline).
10. TBST solution (Tris-buffered saline, with Tween 20).
11. Serum blocking solution. Add 1 mL 10% normal serum (should be the same species as the secondary antibody) to 4.5 mL 5% BSA (in PBS) and 4.5 mL superblock diluent (Thermo Fisher).

2.5 Immunohistochemistry Using Biotinylated Secondary Antibody

1. Antigen retrieval solution (citrate based, pH 6.0 or Tris based, pH 9.0).
2. Steamer or rice cooker (*see* **Note 1**).
3. Pap-pen.
4. Avidin blocking solution.
5. Biotin blocking solution.
6. Primary antibody.
7. Isotype antibody for negative control to primary antibody.
8. Biotinylated secondary antibody.

9. Streptavidin-horseradish peroxidase (SA-HRP).

10. DAB chromogen kit, containing DAB (3,3′-diaminobenzi-dine) chromogen solution, and DAB substrate.

11. PBS solution (phosphate-buffered saline).

12. TBST solution (Tris-buffered saline, with Tween 20).

13. Serum blocking solution. Add 1 mL 10% normal serum (should be the same species as the secondary antibody) to 4.5 mL 5% BSA (in PBS) and 4.5 mL superblock diluent (Thermo Fisher).

2.6 Laser Capture Microdissection (LCM)

1. Cell Cut Machine (Molecular Machines and Industries).

2. Cryostat.

3. OCT (Optimal Cutting Temperature) Compound.

4. Membrane slides—mmi slides RNAse-free.

5. Collecting tubes—0.5 mL with transparent caps and adhesive lids (Molecular Machine and Industries, MMI).

6. H&E MMI staining kit for LCM (Laser Capture Microdissection).

7. Crystal violet stain—Arcturus HistoGene Frozen Section Staining kit.

8. PicoPure RNA Isolation Kit (Arcturus, Molecular Devices).

9. Pathway-Focused, gene expression profiling PCR array (SABiosciences; Qiagen).

10. PicoPure DNA extraction kit (Arcturus, Molecular Device).

---

3 Methods

Carry out all procedures at room temperature unless otherwise specified.

3.1 Tissue Processing, Embedding, and Microtome Sectioning

3.1.1 Tissue Processing: Regular Overnight Processing

Tissue processing includes three major steps—dehydration, clearing, and infiltration. Fixative and the method of processing should be chosen before obtaining the tissue, and those preferences must be compliant with the requirements of any special technique. The dehydrating reagent in the protocol below is ethanol, and the clearing reagent is xylene. The purpose of the clearing reagent is to remove the alcohol used for dehydration and to make the tissue receptive to the infiltration medium: paraffin.

The following protocol can be used either manually or with an automated system; no vacuum or pressure is necessary to complete the process.

1. 10% buffered formalin for 30 min.

2. 70% ethanol for 30 min.

3. 95% ethanol—at room temperature 21 °C for 30 min.

4. 100% ethanol—at room temperature 21 °C for 30 min.

5. 100% ethanol—at room temperature 21 °C for 1 h.

6. 100% ethanol—at room temperature 21 °C for 1 h.

7. Xylene—at room temperature 21 °C for 30 min.

8. Xylene—at room temperature 21 °C for 30 min.

9. Xylene—at room temperature 21 °C for 1 h.

10. Paraffin—at 63 °C for 30 min.

11. Paraffin—at 63 °C for 30 min.

12. Paraffin—at 63 °C for 30 min.

13. Paraffin—at 63 °C for 30 min.

14. Embed all processed samples.

15. Microtome section the samples in a manner compliant with experimental needs.

3.2 H&E Staining

H&E (hematoxylin and eosin) is the most commonly used staining system for cell-type differentiation. Eosin is an acidic dye: it is negatively charged and therefore binds to basic (or acidophilic) structures. Most proteins in the cytoplasm are basic, and so eosin binds to these proteins and stains them pink. This includes cytoplasmic filaments in muscle cells, intracellular membranes, and extracellular fibers. Hematoxylin can be considered a basic dye and is used to stain acidic (or basophilic) structures a purplish blue. The nucleus, parts of the cytoplasm that contain RNA in ribosomes, and the rough endoplasmic reticulum are all acidic, and so hematoxylin binds to them and stains them purple.

The H&E stain can be automated easily; a wide variety of autostainers can expedite the staining process with good reproducibility. The following protocol can be performed manually or can be automated.

1. Deparaffinize the tissue by dunking the slide in xylene for 3 min. Do this a total of three times.

2. Rehydrate tissue (**steps 3–7**).

3. Submerge in 100% ethanol for 1 min.

4. Remove the slide and then re-submerge in 100% ethanol for 1 min.

5. Remove the slide and re-submerge in 100% ethanol for 30 s.

6. Submerge tissue in 80% ethanol for 30 s.

7. Submerge in deionized water for 30 s.

8. Stain with hematoxylin and eosin (**steps 9–16**).

9. Submerge the tissue in hematoxylin for 4 min.

10. Wash in water for 1 min.

11. Submerge in clarifier for 20 s.

12. Wash in water for 1 min.

13. Bluing solution for 30 s.

14. Wash in water for 30 s.

15. Submerge in ethanol 95% for 30 s.

16. Submerge in eosin for 15 s.

17. Dehydrate tissue (**steps 18–25**).

18. Ethanol 95%—15 s.

19. Ethanol 100%—15 s.

20. Ethanol 100%—15 s.

21. Ethanol 100%—30 s.

22. Ethanol 100%—30 s.

23. Xylene—30 s.

24. Xylene—30 s.

25. Xylene—30 s.

26. Coverslip.

27. Results: nuclei in blue and cytoplasm in pale pink (Fig. 1).

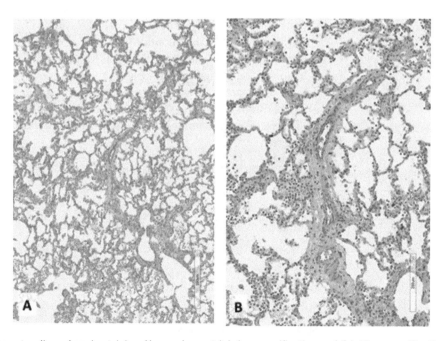

Fig. 1 Hematoxylin and eosin staining. Human lung at (**a**) 4× magnification and (**b**) 10× magnification

3.3 Movat's Pentachrome Staining

1. Deparaffinize the tissue (*see* Subheading 3.2, **step 1**).

2. Hydrate in dH$_2$O.

3. Preheat alkaline alcohol (in Coplin jar) in a 60 °C oven or water bath.

4. Submerge tissue in Bouin's mordant at 60 °C in plastic Coplin jar with lid (heat in microwave for 30 s), and then leave the slides for 30 min. Do this under a fume hood, and wear appropriate safety gear (lab coat and goggles)!

5. Cool 10 min at room temperature with open lid.

6. Wash in running water for 10 min.

7. Submerge tissue in 1% Alcian blue—25 min.

8. Wash in running water—5 min.

9. Submerge in alkaline alcohol at 60 °C for 10 min.

10. Wash in running water—5 min.

11. Submerge in Weigert's for 15 min.

12. Differentiate in 2% ferric chloride for—3 min.

13. Wash slides in water for 10 s.

14. Submerge in sodium thiosulfate for 1 min.

15. Wash slides in running water for 5 min.

16. Submerge in Crocein scarlet/acid fuchsin for 1.5 min.

17. Wash slides in running water for 5 min.

18. Submerge in 5% phosphotungstic acid for 5 min.

19. Transfer slides directly to 1% acetic acid for 5 min.

20. Wash slides in running water for 5 min.

21. Dehydrate slides quickly in 95% ethanol for 1 min, then two times in 100% ethanol for 1 min.

22. Submerge slides in alcoholic saffron for 60 min (reuse).

23. Wash slides in two changes of 100% ethanol, 1 min each.

24. Clear slides in xylene and mount.

25. Results: Your staining will show nuclei and elastic fibers in black, collagen in yellow, ground substance and mucin in blue, fibrinoid and fibrin in intense red, and muscle in red (Fig. 2).

3.4 Immunohistochemistry Using HRP Conjugated Secondary Antibody

1. Deparaffinize in xylene (three times, 3 min each) or use an autostainer.

2. Hydrate tissues in graded alcohols (ethanol 100%, 3× for 1 min; ethanol 80%, 1 min; and ethanol 75%, 30 s).

3. Hydrate in dH$_2$O.

4. Submerge slides in a plastic jar with antigen retrieval (citrate pH 6.0, Tris based pH 9.0, or EDTA, etc.) for 20 min boiling

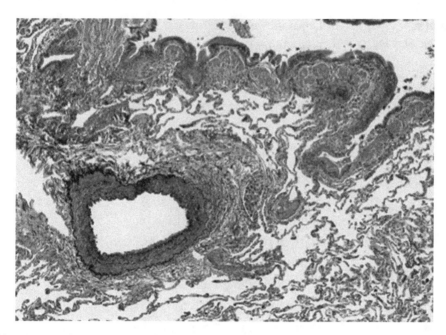

Fig. 2 Movat's pentachrome staining. Human lung at 4× magnification. Nuclei and elastic fibers appear in black, collagen in yellow, ground substance and mucin in blue. Muscle appears red, and fibrin is intense red

in the steamer, and then cool slides with an open lid outside the steamer for 15 min (*see* **Note 1**).

5. Rinse in 1× PBS and draw with a pap-pen around the tissue section.

6. Apply peroxidase—0.03% hydrogen peroxide containing sodium azide.

7. Block for 1 h at room temperature (21 °C) using serum blocking solution.

8. Rinse slides gently with TBST (do not direct water flow directly at specimen).

9. Apply primary antibody for 1 h at room temperature (21 °C) or overnight at 4 °C.

10. Apply isotype IgG antibody to control slide/tissue at the same time as primary antibody (*see* **Note 2**).

11. Rinse slides in 1× TBST.

12. Apply horseradish peroxidase (HRP) conjugated secondary antibody for 30 min at room temperature.

13. Rinse in 1× PBS.

14. Apply DAB chromogen (2 mL of DAB buffer +1 drop of substrate) for 90 s.

15. Rinse in deionized water.

16. Dehydrate tissue in water, alcohols, and xylene (*see* Subheading 3.2, **steps 17–25**).

17. Apply counterstain if desired (e.g., hematoxylin).

18. Coverslip and analyze.

3.5 Immuno-histochemistry Using Biotinylated Secondary Antibody

1. Deparaffinize and hydrate tissues (*see* Subheading 3.4, **steps 1–3**).

2. Preheat steamer during deparaffinization.

3. Submerge slides in plastic jar with antigen retrieval agent (*see* Subheading 3.4, **step 4**).

4. Rinse in 1× PBS and draw with a pap-pen around the tissue section.

5. Rinse in 1× PBS.

6. Apply Avidin block (enough to cover section) for 15 min at room temperature (21 °C).

7. Rinse in 1× PBS.

8. Apply Biotin block (enough to cover section) for 15 min at room temperature (21 °C).

9. Rinse slides in 1× PBS.

10. Apply serum blocking solution for 1 h at room temperature.

11. Do not rinse slides! Flick off excess buffer.

12. Apply primary antibody for 1 h at room temperature or overnight at 4 °C. Apply isotype IgG antibody to control tissue. The isotype serves as a negative control. It should be made from the same species at the primary antibody and added in the same concentration. The control tissue should be processed identically to other tissue specimens.

13. Rinse in 1× PBS.

14. Apply secondary antibody (biotinylated antibody that recognizes your primary antibody) for 30 min at room temperature.

15. Rinse in 1× PBS.

16. Apply RTU SA-HRP for 30 min room temperature.

17. Rinse in 1× PBS.

18. Apply DAB chromogen (2 mL of DAB buffer +1 drop of substrate) for 90 s at room temperature.

19. Rinse in deionized water.

20. Clear the tissue in alcohols (95% and 100% Ethanol) and xylene (*see* Subheading 3.2, **steps 17–25**).

21. Apply counterstain if desired (e.g., hematoxylin).

22. Coverslip and analyze.

3.6 Laser Capture Microdissection (LCM)

Laser capture microdissection (LCM) is a technology for isolating pure cell populations from a heterogeneous tissue specimen. It can precisely target and capture microscopic structures and cells of interest for a wide range of molecular expression assays.

LCM can be applied to a wide range of tissues, including frozen tissue and formalin-fixed, paraffin-embedded (FFPE) tissues. Frozen tissue offers excellent preservation of RNA, DNA, and proteins and is optimal for RNA, DNA, or protein analysis. However, frozen tissue lacks clear histological/morphological details, and the dissection process must be completed within 1 h. FFPE tissue is the standard for preservation of tissue morphology, and it will create cross-links between nucleic acids and proteins and between different proteins. This cross-linking interferes with recovery of RNA and protein from FFPE tissue and is not suitable for RNA or protein-based assays.

1. The microdissected cells can be used for a wide range of downstream analyses, such as qRT-PCR, cDNA microarrays, DNA sequencing, Western blots, RT^{2}Profiler PCR arrays for pathway-focused gene expression profiling, etc. The following protocol for microdissection is appropriate for RNA from fresh frozen tissue. Fresh tissue should be inflated with OCT, then embedded in OCT, and then frozen at −80 °C until use.

2. Warm frozen tissue to −20 °C for cryosectioning.

3. Cryosection the frozen tissue to a thickness of 4–5 μm using membrane slides designed for the LCM (MMI Cell Cut) machine. Use two sections per slide. Each slide should be handled separately to speed the process and to minimize RNA loss. Immediately start the dehydration.

4. 95% Ethanol—30 s.

5. 100% Ethanol—30 s.

6. 100% Ethanol—30 s.

7. Stain with crystal violet 30 s.

8. 100% Ethanol—30 s.

9. 100% Ethanol—30 s.

10. 100% Ethanol—30 s.

11. Xylene—2 min.

12. Xylene—2 min.

13. Xylene—until microdissection.

14. Sandwich the cryosectioned tissue between a standard glass slide and the MMI adhesive membrane slide (Fig. 3).

15. Use MMI Cell Cut Machine for microdissection, and choose desired target cells (Fig. 4).

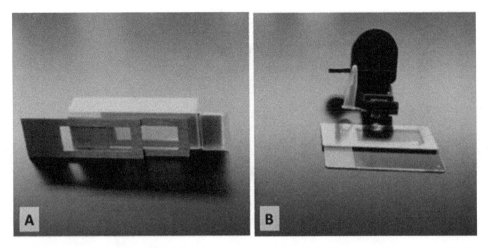

Fig. 3 Slides and caps for processing of cryosectioned material. (**a**) Adhesive membrane slides for MMI Cell Cut machine. (**b**) Normal glass slide underneath adhesive membrane slide with cryosectioned tissue sandwiched in between

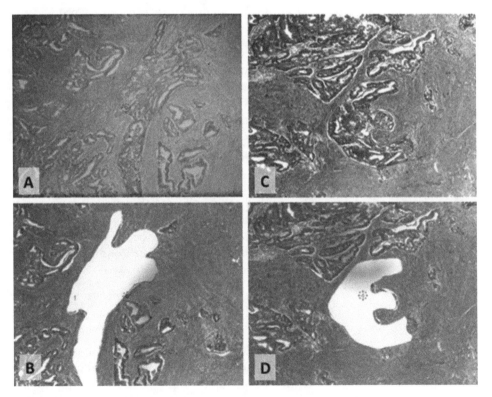

Fig. 4 Selection and removal of target structures. Human lung stained with hematoxylin and eosin. (**a, c**) Selection of targets, red line. (**b, d**) Isolation of selected cells structures from tissue with LCM

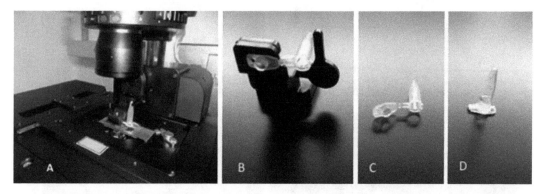

Fig. 5 Collection of microdissected cells. (**a**) Placement of collecting tube cap over tissue. (**b**) Tube holder for collecting dissected target cells and tube. (**c**) Target cells/tissue adhered to the tube cap. (**d**) Tube with adhered cells/tissue and extraction buffer for incubation

16. Collect the desired region from membrane slide in a 0.5 mL tube with adhesive lid, from MMI Cell Cut. Special supply for LCM (Fig. 5).

17. Proceed to RNA extraction using PicoPure RNA extraction kit.

18. Pipette 50 μL of extraction buffer provided in the kit into the microfuge tube with microdissected tissue (tissue is stuck to the adhesive lid of the collecting tube). Place the tube upside down, so the extraction buffer will cover the microdissected tissue/tube lid and incubate at 42 °C for 30 min. Briefly spin at 500 rpm (20 × g), and store at −80 °C until ready for RNA isolation.

3.7 RNA Isolation from LCM Dissected Tissue

The steps provided below use the Arcturus PicoPure RNA isolation kit (Thermo Fisher). Kits from other manufacturers may also be used, and in general, follow similar protocols. Irrespective of the kit, it is important that all steps below be performed on ice (unless otherwise indicated) and in RNAse-free conditions, to minimize RNA degradation (*see* **Note 3**).

1. Into the tube with tissue for RNA extraction (*see* Subheading 3.6, **steps 17** and **18**), add 100 μL of 70% ethanol to the lysate and mix thoroughly by pipetting.

2. Load the 200 μL of lysate onto RNA isolation columns, centrifuge at 1250 × g) for 15 s, and then change the speed to 5000 × g for 15 s.

3. Wash and dry the column, and elute RNA (10 μL) following the manufacturer's protocol.

4. Keep RNA on ice and process with PCR (Subheading 3.9) or freeze at −80 °C for future use.

3.8 DNA Extraction and Isolation from LCM Dissected Tissue

DNA can be extracted from tissue using a wide variety of preparation methods. Superior results can be obtained from formalin-fixed, paraffin-embedded tissue (FFPE) sections that are then LCM microdissected. The following procedure and protocol are designed for FFPE using the Arcturus PicoPure DNA extraction kit.

1. Microtome the FFPE tissue to a thickness of 4–5 μm using membrane slides designed for the LCM (MMI Cell Cut) machine. Use two sections per slide. Proceed with H&E stain either manually or using automated stainer.

2. Deparaffinize the tissue by dunking the slide in xylene for 3 min. Do this a total of three times.

3. Rehydrate tissue (see Subheading 3.2, step 2).

4. Submerge in 100% ethanol for 1 min.

5. Remove the slide and then re-submerge 100% ethanol for 1 min.

6. Remove the slide and re-submerge in 100% ethanol for 30 s.

7. Submerge tissue in 80% ethanol for 30 s.

8. Submerge in deionized water for 30 s.

9. Stain with hematoxylin and eosin (see Subheading 3.2, **steps 8**).

10. Submerge the tissue in hematoxylin for 4 min.

11. Wash in water for 1 min.

12. Submerge in clarifier for 20 s.

13. Wash in water for 1 min.

14. Add bluing reagent for 30 s.

15. Wash in water for 30 s.

16. Submerge in ethanol 95% for 30 s.

17. Submerge in eosin (alcoholic) for 15 s.

18. Dehydrate tissue (see Subheading 3.2, **steps 17**).

19. Ethanol 95%—15 s.

20. Ethanol 100%—15 s.

21. Ethanol 100%—15 s.

22. Ethanol 100%—30 s.

23. Ethanol 100%—30 s.

24. Xylene—30 s. Repeat for a total of three times.

25. Xylene—30 s.

26. Air-dry the slides with sections.

27. Proceed with laser capture microdissection (LCM) as described in Subheading 3.6.

28. Preheat incubator to 65 °C.

29. Pipette 155 μL of Reconstitution Buffer into one vial of Proteinase K and vortex to make the Extraction Solution. Then immediately place the solution on ice and use it as soon as possible.

30. Pipette 50 μL of the Extraction Solution into the LCM collection tube with laser microdissected tissue and invert the tube with the cap down and incubate at 65 °C for 16 h.

31. Remove the tube from incubator and centrifuge for 1 min at 1000 × *g* at room temperature.

32. Heat the extract to 95 °C for 10 min in a heating block to inactivate the Proteinase K. Cool the sample to room temperature. The samples are now ready for PCR analysis. No additional purification is required.

3.9 PCR Array Pathway-Focused Gene Expression Profiling

RT-PCR arrays are a reliable tool for analyzing the expression of a focused panel of genes. RT-PCR arrays are optimized primer assays for panels of pathway or disease-focused genes. PCR arrays can also be customized to contain a panel of genes designed for a specific area of research. Most of the PCR Arrays are very sensitive and require as little as 1 ng total RNA, and results can be obtained on any real-time PCR instrument.

1. Isolated RNA from LCM tissue (*see* Subheading 3.7) should be converted to cDNA by reverse transcription iScript cDNA Synthesis.

2. Prepare reaction mixture for cDNA synthesis by combining the following: 5× iScript reaction mix (4 μL), iScript reverse transcriptase (1 μL), nuclease-free water (10 μL), and RNA template (5 μL).

3. Incubate complete reaction mix, 5 min at 25 °C (room temperature), 45 min at 42 °C, 5 min at 85 °C, and then hold at −20 °C for no more than 1 week. cDNA is ready for downstream analyses.

4. One example for downstream analysis is to use the QIAGEN RT² PCR Array system. These 96-well plates contain primer assays for 84 pathway or disease-focused genes and five housekeeping genes. In addition, one well contains a genomic DNA control, three wells contain reverse transcription controls, and three wells contain positive PCR controls.

5. Loading components for the Array format 96-well RT² PCR reaction are 2× RT² SYBER GREEN (1350 μL), cDNA synthesis reaction (102 μL), and RNAse-free water (1248 μL) for a total volume of 2700 μL. Use 20 μL of this reaction mix in each well.

6. Perform PCR, and determine relative expression using data from the real-time cycler and the $\Delta\Delta CT$ method.

7. To analyze your results use SA Biosciences Pathway-Focused RT^2 Profiler™ PCR Array Service online tools. The GeneGlobe Data Analysis Center is a web resource for scientists analyzing their real-time PCR.

4 Notes

1. Antigen retrieval is used to restore antibody access to an epitope. Most of the methods utilize heat (from a variety of sources such as a microwave, pressure cooker, steamer, water bath, or autoclave) to unmask epitopes in conjunction with unmasking solutions (Citrate pH 6.0, Tris-basedpH 9.0, EDTA, etc.). The preferred technique for optimal retrieval is dependent on the tissue, the fixation method and/or primary antibody, and must be optimized for each staining.

2. The isotype antibody serves as a negative control. It should be made from the same species as the primary antibody and added in the same concentration. The control tissue should be processed identically to other tissue specimens, and if a positive tissue/control is presented, identical slides should be used for the negative control.

3. RNase contamination will cause RNA to degrade and may result in experimental failure. To minimize RNase contamination, clean all work surfaces with RNase decontamination solutions prior to starting RNA isolation steps. In addition, it is essential to wear disposable gloves and change them frequently to prevent the introduction of RNases from skin surfaces. After putting on gloves, avoid touching surfaces that may introduce RNases onto glove surfaces. Finally, only use new plasticware that is certified nucleic acid-free and nuclease-free, and only use new, sterile, RNase-free pipette tips and micro-centrifuge tubes. We prefer to use the PicoPure RNA Isolation Kit; however, kits from other manufacturers will also work. Note that substitution of reagents or kit components may affect yields or introduce RNases.

References

1. Weibel ER, Gomez DM (1962) Architecture of the human lung. Use of quantitative methods establishes fundamental relations between size and number of lung structures. Science 137:577–585

2. Crapo JD, Young SL, Fram EK, Pinkerton KE, Barry BE, Crapo RO (1983) Morphometric characteristics of cells in the alveolar region of mammalian lungs. Am Rev Respir Dis 128:S42–S46

Chapter 21

Intravital Microscopy in the Mouse Lung

Yimu Yang, Joseph A. Hippensteel, and Eric P. Schmidt

Abstract

While reductionist in vitro approaches have allowed for careful interrogation of cellular pathways that underlie innate immune responses, they often fail to capture the complex multicellular interactions characteristic of acute inflammation. Intravital microscopy, by directly observing alveolar cell-cell interactions, provides unique insight into the complex intercellular mechanisms responsible for alveolar inflammation. This review discusses multiple potential approaches to intravital pulmonary imaging, with specific attention to in vivo microscopy of the freely moving mouse lung.

Key words Microscopy, Intravital, Neutrophil, Inflammation

1 Introduction

Reflecting its continuous exposure to the external atmosphere, the lung demonstrates unique innate immune responses to external irritants and/or pathogens. While neutrophil extravasation from non-pulmonary vascular beds occurs by virtue of coordinated processes of neutrophil activation, rolling, and adhesion within postcapillary venules, alveolar neutrophil diapedesis occurs in a rolling-independent fashion within the pulmonary capillaries [1, 2]. However, this lung-specific, rolling-independent extravasation is difficult to model in vitro, and even the most sophisticated multicellular approaches [3] fail to capture the structural and spatial dynamics known to influence pulmonary neutrophil extravasation within the ventilating lung [1]. Furthermore, in vitro preparations may often lose critical lung structures (e.g., the endothelial glycocalyx) that regulate chemokine and adhesion molecule availability [4].

Accordingly, acute pulmonary inflammatory responses may be best observed via intravital microscopy, which allows real-time visualization of neutrophil adhesion and extravasation in a multicellular environment that maintains the structural characteristics of a moving lung. Pulmonary intravital microscopy may be performed via several different approaches [5], each of which has

Scott Alper and William J. Janssen (eds.), *Lung Innate Immunity and Inflammation: Methods and Protocols*,
Methods in Molecular Biology, vol. 1809, https://doi.org/10.1007/978-1-4939-8570-8_21,
© Springer Science+Business Media, LLC, part of Springer Nature 2018

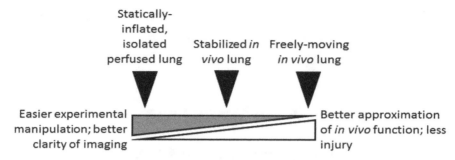

Fig. 1 Relative benefits of different in vivo microscopy approaches

specific advantages and disadvantages (Fig. 1). Intravital microscopy may be performed using an ex vivo, statically inflated isolated blood-perfused mouse lung [6]. This approach allows for precise manipulations of alveolar compartments using experimental agents that are either rapidly cleared by the systemic circulation (e.g., fluorescent reporters) or have systemic toxicity (e.g., cardiotoxic effects of gap junction inhibitors). However, the lack of ventilation of these lungs may influence signaling pathways (e.g., NO-cGMP signaling) relevant to lung inflammatory responses. Alternatively, intravital microscopy may be performed in vivo, using a lung immobilization device [7, 8]. This device stabilizes a small window of the peripheral lung, allowing for tracking of individual leukocytes as they patrol the lung vasculature. This local fixation, however, may potentially induce lung injury at the interface between the immobilized window and the neighboring lung, via shear forces arising from alveolar interdependence. Finally, other intravital approaches allow for imaging of a freely moving lung [9, 10]. While such an approach best approximates true in vivo lung function (while minimizing lung injury), imaging requires high-speed microscopy (ideally with gating to lung motion) and has difficulty tracking a single leukocyte.

The ideal intravital microscopy approach to lung imaging, therefore, is determined by the specific experimental question. In the following protocol, we describe the approach to imaging neutrophil dynamics within the freely moving in vivo lung, which provides real-time lung imaging while minimizing pulmonary manipulation. This approach serves as an optimization of previously published protocols measuring pulmonary endothelial surface layer thickness in vivo [9, 10].

2   Materials

1. Transparent polyvinylidene membrane.

2. 5 mm #1 circular glass coverslip.

3. Alpha-cyanoacrylate or optical glue.

4. PE50 tubing (inner diameter 0.58 mm, outer diameter 0.956 mm).

5. PE10 tubing (inner diameter 0.28 mm, outer diameter 0.61 mm).

6. Surgical equipment (dissection scissors, microdissection scissors, forceps, etc.).

7. 4-0 silk suture.

8. 23 gauge, 3/8 circle suture needle. (This may be approximated by manually bending a 23 gauge needle by 30°).

9. Metal tracheostomy catheter and mouse ventilator (Harvard Apparatus).

10. Syringe pump.

11. Dissecting microscope.

12. TRITC-150 kDa dextran (Sigma), 1% solution in saline.

13. 150 kDa unlabeled dextran (Sigma), 6% solution in saline.

14. Electrocautery apparatus.

15. Vascular clips (removable).

16. C57BL/6 mice (or other strain).

17. Bone marrow-derived neutrophils isolated (as per Chapter 4 and reference [11]) from C57BL/6-Tg(UBC-GFP)30Scha/J mice (Jackson Laboratories), which express GFP in all cells.

3 Methods

3.1 Preoperative Preparations

1. Using scissors, cut a 6 × 4 cm oval from transparent polyvinylidene membrane. Glue a 5 mm coverslip to the center of the oval, using clear alpha-cyanoacrylate glue. The choice of these materials is designed to minimize degradation of fluorescent light passing through plastic components of the thoracic window (*see* **Note 1** for an alternative approach that avoids any plastic).

2. To create a mouse thoracostomy tube, cut a 10 cm length of PE50 tubing. Wrap a 4-0 silk suture several times around the tubing 1 cm from its end, and tie off. The tie should be tight; the rigidity of the PE50 tubing will prevent collapse. This suture will serve as an anchor, preventing the thoracostomy tube from being pulled completely out of the chest wall once inserted. Using a 30 gauge needle, make multiple punctures along the 1 cm length of tubing distal to the suture anchor, creating fenestrations to improve air aspiration. At the proximal end of the tubing, insert the blunt end of a 23 gauge 3/8 curved needle, with the sharp end pointed outward. This curve facilitates manipulation of the needle (and trailing tube) within the chest cavity, allowing for easier inside-out puncturing of the chest wall.

3.2 Anesthesia, Venous Catheterization, and Tracheostomy Placement

1. Weigh the mouse, and then anesthetize it with intraperitoneal ketamine (100 mg/kg) and xylazine (20 mg/kg). The depth of anesthesia may be confirmed via paw pinch.

2. Once anesthesia is confirmed, place the mouse in a supine position upon a heating pad, with the mouse's nose pointing toward the surgeon. Use adhesive tape to secure all four limbs.

3. Using surgical scissors (and a dissecting microscope), make a 2 cm vertical incision in the midline of the neck, exposing the submaxillary glands (Fig. 2a). Separate the right submaxillary gland at midline and then reflect it rostrally.

4. Using blunt dissection with forceps, expose the external jugular vein and its superficial branches. Pass 4-0 sutures under the two superficial veins rostral to the external jugular vein and tie. With light tension maintained, use adhesive tape to secure these sutures in place. These stay sutures provide exposure of the external jugular vein and prevent bleeding. Identify the

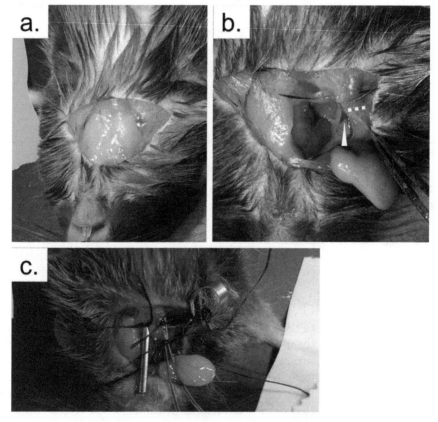

Fig. 2 Jugular catheter placement. (**a**) A midline neck incision reveals submaxillary glands in an anesthetized mouse. (**b**) After retraction of the right submaxillary gland, the bifurcation of the external jugular is visualized (arrowhead). This carina will be the site of catheter insertion. Prior to insertion, deep perforating vessels (dashed line) must be exposed and clamped with a vascular clamp. (**c**) Final placement of two right jugular catheters, with vascular clamp still in position. Note tracheostomy placement (left)

deep branches from the external jugular vein, which can be exposed laterally (dashed line, Fig. 2b), and apply a clip to prevent bleeding. Pass two 4-0 sutures under the external jugular vein, which will be used to stabilize the catheters after insertion.

5. While holding the roof of the external jugular vein, use microdissection scissors to cut a horizontal hole in the carina (arrowhead, Fig. 2b) of the external jugular vein's bifurcation with the previously tied superficial veins. Advance two PE 10 catheters sequentially through this incision and to a depth of approximately 1 cm. The 4-0 sutures that were previously passed under the external jugular vein should then be gently tightened to seal the vessel around the catheters. Aspirate each venous catheter to verify blood return, and then administer 200 μL 6% dextran solution for volume resuscitation. If there is no blood return, the catheter should be gently withdrawn until blood flows freely. Firmly tighten the 4-0 sutures to maintain the catheters in position (Fig. 2c). Cut the stay sutures on the superficial branches and release the clip from the deep branches.

6. Attach one venous catheter to a syringe pump for continuous infusion of anesthesia (ketamine 10 mg/kg/h, xylazine 2 mg/kg/h). If the mouse exhibits signs of decreased anesthesia on the continuous infusion (specifically showing spontaneous movements or painful response to tail pinch), administer a bolus of 25 mg/kg ketamine and 2 mg/kg xylazine. This may be further titrated as necessary. The other venous catheter is reserved for experimental agent injection.

7. Expose and anteriorly hemisect the trachea. Place a tracheostomy catheter through the incision and secure with a 4-0 suture (Fig. 2c). Begin ventilation at 120 breaths/min (8–10 μL/mg body weight) using a small animal ventilator (*see* **Note 2**). Positive end-expiratory pressure (PEEP) is not started at this time, minimizing lung volume during chest wall surgery.

3.3 Engraftment of Chest Wall Window

1. Place the anesthetized mouse left side down (i.e., left lateral decubitus position) on a heating pad and make an incision into the right chest wall. Dissect the skin and underlying intercostal muscle, exposing the rib cage below.

2. Using blunt dissection, puncture the pleura to induce a pneumothorax. Avoid damaging the underlying lung.

3. Partially resect the right third and fifth ribs using electrocautery forceps (*see* **Note 3**), leaving a 3 mm non-bleeding circular hole in the chest wall (Fig. 3a).

4. Place the thoracostomy tube "inside out" by introducing the catheter into the thoracic cavity through the chest wall hole, then

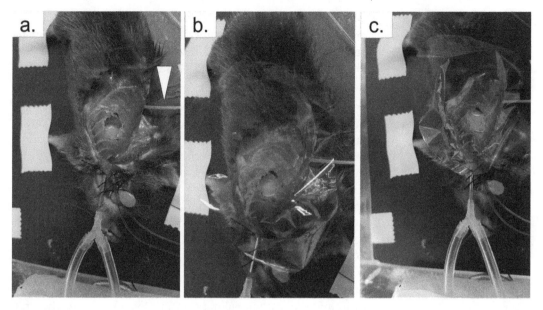

Fig. 3 Placement of thoracic window. (**a**) Mouse after creation of chest wall hole using electrocautery. Note the absence of bleeding (*see* **Note 3**) and the presence of a chest tube (arrowhead). (**b**) Mouse after sealing chest wall hole with window. In panels (**a**, **b**), the mouse is not ventilated for purposes of demonstrating lung collapse. (**c**) Mouse after lung recruitment and suction applied to chest tube. The lung approximates the window, enabling imaging

using the curved needle end to puncture the chest wall from the inside, creating a thoracostomy exit site caudal to the chest wall hole. The exit site should be caudal in the chest cavity (near the diaphragm) so that it does not interfere with imaging through the chest window. The chest tube is slowly pulled out of the body until the suture anchor is pulled against the chest wall, leaving the 1 cm fenestrated end of the chest tube remaining within the thorax. The curved needle is detached, and the distal end of the thoracostomy tube is connected to a syringe for suction.

5. Place a ring of alpha-cyanoacrylate glue around the chest wall hole. Glue the previously prepared polyvinylidene membrane (with glass coverslip on the outside aspect of the membrane), sealing the chest wall hole. During an inspiratory breath-hold (accomplished by briefly using your fingers to clamp the ventilator outflow), aspirate the trapped intrapleural air (i.e., the pneumothorax that accumulated during the surgery) via the transdiaphragmatic chest tube, allowing the lung to expand against the newly created thoracic wall window (Fig. 3b).

6. Begin positive end-expiratory pressure (PEEP) at 3 cm H_2O.

3.4 Intravital Microscopy

1. Transfer the mouse to an upright intravital microscope (*see* **Note 4**). A drop of deionized water is added to the coverslip, allowing for interface with a water immersion objective.

After injecting 100 μL of 1% 150 kDa TRITC-dextran through the jugular venous catheter, the subpleural microvasculature (vessel diameter <30 μm) of the inferior aspect of the right upper lobe can be easily observed.

2. GFP-expressing neutrophils (and/or additional experimental agents) may be subsequently injected via the available jugular venous catheter.

3. After completion of the protocol (*see* **Note 5**), euthanize the terminally sedated mouse by performing direct cardiac puncture (easily performed by virtue of the thoracic window) and creating bilateral pneumothoraces.

4 Notes

1. *Window alternatives.* Fluorescent light is significantly degraded by plastic; hence, glass is typically used as the only interface (other than water applied to the coverslip) for imaging. Thus, the choice of membrane material and glue is critical; the options discussed above have minimal degradation of signal. An alternative approach is to manufacture a window that only has glass in the light source. Briefly, purchase silicone membranes (<0.5 mm thickness), and cut a 6 × 4 cm oval. Punch a hole in the center of the oval with a 2 mm hole punch. Place the coverslip over the hole, and add liquid silicone to fix the coverslip in place (Fig. 4b).

2. *Optimization of mouse supportive care.* To ensure mice maintain proper oxygenation, ventilation, and perfusion during the experimental protocol, we recommend performing preliminary experiments in which arterial blood gas indices (pH, $PaCO_2$, PaO_2, as measured via a clinical or portable analyzer) are optimized via adjustment of respiratory rate, tidal volume, and fraction of inhaled oxygen. Blood pressure may be measured via direct arterial catheterization or via tail cuff measurement.

3. *Electrocautery.* The blood on the chest wall window prevents clear imaging of the lung parenchyma. Accordingly, the chest wall hole must not be bleeding at the time of window engraftment. This requirement necessitates the use of electrocautery to perform the rib resection (thereby preventing intercostal artery bleeding). Take care to avoid lung damage with the electrocautery probe.

4. *Tips for microscopy.* As described before, the major drawback from intravital imaging of the freely moving mouse lung is motion artifact. This can be negated by the use of high-speed imaging, such as a confocal resonant or spinning disk microscope [10]. Additionally, some mouse ventilators can be synchronized

a.

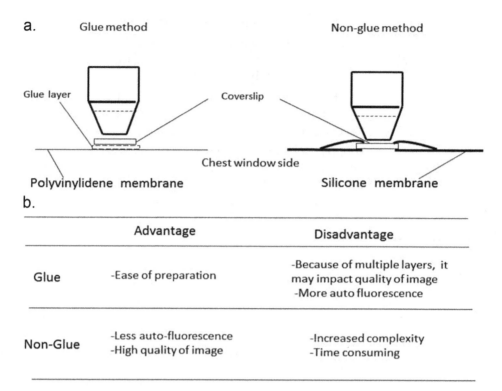

b.

	Advantage	Disadvantage
Glue	-Ease of preparation	-Because of multiple layers, it may impact quality of image -More auto fluorescence
Non-Glue	-Less auto-fluorescence -High quality of image	-Increased complexity -Time consuming

Fig. 4 Approaches to chest window design. (**a**) The glue method (left), as previously published [9–11], adheres a glass coverslip to the polyvinylidene membrane using clear (optical) cyanoacrylate glue. This presence of glue (and membrane) within the light path may decrease the clarity of fluorescent imaging. An alternative approach (right) uses as silicone membrane, in which the coverslip is placed over a membrane fenestration (punched-out hole). (**b**) Relative advantages of individual approaches. The alternative approach adds additional complexity but improves fluorescent imaging, as light only passes through the glass

to microscopy software, allowing for imaging to be coordinated with a respiratory hold. We have found minimal benefit from deconvolution software, due to the complex (and unpredictable) combined artifacts of respiratory and cardiac motion (i.e., pulsatile motion). In our experience, there is minimal additional benefit of two-photon imaging (over high-speed confocal), as the major benefit of two-photon imaging (e.g., depth of penetration) in the lung is largely limited by signal degradation from light passing through multiple alveolar air-fluid interfaces. A better investment of resources is to purchase a long working distance, high numerical aperture objective, which allows for high-resolution lung imaging.

5 *The duration of imaging.* It is important to ensure that your surgical approach does not damage the lung; furthermore, it is critical to determine how long the mouse remains stable on the preparation. While this may be accomplished via a number of approaches, a simple measure is comparing the wet-dry ratio (an index of edema) of the right (imaged) lung with the

left (un-imaged) lung [11]. The wet-dry ratios should be equal between lungs and unchanged from control mice that do not undergo intravital surgery. The duration of preparation viability is defined as the time in which wet-dry ratios remain stable; in our experience, by 2–3 h there is a symmetric increase in lung edema, negating further utility of imaging. This should be optimized for your laboratory.

References

1. Schmidt EP, Kuebler WM, Lee WL, Downey GP (2016) Adhesion molecules: master controllers of the circulatory system. Compr Physiol 6(2):945–973. https://doi.org/10.1002/cphy.c150020

2. Doerschuk CM (2001) Mechanisms of leukocyte sequestration in inflamed lungs. Microcirculation 8(2):71–88

3. Huh D, Matthews BD, Mammoto A, Montoya-Zavala M, Hsin HY, Ingber DE (2010) Reconstituting organ-level lung functions on a chip. Science 328(5986):1662–1668. https://doi.org/10.1126/science.1188302

4. Potter DR, Damiano ER (2008) The hydrodynamically relevant endothelial cell glycocalyx observed in vivo is absent in vitro. Circ Res 102(7):770–776. https://doi.org/10.1161/CIRCRESAHA.107.160226

5. Kuebler WM, Parthasarathi K, Lindert J, Bhattacharya J (2007) Real-time lung microscopy. J Appl Physiol (1985) 102(3):1255–1264. https://doi.org/10.1152/japplphysiol.00786.2006

6. Islam MN, Das SR, Emin MT, Wei M, Sun L, Westphalen K, Rowlands DJ, Quadri SK, Bhattacharya S, Bhattacharya J (2012) Mitochondrial transfer from bone-marrow-derived stromal cells to pulmonary alveoli protects against acute lung injury. Nat Med 18(5):759–765. https://doi.org/10.1038/nm.2736

7. Presson RG Jr, Brown MB, Fisher AJ, Sandoval RM, Dunn KW, Lorenz KS, Delp EJ, Salama P, Molitoris BA, Petrache I (2011) Two-photon imaging within the murine thorax without respiratory and cardiac motion artifact. Am J Pathol 179(1):75–82. https://doi.org/10.1016/j.ajpath.2011.03.048

8. Looney MR, Thornton EE, Sen D, Lamm WJ, Glenny RW, Krummel MF (2011) Stabilized imaging of immune surveillance in the mouse lung. Nat Methods 8(1):91–96. https://doi.org/10.1038/nmeth.1543

9. Tabuchi A, Mertens M, Kuppe H, Pries AR, Kuebler WM (2008) Intravital microscopy of the murine pulmonary microcirculation. J Appl Physiol (1985) 104(2):338–346. https://doi.org/10.1152/japplphysiol.00348.2007

10. Yang Y, Yang G, Schmidt EP (2013) In vivo measurement of the mouse pulmonary endothelial surface layer. J Vis Exp 72:e50322. https://doi.org/10.3791/50322

11. Schmidt EP, Yang Y, Janssen WJ, Gandjeva A, Perez MJ, Barthel L, Zemans RL, Bowman JC, Koyanagi DE, Yunt ZX, Smith LP, Cheng SS, Overdier KH, Thompson KR, Geraci MW, Douglas IS, Pearse DB, Tuder RM (2012) The pulmonary endothelial glycocalyx regulates neutrophil adhesion and lung injury during experimental sepsis. Nat Med 18(8):1217–1223. https://doi.org/10.1038/nm.2843

Chapter 22

Mouse Models of Acute Lung Injury and ARDS

Franco R. D'Alessio

Abstract

The acute respiratory distress syndrome (ARDS) is a devastating illness characterized by severe hypoxemia and diffuse alveolar damage. Direct lung infection is the leading cause of ARDS and can be modeled in mice using sterile models of inflammation or live pathogens. In this chapter, two mouse models for ARDS are defined. These include an infectious model of ARDS driven by direct administration of *Streptococcus pneumoniae* and a sterile inflammatory model mediated by intratracheal administration of lipopolysaccharide. Methods for growth and preparation of *Streptococcus pneumoniae* are provided as methods to assess lung inflammation and injury.

Key words Acute lung injury, ARDS

1 Introduction

Acute respiratory distress syndrome (ARDS) is a devastating inflammatory lung disease that has an overall mortality of 40% [1]. ARDS is characterized by its acute onset (occurs within 1 week of the precipitating event) and the development of diffuse alveolar damage, which results in severe hypoxemia ($P_aO_2/FiO_2 \leq 300$) and bilateral pulmonary infiltrates on chest radiography [2]. Bacterial and viral pneumonia are among the most common causes of ARDS [3], although non-pulmonary sepsis, aspiration of gastric contents, trauma, blood transfusions, and pancreatitis also represent risk factors for the development of ARDS.

In humans, the cardinal pathophysiologic features of ARDS include dysregulated inflammation and alveolar-capillary barrier disruption [4]. These are represented histologically by (a) neutrophilic alveolitis, (b) deposition of hyaline membranes in the alveoli, and (c) formation of microvascular thrombi, indicating activation of the coagulation cascade and endothelial damage [5]. Although the term acute lung injury (ALI) has fallen out of favor as a descriptor for human disease, it remains applicable to preclinical animal models, since no single model fully recapitulates all of the features of ARDS.

Scott Alper and William J. Janssen (eds.), *Lung Innate Immunity and Inflammation: Methods and Protocols*,
Methods in Molecular Biology, vol. 1809, https://doi.org/10.1007/978-1-4939-8570-8_22,
© Springer Science+Business Media, LLC, part of Springer Nature 2018

In order to maximize the value of animal studies, it is imperative that the investigator chooses the model that most closely reproduces the cardinal features of clinical ARDS while enabling testing of the study's specific hypothesis. In this context, several variables must be considered:

1. *Sex.* Male mice have been reported to be more susceptible to hyperoxia-induced lung injury [6, 7] and bleomycin-induced lung injury [8]. Similarly, studies have shown that estrogen imparts protective effects on lipopolysaccharide (LPS)-induced acute lung inflammation [9] and hemorrhagic shock-induced lung injury [10]. Conversely, in infectious models that employed *Pseudomonas aeruginosa*, females displayed a more robust lung inflammatory response while displaying higher bacterial loads [11]. This underscores the complexities underlying the sex differences in lung immune responses to diverse injuries [12].

2. *Strain.* Considerable differences in susceptibility to lung injury exist between strains. For example, in ozone-induced lung injury models, inbred A/J mice were highly sensitive, while C57BL/6J displayed a resistant pattern [13]. In contrast, C57BL/6 and BALB/c mice were found to be more sensitive than A/J mice in a model of lung ischemia-reperfusion injury [14].

3. *Age.* Aging increases the incidence and mortality of ARDS in humans [1]. Animal models parallel this, with aged mice displaying increases in neutrophilic infiltration and alveolar-capillary permeability after intranasal LPS challenge [15]. Moreover, we have determined that neonatal mice have delayed resolution of lung inflammation as compared to their juvenile counterparts [16].

4. *Phases of ALI.* The field has mostly focused in studying underlying mechanisms of early lung injury. In contrast, fewer studies have evaluated determinants of resolution of lung injury, a phase with distinct active pathways that could offer novel therapeutic avenues. Our group and others have described models of intratracheal LPS or influenza-induced lung injury that allow for the investigation of both early and resolution phases of lung injury [17–19].

5. *Route of injury.* ARDS can result from direct injury to the lungs (e.g., pneumonia, acid aspiration) or from indirect causes (e.g., pancreatitis, non-pulmonary sepsis). In this regard, murine models of direct ALI include ones in which agents are delivered intratracheally or intranasally. Common injury agents include LPS, hydrochloric acid, live (or heat killed) bacteria, and viruses. Hyperoxia can also drive direct lung injury. In comparison, indirect lung injury can be induced by intravenous injection of LPS, cecal ligation and puncture,

hemorrhagic shock, and mesenteric ischemia and reperfusion. As mentioned above, no model fully recapitulates the complex findings seen in human ARDS. For instance, intratracheal LPS administration causes robust neutrophilic alveolitis, while models of lung ischemia-reperfusion cause disruption of capillary membranes with intra-alveolar protein accumulation. In contrast, indirect models of sepsis produce minimal or no neutrophilic alveolitis, but are characterized by profound accumulation of neutrophils and monocytes in the pulmonary vasculature.

This chapter provides methods for two preclinical mouse models of ARDS that arise from direct lung injury. These use intratracheal instillation of LPS (sterile inflammation) or *Streptococcus pneumoniae* to mimic human pneumonia. The models are chosen because they are highly reproducible, elicit robust neutrophilic alveolitis and disruption of the alveolar-capillary membrane, are easily titratable (degrees of pulmonary inflammation), and allow for evaluation of both the early and resolution phases of ALI.

2 Materials

1. Animals: C57BL/6, BALB/c, or any transgenic mouse that will help address your hypothesis.

2. Lipopolysaccharide from *Escherichia coli* (O55:B5) (*see* **Note 1**).

3. *Streptococcus pneumoniae* culture and preparation.

   (a) *Streptococcus pneumoniae* serotype 19 (ATCC 49619) (*see* **Note 2**).

   (b) Trypticase soy agar with 5% sheep blood plates (100 × 15 mm).

   (c) Todd-Hewitt broth (DIFCO).

   (d) Fetal bovine serum (FBS from Gibco).

   (e) Inoculating loops.

   (f) Incubator shaker.

   (g) Personal protective equipment including gloves, mask, and eye goggles

4. Oropharyngeal or intratracheal administration of LPS or *S. pneumoniae*

   (a) Induction chamber or bell jar.

   (b) Sterile curved forceps.

   (c) Sterile scissors.

   (d) Surgery board or wedge (we use an old cover for a laboratory water bath).

(e) 200 μL pipette.

(f) 200 μL pipette tips (for oropharyngeal administration).

(g) Gel loading tips (for intratracheal administration).

(h) Isoflurane (Henry Schein, USP).

(i) Ketamine (for intratracheal administration).

(j) Acetylpromazine (for intratracheal administration).

(k) Vetbond or Superglue.

5. Protein assay (BCA assay) for measurement of bronchoalveolar lavage (BAL) fluid

(a) Microplate.

(b) Spectrophotometer (microplate reader).

(c) BCA protein assay kit.

6. Bronchoalveolar cell count and differential

(a) Microfuge tubes.

(b) Red blood cell lysis buffer.

(c) Microcentrifuge.

(d) Hemocytometer.

(e) Trypan blue solution.

(f) Frosted slides.

(g) Microscope.

(h) Tabletop cytocentrifuge (Thermo Shandon).

(i) Phosphate-buffered   saline   (PBS),   calcium,   and magnesium-free.

3   Methods

Oropharyngeal aspiration is a rapid and simple way to directly deliver agents into the lungs. We compared oropharyngeal aspiration (*see* Subheading 3.2) or intratracheal (*see* Subheading 3.3) routes of instillation for LPS and found that both delivery methods produced robust early injury [20]. However, inflammation resolved more rapidly in mice treated with the oropharyngeal method (Fig. 1).

3.1   Bacterial Growth and Quantification

3.1.1   *Todd-Hewitt (TH) Broth Preparation*

1. Dissolve 30 g of Todd-Hewitt powder in 1 L of distilled water, and put in 2000 mL autoclavable flask.

2. Use heated magnet spinner to completely dissolve.

3. Autoclave at 121 °C for 15 min (can be stored *without* FBS in 4 °C for 2 months).

3.1.2   *Bacterial Growth (Always Work Near a Bunsen Burner)*

1. Streak blood agar plate with bacterial loop.

2. Grow overnight (18 h).

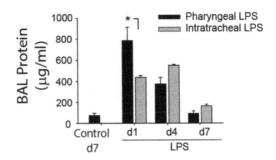

Fig. 1 Bronchoalveolar lavage protein levels in C57BL/6 mice challenged with LPS using either oropharyngeal aspiration or intratracheal delivery

3. Pick ten colonies, and add to a mixture of 90 mL Todd-Hewitt broth and 18 mL filtered FBS in 250 mL autoclaved flask.

4. Incubate for 8 h at 37 °C in incubator shaker at 225 rpm.

5. After 8–10 h of incubation, begin taking aliquots of broth every hour to estimate bacterial concentration with a spectrophotometer. We use 6 mL cuvettes for measurement and therefore find it easiest to remove 3 mL of broth with bacteria and mix it with 3 mL of stock broth for measurement. Bacterial concentrations are adequate when an OD of 0.25 at 620 nm is reached.

6. Immediately aliquot bacteria into 1 mL aliquots and flash freeze for 5 min.

7. Store aliquots of bacteria at −80 °C (good up to 4 years).

8. After 5 days of freezing, aliquots can be used for animal challenges. Stock should have 1×10^8 CFU/mL, although this will need to be determined by streaking bacteria on blood agar plates (*see* **step 11** below).

9. Rapidly thaw frozen vials in warm water and mix well. Bacteria should be used within 1 h to avoid declines in CFU.

10. Instill bacteria into mice as described below (Subheadings 3.2 and 3.3).

11. Quantify the bacteria load injected. Make serial tenfold dilutions of the thawed bacteria. Use 200 μL per plate, and spread using bacterial loops across the surface of a blood agar plate. Incubate (incubator shaker) at 37 °C overnight and count colonies. Apply the following formula to determine bacterial counts:

CFU/mL = (no. of colonies × dilution factor)/volume of culture plate

3.2 Oropharyngeal Aspiration of LPS or S. pneumoniae

1. Use either this method or intratracheal instillation (Subheading 3.3).

2. Prepare LPS or *S. pneumoniae* by diluting stock solution in sterile PBS to achieve the desired concentration in a volume of 50 μL (*see* **Notes 1** and **2**).

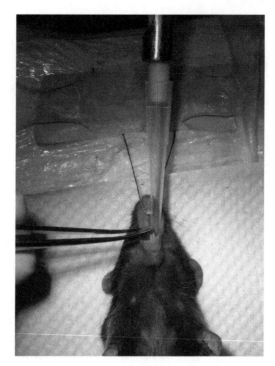

Fig. 2 Technique showing oropharyngeal method of delivery of injury agents directly into the lungs

3. Fast mice for at least 1 h before induction of anesthesia.

4. Place mouse in induction chamber with isoflurane. The duration of anesthetic exposure will vary between strains and by age. A significant decrease in respiratory rate and deep respirations is desirable in conjunction with complete unresponsiveness.

5. Suspend anesthetized mouse by the front incisors on a surgery board.

6. Using a curved forceps, gently pull the tongue forward, and using a pipette, instill 50 μL of instillation into the posterior oropharynx. Keep the tongue extended until instillate is completely aspirated (Fig. 2).

7. Mice randomized to a control group should receive 50 μL of sterile PBS or water using the same approach.

8. Allow animals to recover from anesthesia.

3.3 Intratracheal Instillation of LPS or S. pneumoniae

1. Use either this method or oropharyngeal aspiration (Subheading 3.2).

2. Place mice in induction chamber with isoflurane 5% until immobility and decreased respiratory rate are achieved.

3. Remove mice from chamber, and inject intraperitoneally with ketamine (150 mg/kg) and acetylpromazine (13.5 mg/kg).

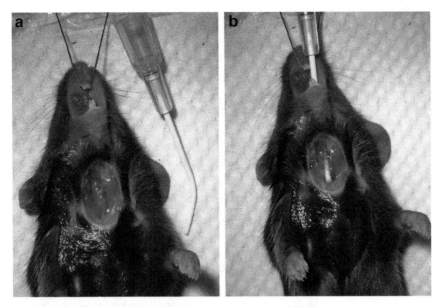

Fig. 3 Technique showing intratracheal delivery of injury agents directly to the lungs. (**a**) A small midline incision in the neck is used to visualize the trachea. (**b**) The angiocatheter is advanced through the mouth. Its distal tip can be seen in the trachea

4. Place anesthetized mouse on a surgical board or wedge, and hang by the incisors. Gently tape forelimbs to the board to stabilize the mouse.

5. Wet the skin with 80% ethanol. Using scissors, open a small superficial midline neckline incision to visualize the trachea (Fig. 3a).

6. Gently pull the tongue outward using curved forceps while introducing a 20 gauge angiocatheter into the trachea. The angiocatheter should be slightly angled and almost parallel to the tracheal plane. Advance the angiocatheter through the mouth and into the trachea (Fig. 3b). Can also apply gentle pressure over trachea to facilitate intubation.

7. Using a 200 μL pipettor and gel loading tips, carefully instill the injury agent into the angiocatheter (*see* **Note 3**).

8. Close incision by adding one small drop of glue (Vetbond or Superglue works well for this purpose). Bring skin folds together, and hold light pressure for 3–5 s.

9. Place animals on their sides on a heating pad in their cage until they recover.

3.4 Bronchoalveolar Lavage (BAL)

Bronchoalveolar lavage is used to isolate constituents from the alveolar space. These include cells (namely, leukocytes) that can be enumerated and identified to quantify inflammation and fluid that will contain protein (commonly used as a marker of injury) and cytokines.

1. Euthanize the mouse (*see* **Note 4**).

2. Cannulate the trachea with a 20 gauge catheter.

3. Lavage the right lung twice using calcium-free PBS and a 1 mL syringe. Each aliquot should be 0.7 mL. The total return typically averages 0.9–1.1 mL. The left lung can be used for histology (*see* Chapter 20). We found that when using the bacterial model, the injury often localizes to the left lung. Hence, for bacterial infection models, we often perform lavage of both lungs. Two 1 mL aliquots are used when lavaging both lungs.

4. Centrifuge BAL fluid at $500 \times g$ for 8 min at 4 °C.

5. Remove the cell-free supernatant, and store at −80 °C for future analysis of protein content (*see* **Note 5**). The supernatant can also be used to measure other biomarkers (e.g., albumin, RAGE) or specific cytokines (ELISA).

6. Dilute the cell pellet in PBS. If the cell pellet is bloody, red blood cell lysis can be performed (*see* **Note 6**). The volume of PBS used will depend on the size of the pellet. For small pellets we use 100 μL; for large pellets we use 300 μL.

7. Remove an aliquot of cells. Add trypan blue stain to enhance their visualization and determine cell numbers using a hemacytometer.

8. Differential cell counts are performed by placing 50 μL of resuspended BAL on cytocentrifuge preparations (Cytospin 3, Shandon Scientific).

9. Stain cytospins with Diff-Quik stain. Count 300 cells per sample.

4   Notes

1. We use LPS generated from *E. coli* (strain O55:B5, Sigma-Aldrich). However, LPS from other strains of *E. coli* or from other Gram-negative bacteria also produce inflammation. Inflammation will vary depending on the source of LPS and will even vary between lot numbers. We recommend titrating the dose of LPS based on the hypothesis to be evaluated. Doses from 1 to 5 mg/kg are typically sufficient and will provide various degrees of mortality, early inflammation, and reproducible resolution patterns.

2. Other strains of *Pneumococcus* can be used. We use *S. pneumoniae* serotype 19 (ATCC) since it remains pneumonic without causing overt systemic involvement. Notably, strains that become systemic may provide a better model for sepsis.

3. Often, mice will develop a vasovagal reflex and become apneic. In this case a small mouse ventilator can be used to briefly provide ventilatory support. Usually only 10–15 s is required.

4. We typically use a lethal dose of pentobarbital (200 mg/kg intraperitoneally). Other methods approved by your Institutional Animal Care and Use Committee (IACUC) are acceptable. Cervical dislocation should be avoided since it can cause trauma to the trachea and lead to hemorrhage in the airways.

5. Protein concentrations in the BAL fluid correspond with the degree of inflammation and alveolar-capillary permeability. We use the BCA assay; however, any other method of protein determination, such as Lowry, can be used. Other methods use specific reagents, although invariably albumin is utilized to create the standard curve. Moreover, every method has a specific spectrophotometer wavelength to detect the signal.

6. RBC lysis buffer is used for BAL pellets that are visually bloody. We add 100–300 μL RBC lysis buffer for 1–3 min (depending on pellet size and degree of bleeding). RBC lysis buffer is neutralized by 2–5× volumes of PBS and then centrifuge to obtain a clear BAL cell pellet. Some investigators use red blood cell (RBC) counts as a measure of lung injury for which the lysis step would be skipped.

Acknowledgments

This work was supported by NIH grant R56HL131812 and DoD W81XWH-16-1-0510.

References

1. Rubenfeld GD, Caldwell E, Peabody E et al (2005) Incidence and outcomes of acute lung injury. N Engl J Med 353(16):1685–1693

2. Force ADT, Ranieri VM, Rubenfeld GD et al (2012) Acute respiratory distress syndrome: the Berlin definition. JAMA 307(23):2526–2533

3. Matthay MA, Ware LB, Zimmerman GA (2012) The acute respiratory distress syndrome. J Clin Invest 122(8):2731–2740

4. Ware LB, Matthay MA (2000) The acute respiratory distress syndrome. N Engl J Med 342(18):1334–1349

5. Matute-Bello G, Frevert CW, Martin TR (2008) Animal models of acute lung injury. Am J Physiol Lung Cell Mol Physiol 295(3):L379–L399

6. Lingappan K, Jiang W, Wang L, Couroucli XI, Barrios R, Moorthy B (2013) Sex-specific differences in hyperoxic lung injury in mice: implications for acute and chronic lung disease in humans. Toxicol Appl Pharmacol 272(2):281–290

7. Lingappan K, Jiang W, Wang L, Moorthy B (2016) Sex-specific differences in neonatal hyperoxic lung injury. Am J Physiol Lung Cell Mol Physiol 311(2):L481–L493

8. Redente EF, Jacobsen KM, Solomon JJ et al (2011) Age and sex dimorphisms contribute to the severity of bleomycin-induced lung injury and fibrosis. Am J Physiol Lung Cell Mol Physiol 301(4):L510–L518

9. Speyer CL, Rancilio NJ, McClintock SD et al (2005) Regulatory effects of estrogen on acute lung inflammation in mice. Am J Physiol Cell Physiol 288(4):C881–C890

10. Suzuki T, Shimizu T, Yu HP, Hsieh YC, Choudhry MA, Chaudry IH (2007) Salutary effects of 17beta-estradiol on T-cell

signaling and cytokine production after trauma-hemorrhage are mediated primarily via estrogen receptor-alpha. Am J Physiol Cell Physiol 292(6):C2103–C2111

11. Guilbault C, Stotland P, Lachance C et al (2002) Influence of gender and interleukin-10 deficiency on the inflammatory response during lung infection with Pseudomonas aeruginosa in mice. Immunology 107(3):297–305

12. Carey MA, Card JW, Voltz JW, Germolec DR, Korach KS, Zeldin DC (2007) The impact of sex and sex hormones on lung physiology and disease: lessons from animal studies. Am J Physiol Lung Cell Mol Physiol 293(2):L272–L278

13. Prows DR, Shertzer HG, Daly MJ, Sidman CL, Leikauf GD (1997) Genetic analysis of ozone-induced acute lung injury in sensitive and resistant strains of mice. Nat Genet 17(4):471–474

14. Dodd-o JM, Hristopoulos ML, Welsh-Servinsky LE, Tankersley CG, Pearse DB (2006) Strain-specific differences in sensitivity to ischemia-reperfusion lung injury in mice. J Appl Physiol (1985) 100(5):1590–1595

15. Kling KM, Lopez-Rodriguez E, Pfarrer C, Muhlfeld C, Brandenberger C (2017) Aging exacerbates acute lung injury-induced changes of the air-blood barrier, lung function, and inflammation in the mouse. Am J Physiol Lung Cell Mol Physiol 312(1):L1–L12

16. McGrath-Morrow SA, Lee S, Gibbs K et al (2015) Immune response to intrapharyngeal LPS in neonatal and juvenile mice. Am J Respir Cell Mol Biol 52(3):323–331

17. D'Alessio FR, Tsushima K, Aggarwal NR et al (2009) CD4+CD25+Foxp3+ Tregs resolve experimental lung injury in mice and are present in humans with acute lung injury. J Clin Invest 119(10):2898–2913

18. D'Alessio FR, Tsushima K, Aggarwal NR et al (2012) Resolution of experimental lung injury by monocyte-derived inducible nitric oxide synthase. J Immunol 189(5):2234–2245

19. Singer BD, Mock JR, Aggarwal NR et al (2015) Regulatory T cell DNA methyltransferase inhibition accelerates resolution of lung inflammation. Am J Respir Cell Mol Biol 52:641–652

20. Aggarwal NR, Tsushima K, Eto Y et al (2014) Immunological priming requires regulatory T cells and IL-10-producing macrophages to accelerate resolution from severe lung inflammation. J Immunol 192(9):4453–4464

Chapter 23

Mouse Models of Asthma

Magdalena M. Gorska

Abstract

Mouse models are critical for delineating the mechanisms that underlie asthma pathogenesis and developing new treatments. In this chapter we describe four different asthma models that offer unique benefits and allow investigators to answer distinct research questions. We also describe key surgical procedures that are necessary for assessing experimental asthma.

Key words Mouse model, Experimental asthma, Allergen, Immunization, Lung, Bronchoalveolar lavage

1 Introduction

Asthma is a common, chronic, and heterogeneous disease of the airways that is characterized by airway hyperresponsiveness (AHR), airflow obstruction, and airway inflammation and remodeling. In the majority of asthma endotypes, the inflammation has a master regulatory role that drives the other features of asthma. Based on the nature of the inflammation, asthma is classified into eosinophilic (most common), neutrophilic, mixed granulocytic, or pauci-inflammatory [1]. Eosinophilic asthma is frequently a result of exaggerated type 2 immune responses to airborne allergens. The type 2 responses are driven by adaptive and innate lymphocytes. Most studies have focused on the importance of CD4 T cells and, more recently, innate lymphoid type 2 cells (ILC2s) [2]. CD4 T cells promote the emergence of allergen-specific IgE which is a classical indicator of the allergic response.

In addition to cells of the immune system, airway structural cells are activated and contribute to the pathogenesis of asthma [3]. Activation of airway structural cells is manifested by goblet cell hyperplasia and mucus overproduction, subepithelial fibrosis, smooth muscle hypertrophy, and an increased number of blood vessels. These structural changes are collectively called "airway remodeling" [4].

Scott Alper and William J. Janssen (eds.), *Lung Innate Immunity and Inflammation: Methods and Protocols*,
Methods in Molecular Biology, vol. 1809, https://doi.org/10.1007/978-1-4939-8570-8_23,
© Springer Science+Business Media, LLC, part of Springer Nature 2018

Mouse models are instrumental in mechanistic studies of asthma. Existing mouse models recapitulate many features of human asthma. Over the years, several models have been developed. The unifying feature of all these models is the use of an environmental trigger (usually an allergen) to elicit an asthma-like reaction in the lung. In this chapter we describe four models. These include the classic ovalbumin sensitization model and models based on exposures to house dust mite extract, *Alternaria alternata* extract, and a combination of dust mite, ragweed, and aspergillus extracts. Each one has unique advantages and disadvantages, as described below.

The classic ovalbumin (OVA)-induced model involves sensitization with OVA in the presence of alum as an adjuvant, followed by airway challenge with OVA to elicit features of asthma [5–7]. These features are then typically studied 24–72 h after the challenge. Multiple variants of this model have been described. In this chapter we describe a variant with two subcutaneous sensitizations followed by five consecutive challenges through the intranasal route [7]. The OVA-induced model generates robust adaptive type 2 immune response (including emergence of OVA-specific IgE), eosinophilic inflammation of airways, goblet cell hyperplasia, and AHR. The key advantage of this model is its compatibility with OVA-related reagents and mouse strains. The OT2 and DO11.10 mouse strains are transgenic for OVA peptide-specific T-cell receptors (TCRs) that pair with the CD4 receptor. The OT2 and DO11.10 strains are extremely useful in studies of CD4 T-cell responses. The added bonus is that the OT2 and DO11.10 strains were developed on C57BL/6 and BALB/c backgrounds, respectively. C57BL/6 and BALB/c represent the two most commonly used genetic backgrounds in mouse models of experimental asthma. The disadvantages of the OVA-induced model include the use of an artificial antigen, use of an adjuvant, and an immunization route that are not directly relevant to human asthma. The OVA-induced model is an acute model; features of asthma resolve within 1–2 weeks after cessation of the challenge procedure.

To overcome some of the problems of the OVA-induced model, the house dust mite (HDM) extract-induced model has been developed [8, 9]. The HDM extract-induced model utilizes a human asthma-relevant allergen and immunization method (administration of the allergen directly onto airway mucosal surfaces, no artificial adjuvant). The natural immunization is possible due to intrinsic adjuvant-like properties of HDM, including its protease activity. In general, the HDM extract-induced model reiterates most of the asthma features that are characteristic of the OVA-induced model. However, the generation of allergen-specific IgE is not as robust as in the OVA-induced model [10]. Similarly to the OVA-induced model, features of asthma in the HDM extract-induced model resolve over a very short period of time.

In recent years, two new valuable mouse asthma models have been developed. These models address two important aspects of asthma—involvement of innate lymphoid cells and persistence of asthma in the absence of an environmental trigger. The first model involves three to four consecutive intranasal challenges with an extract of the fungus *Alternaria alternata* [11]. The features of asthma are studied shortly after the last challenge, typically at 24 h. Understanding of the responses to *Alternaria* is relevant to human pathology since exposure to this fungus is associated with severe asthma [12]. The *Alternaria* extract-induced model generates strong responses of the innate immune system, including responses of innate lymphoid type 2 cells (ILC2s), which induce airway eosinophilia and AHR [11, 13]. The role of T cells and B cells in this model is uncertain at this moment and requires further investigation. The second model involves intranasal administration of three allergens—extracts of dust mite, ragweed, and *Aspergillus* (DRA) without adjuvant—twice a week for 6 weeks [7, 14]. In the DRA extract-induced model, high-level AHR persists for at least 6 months after cessation of the allergen exposure. Airway inflammation and remodeling peak 1–2 months postexposure, but remain significantly increased at 6 months. The persistence of asthma is independent of T cells. Instead, it relies on ILC2s. T cells contribute to the severity (magnitude) of asthma features. This is the first model, to our knowledge, that truly reflects the chronic aspect of human asthma, i.e., the long-term persistence of airway pathology in the absence of an environmental trigger.

Protocols for each of the four animal models of asthma are provided below.

2 Materials

2.1 OVA-Induced Model of Acute Asthma

1. Mixture of OVA and alum: Dissolve OVA grade V (*see* **Note 1**) to a final concentration of 1.2 mg/mL in sterile phosphate-buffered saline (PBS) to prepare OVA solution # 1. For immunization, you will need 50 μL (60 μg) of solution # 1 per mouse per day. The solution can be distributed into single-use aliquots and frozen at −80 °C. Use the aliquots within 1 year of preparation. To prepare the immunizing mixture, add an equal volume of Imject Alum (Pierce) dropwise to OVA solution #1. Imject Alum is an aqueous suspension of aluminum hydroxide (40 mg/ml) and magnesium hydroxide (40 mg/mL) (*see* **Note 2**). You will need 50 μL (2 mg of aluminum hydroxide and 2 mg of magnesium hydroxide) of Imject Alum per mouse per day. Work under sterile conditions. Mix (vortex mixer) the OVA-alum suspension for 30 min. Mixing allows the Imject Alum to adsorb OVA. Use immediately after preparation.

2. OVA solution # 2: Dissolve OVA grade V (*see* **Note 1**) to a final concentration of 4 mg/mL in sterile PBS. OVA solution # 2 is used for challenge by intranasal instillation. You will need 15 μL (60 μg) of this solution per mouse per challenge day. The solution can be distributed into single-use aliquots and frozen at −80 °C. Use the aliquots within 1 year of preparation. Work under sterile conditions.

3. 70% (v/v) ethanol.

4. Isoflurane.

5. Isoflurane vaporizer connected to an anesthesia induction chamber.

6. 1 mL syringes.

7. 25 gauge needles.

8. 6-week-old C57BL/6 or BALB/C mice (*see* **Note 3**).

2.2 HDM Extract-Induced Model of Acute Asthma

1. House dust mite (HDM) solution # 1: Dissolve the HDM powder (lyophilized extract of *Dermatophagoides pteronyssinus*, Greer Laboratories) to a final concentration of 0.67 mg/mL (*see* **Notes 1, 4,** and **5**) in sterile PBS. Prepare single-use aliquots, and freeze at −20 °C. Use the aliquots within 1 year of preparation. Each aliquot should be thawed only once. Do not refreeze. HDM solution # 1 is used for challenge by intranasal instillation. You will need 15 μL (10 μg) of solution # 1 per mouse per challenge day. HDM solution # 1 also serves as a stock solution for preparation of HDM solution # 2.

2. HDM solution # 2: Use HDM solution # 1 to prepare solution # 2 at a final concentration of 20 μg/mL. Use sterile PBS for dilution. HDM solution # 2 is used for immunization by intratracheal instillation. You will need 50 μL (1 μg) of solution # 2 per mouse.

3. Isoflurane.

4. Isoflurane vaporizer connected to an anesthesia induction chamber.

5. A stand for intratracheal instillation: An angled (60° incline) platform with a rubber band stretched between two bolts.

6. Forceps.

7. 6–8-week-old C57BL/6 or BALB/C mice (*see* **Note 3**).

2.3 Alternaria Extract-Induced Model of ILC2-Mediated Asthma

1. *Alternaria* solution: Dissolve the lyophilized *Alternaria alternata* extract (Greer Laboratories) to a final concentration of 0.33 mg/mL in sterile PBS (*see* **Notes 1, 4,** and **5**). You will need 15 μL (5 μg) of this solution per mouse per challenge day. The solution should be distributed into single-use aliquots and frozen at −20 °C. Use the aliquots within 1 year of preparation. Each aliquot can be thawed only once. Do not refreeze.

2. Isoflurane.

3. Isoflurane vaporizer connected to an anesthesia induction chamber.

4. 6–8-week-old C57BL/6 or BALB/C mice (*see* **Note 3**).

2.4 DRA Extract-Induced Model of Chronic Asthma

1. DRA solution: Dissolve lyophilized extracts of HDM (*Dermatophagoides pteronyssinus*), ragweed (*Ambrosia artemisiifolia*), and *Aspergillus fumigatus* (all extracts from Greer Laboratories) in sterile PBS to a final concentration of 0.33 mg/mL, 1 mg/mL, and 0.33 mg/mL, respectively (*see* **Notes 1, 4,** and **5**). You will need 15 μL of this solution (5 μg of HDM, 15 μg of ragweed, and 5 μg of *Aspergillus*) per mouse per challenge day. The solution should be distributed into single-use aliquots and frozen in −20 °C. Use the aliquots within 1 year of preparation. Each aliquot can be thawed only once. Do not refreeze.

2. Isoflurane.

3. Isoflurane vaporizer connected to an anesthesia induction chamber.

4. 6–8-week-old C57BL/6 or BALB/C mice (*see* **Note 3**).

2.5 Tracheotomy and Bronchoalveolar Lavage (BAL)

1. PBS-EDTA solution: Prepare sterile PBS containing 0.5 mM EDTA.

2. Mixture of xylazine at 30 mg/kg mouse weight and ketamine at 500 mg/kg mouse weight.

3. 70% (v/v) ethanol.

4. Intratracheal cannula: 18 or 21 gauge needle with polyethylene tubing around the metal part of the needle. The tubing protects the trachea from laceration by the sharp end of the needle.

5. 25 gauge needles.

6. 1 mL syringe.

7. Scissors and forceps.

8. 15 mL conical tubes.

9. Refrigerated centrifuge with a swinging bucket rotor.

10. Upright microscope.

11. Hemacytometer.

12. C57BL/6 or BALB/c mice (*see* **Note 3**).

2.6 Lung Harvest

1. Sterile PBS.

2. Mixture of xylazine at 30 mg/kg mouse weight and ketamine at 500 mg/kg mouse weight.

3. 70% (v/v) ethanol.

4. 25 gauge needles.

5. 1 and 10 mL syringes.

6. Scissors and forceps.

7. 60 mm plastic dishes.

8. C57BL/6 or BALB/c mice (*see* **Note 3**).

3 Methods

3.1 OVA-Induced Model of Acute Asthma (Fig. 1a)

3.1.1 Immunization

Day 0

1. Disinfect the dorsal skin of the mouse neck with 70% ethanol.

2. Grasp the dorsal skin of the neck using your thumb and your forefinger. The skin in the neck area is loose and can be gently lifted between your fingers.

3. Insert a 25 gauge needle under the lifted skin and subcutaneously inject 100 μL of the OVA-alum mixture.

4. Release the mouse (*see* **Note 6**).

Day 7

Repeat the immunization procedure as described in **steps 1–4** for Day 0.

3.1.2 Challenge

Day 14

1. Place the mouse in an anesthesia induction chamber connected to an isoflurane vaporizer. Switch on the vaporizer. Anesthetize the mouse with 4–5% isoflurane in oxygen or air. Anesthesia is considered adequate when an animal stays still quietly, loses a toe pinch reflex, and its respiratory rate has decreased and is noticeably deeper.

2. Remove the mouse from the anesthesia induction chamber, and immediately (*see* **Note 7**) proceed with intranasal instillation.

3. Intranasal instillation: Hold the mouse vertically head upward and rear legs downward. Slowly administer 15 μL of OVA solution # 2 into one nostril using a sterile pipet tip. Maintain a vertical position of the mouse with the head upward for 1 min to permit the OVA solution to distribute downward into the lung.

4. Place the mouse into its cage. Switch off the isoflurane vaporizer.

5. Monitor the mouse during recovery from anesthesia.

Days 15, 16, 17, and 18

Repeat the challenge procedure on each day as described in **steps 1–5** for Day 14.

A.

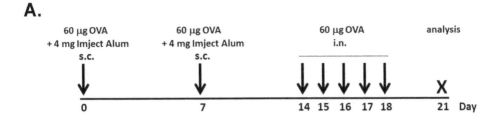

B.

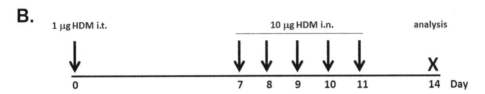

C.

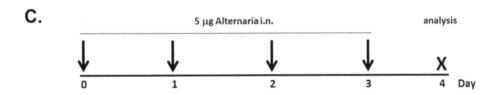

D.

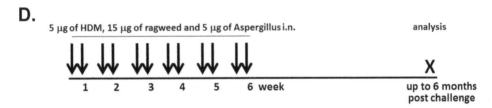

Fig. 1 Four models of experimental asthma. (**a**) Ovalbumin (OVA)-induced model of acute asthma, (**b**) house dust mite (HDM) extract-induced model of acute asthma, (**c**) *Alternaria alternata* extract-induced model of ILC2-mediated asthma, (**d**) dust mite–ragweed–*Aspergillus* (DRA) extract-induced model of chronic asthma. *i.n.* intranasal route, *i.t.* intratracheal route, *s.c.* subcutaneous route; *see* the text for detailed description of models

3.1.3 Analysis	*Day 21*
	Measure features of asthma (*see* **Note 8**).
3.2 HDM Extract-Induced Model of Acute Asthma (Fig. 1b)	*Day 0*
	1. Anesthetize a mouse with isoflurane as described in Subheading 3.1.2.
3.2.1 Immunization	2. Remove the mouse from the anesthesia chamber, and immediately (*see* **Note 7**) proceed with the intratracheal instillation procedure.
	3. Intratracheal instillation: Gently suspend the mouse by its cranial incisors on a rubber band attached to the angled instil-

lation stand. Use forceps to pull the tongue forward and then downward. A full extension of the tongue is needed to prevent the animal from swallowing. Place the pipet tip containing 50 μL of HDM solution # 2 over the tongue. Inject the solution toward the back of the throat. Keep the tongue fully extended, and monitor mouse's breathing to ensure that the fluid is aspirated into the lungs. The common practice is to maintain the tongue extended for five breaths after the liquid disappears from the throat. Next, release the tongue.

4. Place the mouse into its cage. Switch off the isoflurane vaporizer.

5. Monitor the mouse during recovery from anesthesia.

3.2.2   Challenge

Day 7

1. Anesthetize the mouse with isoflurane as described in the Subheading 3.1.2.

2. Perform the intranasal instillation procedure as described in the Subheading 3.1.2 using 15 μL of HDM solution # 1 per mouse.

Days 8, 9, 10, and 11
Repeat the challenge procedure on each day as described in **steps 1** and **2** for Day 7

3.2.3   Analysis

Day 14
Measure features of asthma (*see* **Note 8**).

3.3   Alternaria Extract-Induced Model of ILC2-Mediated Asthma (Fig. 1c)

3.3.1   Challenge

Day 0

1. Anesthetize the mouse with isoflurane as described in the Subheading 3.1.2.

2. Perform intranasal instillation as described in the Subheading 3.1.2 using 15 μL of the *Alternaria* solution per mouse.

Days 1, 2, and 3
Repeat the challenge procedure on each day as described in **steps 1** and **2** for Day 0.

3.3.2   Analysis

Day 4
Measure features of asthma (*see* **Note 8**).

3.4   DRA Extract-Induced Model of Chronic Asthma (Fig. 1d)

3.4.1   Immunization and Challenge

Day 0 (Week 1)

1. Anesthetize the mouse with isoflurane as described in the Subheading 3.1.2.

2. Perform intranasal instillation as described in the Subheading 3.1.2 using 15 μL of the DRA solution per mouse.

Day 3 (Week 1)

Repeat the intranasal instillation procedure as described in **steps 1** and **2** for Day 0.

Weeks 2, 3, 4, 5, and 6

Repeat the intranasal instillation procedure twice a week as described in **steps 1** and **2** for Day 0 of week 1.

3.4.2 Analysis

Up to 6 Months Post-challenge

Measure features of asthma (*see* **Note 8**).

3.5 Tracheotomy and Bronchoalveolar Lavage (BAL)

1. Euthanize the mouse through intraperitoneal injection of xylazine (30 mg/kg mouse weight) and ketamine (500 mg/kg mouse weight). Use a 25 gauge needle and 1 mL syringe for injection (*see* **Note 9**).

2. Disinfect the ventral skin of the neck with 70% ethanol.

3. Cut the ventral skin of the neck in the midline. Carefully separate the salivary glands and the sternohyoid muscle laterally to expose the trachea.

4. Make a small incision in the trachea. The incision should be made between the first and the second cartilage ring proximal to the larynx.

5. Insert the cannula into the trachea. The cannula should be attached to the 1 mL syringe containing the PBS-EDTA solution. Inject 1 mL of this solution into the lungs. Massage the chest for a few seconds. Aspirate and collect the BAL fluid into a 15 mL conical tube.

6. Centrifuge this first milliliter of the BAL fluid for 5 min, $300 \times g$, 4 °C. Collect the supernatant, and freeze it at −80 °C for future measurements (*see* **Notes 10** and **11**). Keep the BAL cell pellet on ice.

7. Repeat the inject-aspirate procedure twice to collect an additional 2 mL of the BAL fluid.

8. Pool these two lavages, and add them to the cell pellet obtained from the first lavage. Centrifuge 5 min, $300 \times g$, 4 °C to obtain the final cell pellet. Discard the supernatant.

9. Perform cell counts using a hemacytometer.

10. Cells can be cytospin and stained with the Diff-Quick stain for morphometric analysis and differential counting. Alternatively, cells can be stained with fluorescently labeled antibodies and analyzed by flow cytometry.

3.6 Lung Harvest

1. Euthanize the mouse as described in the Subheading 3.5.

2. Disinfect the ventral side of the mouse (the neck, chest, and abdomen) with 70% ethanol.

3. Make a midline incision of the ventral skin, separate it aside, and remove it. Lift the xiphoid process of the sternum with forceps. Cut the diaphragm away from the ribs. Cut the entire sternum in the midline from its caudal part to the cranial part (*see* **Note 12**). Open the rib cage, and, using scissors, remove its ventral and lateral parts to expose the lung and the heart.

4. Make a small incision in the left ventricle of the heart. Insert a 25 gauge needle into the right ventricle, and slowly perfuse the pulmonary circulation with 5–10 mL of PBS to remove blood. Continue the perfusion until the lungs are expanded and white and the fluid flowing out of the left ventricle is completely clear.

5. Remove the lungs from the chest cavity. Make sure to separate the lungs from mediastinal organs and tissues including the heart, major blood vessels, thymus, lymph nodes, esophagus, and trachea.

6. Place the lungs in a dish containing PBS on ice.

7. The lungs can be then processed for histology, RNA or protein analysis, and flow cytometry (*see* **Note 13**).

4 Notes

1. It is advisable to measure the endotoxin content of an allergen. A common approach is to use the limulus amebocyte lysate (LAL) assay.

2. Aluminum hydroxide and magnesium hydroxide are not soluble in water. As a result, Imject Alum is a suspension of particles that sediment to the bottom of the bottle over time. Therefore, shake the bottle of Imject Alum thoroughly before each use.

3. BALB/c and C57BL/6 strains differ in the magnitude of pulmonary responses [15]. BALB/c mice develop stronger AHR. C57BL6/mice develop more robust eosinophilia in the BAL fluid (BALF).

4. Allergen extracts are complex mixtures of protein and nonprotein molecules. Allergen batches (lots) differ in composition, including variations in the protein content. It is advisable to use the same lot for all mice in a given project. If another lot is needed to continue the project, make sure to buy a lot with similar protein content.

5. Concentration of an allergen should reflect its protein content.

6. Gently mix the OVA-alum mixture between injections/mice. This will prevent the alum from depositing at the bottom of a tube and ensure a homogenous distribution of the suspension.

7. The application of an experimental agent should be done immediately after removal of a mouse from the anesthesia induction chamber. The isoflurane anesthesia is very short-lasting, and mice recover from anesthesia within a few minutes.

8. The key features of experimental asthma include elevated airway eosinophils and type 2 cytokines, peribronchial inflammation and goblet cell hyperplasia, emergence of allergen-specific IgE, and airway hyperresponsiveness (AHR) to methacholine (*see* Chapter 19). The *Alternaria* extract-induced model is used to analyze responses of ILC2s.

9. Other methods of euthanasia can also be used, including an overdose of pentobarbital (300 mg/kg mouse weight) or CO_2 asphyxiation.

10. The first milliliter of the BAL supernatant has the highest concentration of proteins (e.g., cytokines). It is important not to dilute it with additional lavages.

11. BAL supernatant can be used for measurements of inflammatory mediators (e.g., cytokines and chemokines), growth factors, enzymatic activities (e.g., lactate dehydrogenase), metabolites (e.g., ATP, uric acid), and mucus.

12. Keeping the scissor tips horizontal will ensure that the heart, large blood vessels, and the lung are not punctured.

13. Histological stains (*see* Chapter 20) are used for morphometric measurements of peribronchial inflammation (hematoxylin and eosin—H&E), goblet cell hyperplasia (periodic acid–Schiff—PAS), and deposition of extracellular matrix (trichrome stain). The lung tissue sections can also be stained for individual proteins using specific antibodies (immunohistochemistry and immunofluorescence). The lung can be lysed; proteins and RNAs can be then measured by western blot, ELISA, real-time PCR, or RNA-seq. Finally, the lung can be digested with collagenase, and the cell suspension can be analyzed by flow cytometry.

Acknowledgments

This work was supported by the National Institutes of Health grant R01HL122995 to M.M.G.

References

1. Wenzel SE (2006) Asthma: defining of the persistent adult phenotypes. Lancet 368:804–813

2. Kim HY, Umetsu DT, Dekruyff RH (2016) Innate lymphoid cells in asthma: will they take your breath away? Eur J Immunol 46:795–806

3. Hammad H, Lambrecht BN (2015) Barrier epithelial cells and the control of type 2 immunity. Immunity 43:29–40

4. Saglani S, Lloyd CM (2015) Novel concepts in airway inflammation and remodelling in asthma. Eur Respir J 46:1796–1804

5. Kanehiro A, Lahn M, Mäkelä MJ et al (2001) Tumor necrosis factor-alpha negatively regulates airway hyperresponsiveness through gamma-delta T cells. Am J Respir Crit Care Med 164:2229–2238

6. Stafford S, Li H, Forsythe PA et al (1997) Monocyte chemotactic protein-3 (MCP-3)/fibroblast-induced cytokine (FIC) in eosinophilic inflammation of the airways and the inhibitory effects of an anti-MCP-3/FIC antibody. J Immunol 158:4953–4960

7. Goplen N, Karim MZ, Liang Q et al (2009) Combined sensitization of mice to extracts of dust mite, ragweed, and Aspergillus species breaks through tolerance and establishes chronic features of asthma. J Allergy Clin Immunol 123:925–32.e11

8. Plantinga M, Guilliams M, Vanheerswynghels M et al (2013) Conventional and monocyte-derived CD11b(+) dendritic cells initiate and maintain T helper 2 cell-mediated immunity to house dust mite allergen. Immunity 38:322–335

9. Debeuf N, Haspeslagh E, van Helden M et al (2016) Mouse models of asthma. Curr Protoc Mouse Biol 6:169–184

10. Birrell MA, Van Oosterhout AJ, Belvisi MG (2010) Do the current house dust mite-driven models really mimic allergic asthma? Eur Respir J 36:1220–1221

11. Zhou W, Toki S, Zhang J et al (2016) Prostaglandin I2 signaling and inhibition of group 2 innate lymphoid cell responses. Am J Respir Crit Care Med 193:31–42

12. O'Hollaren MT, Yunginger JW, Offord KP et al (1991) Exposure to an aeroallergen as a possible precipitating factor in respiratory arrest in young patients with asthma. N Engl J Med 324:359–363

13. Bartemes KR, Iijima K, Kobayashi T et al (2012) IL-33-responsive lineage- CD25+ CD44(hi) lymphoid cells mediate innate type 2 immunity and allergic inflammation in the lungs. J Immunol 188:1503–1513

14. Christianson CA, Goplen NP, Zafar I et al (2015) Persistence of asthma requires multiple feedback circuits involving type 2 innate lymphoid cells and IL-33. J Allergy Clin Immunol 136:59–68.e14

15. Takeda K, Haczku A, Lee JJ et al (2001) Strain dependence of airway hyperresponsiveness reflects differences in eosinophil localization in the lung. Am J Physiol Lung Cell Mol Physiol 281:L394–L402

Chapter 24

Animal Models of Pulmonary Fibrosis

David N. O'Dwyer and Bethany B. Moore

Abstract

Pulmonary fibrosis is a debilitating disease and is often fatal. It may be the consequence of direct lung injury or the result of genetic defects and occupational, environmental, or drug-related exposures. In many cases the etiology is unknown. The pathogenesis of all forms of pulmonary fibrosis regardless of type of injury or etiology is incompletely understood. These disorders are characterized by the accumulation of extracellular matrix in the lung interstitium with a loss of lung compliance and impaired gas exchange that ultimately leads to respiratory failure. Animal models of pulmonary fibrosis have become indispensable in the improved understanding of these disorders. Multiple models have been developed each with advantages and disadvantages. In this chapter we discuss the application of two of the most commonly employed direct lung instillation models, namely, the induction of pulmonary fibrosis with bleomycin or fluorescein isothiocyanate (FITC). We provide details on design, materials, and methods and describe how these models can be best undertaken. We also discuss methods to induce fibrosis in aged mice using murine gamma-herpesvirus (γHV-68) and approaches to exacerbate bleomycin- or FITC-induced fibrosis using γHV-68.

Key words Animal models, Pulmonary fibrosis, Bleomycin, Fluorescein isothiocyanate, Gamma-herpesvirus

1 Introduction

Pulmonary fibrosis consists of a collection of heterogeneous pulmonary disorders that are characterized by progressive dyspnea, cough, restrictive physiology, impaired gas exchange, and respiratory failure [1]. Used interchangeably with the term interstitial lung disease (ILD), pulmonary fibrotic disorders can develop from a wide variety of insults including drugs, acute lung injury, occupational and environmental insults, autoimmune disease, and genetic defects. In many cases, the etiology of the fibrotic remodeling is not known and the condition is idiopathic. The idiopathic disorders are further categorized as idiopathic interstitial pneumonias (IIPs), of which the most common is idiopathic pulmonary fibrosis (IPF) [1]. Irrespective of the etiology, the pathogenesis of all forms of pulmonary fibrosis is incompletely understood. The histologic hallmark of IPF is a pattern of usual interstitial pneumonia (UIP) on lung tissue

Scott Alper and William J. Janssen (eds.), *Lung Innate Immunity and Inflammation: Methods and Protocols*, Methods in Molecular Biology, vol. 1809, https://doi.org/10.1007/978-1-4939-8570-8_24,
© Springer Science+Business Media, LLC, part of Springer Nature 2018

from surgical biopsies or at autopsy [1]. Unfortunately, to date no animal model fully recapitulates the histological pattern of UIP. However, this does not mean that animal models of lung fibrosis are not useful. Traditional models of fibrosis have led to important insights into disease pathobiology, including features of inflammation and fibroproliferation [2, 3], and many of the gene expression changes seen in murine models (e.g., the bleomycin model) recapitulate alterations in human disease [4].

Several methods exist for modeling human pulmonary fibrosis. The overexpression of certain cytokines results in pulmonary fibrosis. These overexpression models have been achieved using both gene transfer and transgenic approaches. Transforming growth factor-β (TGF-β) overexpression by adenoviral delivery [5] or doxycycline-regulated transgenic expression in epithelial cells is an example [6]. Transforming growth factor-α (TGF-α) overexpression in animal models results in both pulmonary fibrosis and pulmonary hypertension [7, 8]. Other overexpression models include IL-13, IL-1β, and tumor necrosis factor-α (TNF-α) [9–11]. A recent novel development has been the introduction of models that target type II alveolar epithelial cell (AEC) death. This is achieved through the expression of the diphtheria toxin receptor under the control of a type II AEC promoter [surfactant protein C (SP-C)]. Repetitive delivery (intraperitoneal) of diphtheria toxin injures type II AECs specifically and induces pulmonary fibrosis [12]. Animal models of familial pulmonary fibrosis (FPF) and models of direct lung injury-mediated pulmonary fibrosis have also been developed and are reviewed elsewhere [3]. Humanized models also exist. The best characterized "humanized" model involves the instillation of fibroblasts isolated from patients with IPF into immunodeficient nonobese diabetic/severe combined immunodeficiency (NOD/SCID/beige) mice [13, 14]. Overall there are a wide variety of specialized platforms available for the evaluation of pulmonary fibrosis in animal models.

A NHLBI workshop highlighted the importance of animal models and the need for their further refinement in understanding the pathogenesis of IPF [15]. As mentioned, transgenic approaches offer methods to study cell-cell interactions and the effects of soluble mediators on the pathogenesis of pulmonary fibrosis. These approaches offer opportunities to better understand novel targets and assess and validate these targets in human disease. However, to best take advantage of these tools, it is necessary to study the phenotypes of these mice in well-characterized models of lung fibrosis. This chapter summarizes two of the most commonly employed preclinical models of pulmonary fibrosis, namely, the use of bleomycin and fluorescein isothiocyanate (FITC). We also describe the application of an exacerbation model of pulmonary fibrosis that reflects many of the features common to acute exacerbations of IPF. These acute exacerbations are episodes that are characterized by acute respiratory worsening in patients and are associated with very high mortality rates [16].

The incidence of IPF increases with age [17]. However, most animal models use mice that are 6–8 weeks old. To model the age range for IPF in mice, one needs to use animals that are 18–24 months old [18, 19]. While several studies have utilized bleomycin in older animals [20] and shown worsened pathology or impaired resolution, there is also evidence that infection with γHV-68 can induce fibrosis in aged but not young mice [21–23]. Aged mice infected with a murine γHV-68 develop pulmonary fibrosis with associated endoplasmic reticulum stress within alveolar epithelial cells and increased fibroblast transforming growth factor-β (TGF-β) signaling [24, 25]. Thus aged lungs are uniquely susceptible to fibrosis induced by this viral pathogen, offering a rare opportunity to study a model system with such dramatic age-associated effects. Not surprisingly, this same viral pathogen is able to exacerbate the degree of fibrosis caused by bleomycin or FITC in young animals even though infection alone does not induce fibrosis [26].

Bleomycin is a glycopeptide antibiotic produced by *Streptomyces verticillus* [27]. Bleomycin has been used in the treatment of several different forms of human neoplasms for decades [28–31]. However, its use has been limited by the potential development of pulmonary fibrosis as an adverse effect. The reported rates of this effect vary in studies but range from 6.8 to 46% [32, 33]. The bleomycin injury occurs through cytotoxicity; bleomycin induces single- and double-stranded breaks in DNA and is associated with rapid DNA fragmentation [34, 35]. This results in the generation of free radicals and oxidant-mediated stress [36]. Recent work supports direct scission by DNA H+ abstraction as a major contributor to DNA strand breaks [37, 38]. Intratracheal administration results in an initial lung-specific injury of the AECs. Type I AEC apoptosis occurs with significant loss of this cell type. This critical loss is followed by the development of hyperplastic type II AECs and abnormal epithelial regeneration [39, 40]. There is subsequent pro-inflammatory cytokine release and inflammatory cell influx; peak inflammation occurs on day 7 post-bleomycin administration. At later stages there is increased expression of pro-fibrotic cytokines including transforming growth factor (TGF)-β with fibroblast accumulation, myofibroblast differentiation, collagen deposition, and thickening of the extracellular matrix. When bleomycin is administered systemically, the initial injury occurs at the level of the pulmonary vascular endothelium, which is followed by extensive AEC damage. Following a single intratracheal or trans-oral administration of bleomycin, histological and biochemical evidence of lung fibrosis is noted starting at day 14, with day 21 being the most common time point for analysis. Figure 1 shows the characteristic histological pattern of bleomycin-induced fibrosis at day 21 post-instillation in mice. Figure 2 demonstrates collagen deposition with Masson's trichrome staining

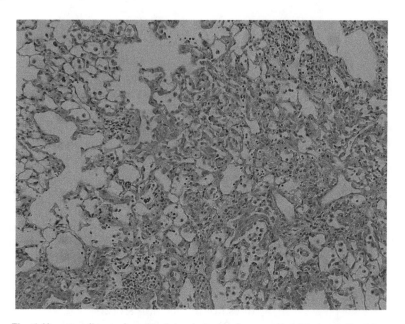

Fig. 1 Hematoxylin- and eosin-stained murine lung section 21 days after bleomycin administration. Diffuse interstitial abnormalities and loss of normal lung architecture are present

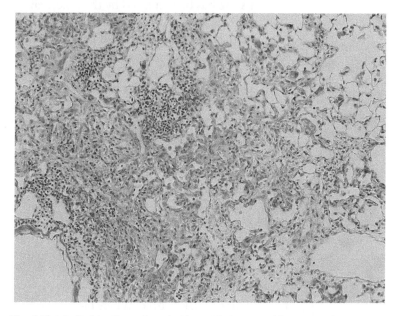

Fig. 2 Histological sections of murine lung 21 days post-bleomycin administration demonstrating interstitial collagen deposition (blue) by Masson's trichrome staining

in bleomycin-treated mice at day 21. For comparison, normal lung architecture is shown in Fig. 3.

FITC is a skin-sensitizing hapten that has been used to model pulmonary fibrosis in rodents [41, 42]. The intratracheal delivery of FITC initiates acute lung injury with inflammatory cell influx,

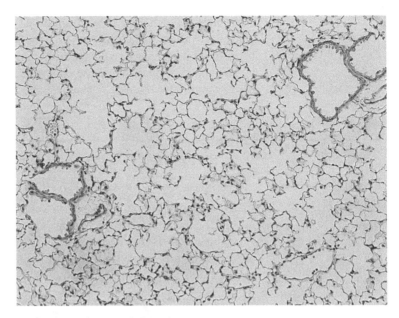

Fig. 3 Hematoxylin- and eosin-stained section of murine lung 21 days after saline instillation. Healthy lung architecture is present

including neutrophils. This is followed by fibrosis that establishes within 14–21 days. FITC conjugates to parenchymal proteins and persists at areas of injury and in areas with fibrosis [2]. Immunofluorescence can be utilized to identify localized injury patterns (*see* images in References 2 and 3). FITC can reliably induce fibrosis in animals generally resistant to the effects of bleomycin (i.e., Balb/c mice) [2].

Several different animal species have been used in fibrosis studies. Bleomycin-induced fibrosis has been described in dogs [43], Australian sheep [44], hamsters [45], and rats [46]. Horses develop fibrosis after experimental gamma-herpesvirus infection [47]. Some species develop spontaneous pulmonary fibrosis including donkeys [48], cats [49], and West Highland white terriers [50]. However, for multiple reasons, mice have become the most commonly used tool. They are relatively less expensive to house, and a vast library of transgenic strains and study reagents exists for this species.

Most published studies of bleomycin-induced fibrosis have utilized the intratracheal (IT) method of administration. The administration of bleomycin in clinical settings is parenteral or intrapleural. Early models applied this method of administration to better reflect some of the clinical context. However, the cumulative dose of bleomycin in parenteral models is in the 100–200 mg (units)/kg range. The IT single injection dose is in the range of 1.5–7.5 mg/kg. The IT model of administration has certain advantages over the parenteral model, including a single and lower dose, clarity regarding the initiation of injury, and reproducible kinetic development of fibrosis. However, the trans-oral approach has recently been developed and has demonstrated efficiency with added advantages over the IT method in that it eliminates the need for surgery for the mice. For this reason, it is preferred by some institutional animal use committees and will be the route of administration we discuss.

2 Materials

2.1 Animals and Reagents

1. Animals: C57Bl/6, CBA/J, or BALB/c mice (*see* **Note 1**).

2. Anesthetic drugs: mixture of ketamine HCL and xylazine HCL. Ketamine (80–120 mg/kg mouse body weight) and xylazine (5–10 mg/kg mouse body weight) are diluted in sterile saline and given via intraperitoneal injection.

3. Bleomycin (*see* **Note 2** for source and dose).

4. FITC (*see* **Note 3** for source and dose).

5. γHV-68: γHV-68 (murid herpesvirus, American Type Culture Collection number: VR-1465) is diluted to deliver a dose of 5×10^4 PFU/mouse generally given by intranasal inoculation in a volume of 10 μL.

2.2 Oropharyngeal Administration of Bleomycin or FITC

1. Sterile forceps.

2. Surgery board (at approximately 70° from horizontal). We often use a Styrofoam tray (used for 15 or 50 mL conical centrifuge tubes) with a thin rubber band stretched around the tray. This rubber band will be used to hold the mouse on the board by tucking the incisor teeth over the rubber band.

3. Syringe.

4. 26 gauge needles or a blunt-tipped gavage needle.

5. Warmer pads and absorbent paper.

6. Mouse bedding.

7. Personal protective equipment including gloves, mask, and eye goggles.

8. Paper towels.

9. Eye lubricant (ophthalmic gel with mineral oil, white petrolatum, and lanolin).

10. Vortexer.

2.3 Hydroxyproline Assay

1. Citrate/acetate buffer: 5% citric acid (5 g monohydrate), 1.2% glacial acetic acid (1.2 mL), 7.24% sodium acetate (7.24 g trihydrate), 3.4% sodium hydroxide (3.4 g) to 100 mL sterile water.

2. Chloramine-T solution: 0.282 g chloramine-T, 2 mL N-propanol, 2 mL sterile water, 16 mL citrate/acetate buffer. Heat in a 37°C water bath until completely dissolved (approx. 15 min). Ensure you wear protective gloves.

3. Ehrlich's solution: 2.5 g 4-DMAB Ehrlich's reagent (*p*-dimethylaminobenzaldehyde), 9.3 mL *N*-propanol, 3.9 mL 70% perchloric acid (add last and add slowly). Ensure you wear protective gloves.

4. Hydroxyproline standard: the standard is *cis*-4-hydroxy-l-proline. It is made up in sterile water to 4 mg/mL. To make the first dilution, dilute the stock 1:10 in sterile water. Use 5 μL of this dilution for the first standard (400 μg/mL), and then set up a series of 1:2 dilutions (400, 200, 100, 50, 25, 12.5, 6.25, 3.12, 1.56, 0.78, 0.39, 0 μg/mL).

2.4 Inflation of Lungs for Histologic Assessment

1. Surgical board with rubber band.

2. Absorbent pads.

3. Scissors.

4. Forceps.

5. Silk suture thread.

6. 1 cc syringe.

7. 16 gauge needle.

8. Tubing to fit over 20 gauge needle.

9. 5 cc syringe with 23 gauge needle filled with saline for perfusion.

10. 10% neutral-buffered formalin.

3 Methods

3.1 Trans-oral Bleomycin Administration

1. Age-, weight-, and gender-matched mice should be randomly allocated to bleomycin-treated or control groups.

2. Anesthetize animals via intraperitoneal injection of ketamine (80–120 mg/kg) and xylazine (5–10 mg/kg) diluted in sterile saline.

3. Once anesthesia is adequate, the animals are placed supine on the surgical board and suspended using thread or a rubber band via the animal's incisors. The mouth is opened using a sterile forceps; the tongue is isolated and pulled forward toward the mandible and lower incisors. Adequate lighting is required and should allow visualization of the vocal cords.

4. The bleomycin dose is then administered through the vocal cords using a 100 μL pipette during inspiration (Fig. 4). *See* **Note 2** for dosing information. In most cases, animals allocated

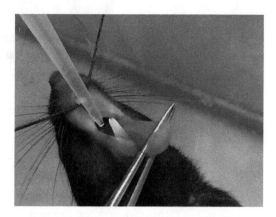

Fig. 4 Trans-oral administration of bleomycin using a pipette tip applied to posterior oropharynx and sterile forceps to carefully manipulate the tongue

to the control group will receive an equal volume of sterile phosphate-buffered saline (PBS) through the same approach.

5. The animals should then be allowed to recover from anesthesia by placing them on warming pads, gently applying eye lubricant to the eye area, and monitoring them until fully awake and recovered from anesthesia. The entire procedure should take approximately 15 min. Animals should fully recover from anesthesia within 1 h of administration.

6. Animals are examined daily, and body weight and survival may be recorded. The period of most significant illness is between days 5 and 10, and wet chow on the cage floor may help keep animals hydrated. Experimental groups are sacrificed at specific predefined time points, with day 21 being the most common time to analyze lung histology and collagen deposition using the hydroxyproline assay. The number of animals needed depends on the outcome variable to be measured, but a good rule of thumb is that 3–5 mice are needed for histologic assessments and 7–10 mice/group are needed for hydroxyproline assessment. These methods are discussed below.

3.2 FITC-Induced Pulmonary Fibrosis

1. Anesthetize and prepare animals as outlined in Subheading 3.1, **step 1**, to Subheading 3.1, **step 3**.

2. FITC is administered by trans-oral injection using the exact approach described in Subheading 3.1, **step 4**, for bleomycin administration. Take care to vortex the FITC slurry prior to administration in each animal. *See* **Note 3** for dosing information.

3. As with bleomycin, animals are monitored closely until they fully recover from anesthesia. Mice are euthanized for histologic assessment and collagen deposition generally on days 14–21. One advantage of the FITC model is that it is persistent;

thus, mice will still show evidence of pathologic fibrosis at later time points (days 35, 42, and 60).

3.3 Viral Exacerbation of Pulmonary Fibrosis

To model an acute exacerbation of lung fibrosis, mice treated with bleomycin or FITC on day 0 can be re-anesthetized on day 14 and given an intranasal administration of γHV-68.

1. Perform bleomycin or FITC installation on day 0 as described in Subheading 3.1 or 3.2.

2. On day 14, re-anesthetize mice as described in Subheading 3.1, **step 2**.

3. Introduce 5×10^4 PFU γHV-68 via intranasal administration. Generally this is delivered in a 10 µL total volume alternating delivery to the mouse nares on both sides of the face by dropping small droplets of liquid delivered by pipette tip to the nose of the mouse and waiting for full inhalation before applying the next droplet. Note that mice are obligate nose breathers and thus should not have both nares covered with droplets of virus suspension at the same time (*see* Fig. 5).

4. We generally follow the 10 µL viral instillation with an additional 10 µL of sterile PBS.

5. Once all liquid has been inhaled, mice should be placed on a warming pad and observed for full recovery from anesthesia.

6. Mice are generally analyzed for exacerbations 7 days post-viral infection which corresponds to day 21 post-bleomycin or FITC.

3.4 Viral-Induced Fibrosis in Aged Mice

1. To induce fibrosis using γHV-68 in aged mice, obtain mice between 15 and 24 months of age, ensuring that you have

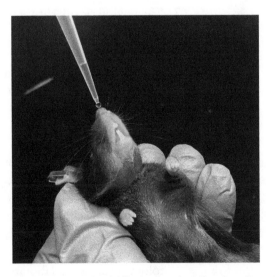

Fig. 5 Intranasal injection in mice (obligate nose breathers) postanesthesia

mice of similar weight, age, and gender in both the control (saline) and virus groups (*see* **Note 1**).

2. Anesthetize mice as described in Subheading 3.1, **step 2**.

3. Deliver a slightly higher dose of virus (5×10^5–1×10^6) PFU via intranasal injection as described in Subheading 3.3, **step 3**.

4. Allow mice to recover from anesthesia.

5. Monitor daily until day 21, when mice can be euthanized for measurements of fibrosis using hydroxyproline assays and histologic assessments as described below.

3.5 Preparation of Lungs for Histologic Assessment

1. Following euthanasia, open the chest cavity of the mouse via a horizontal incision across the upper abdominal cavity and a second incision through the chest wall, rib cage, and diaphragm. Caution must be taken not to damage the underlying lung parenchyma.

2. Remove the rib cage using sterile scissors; the contents of the thoracic cage are now exposed.

3. Isolate the right ventricle, and inject sterile saline slowly into the right ventricle using a sterile 21 gauge needle attached to a sterile 10 mL syringe. Infuse saline slowly until the lung lobes are blanched.

4. Bronchoalveolar lavage (BAL) can be performed if required at this point.

5. The lung tissue can then be carefully extracted from the thoracic cage and harvested for cell, RNA, and/or protein analysis.

6. In a separate procedure to prepare the lungs for histology, perfuse the mouse with 5 mL PBS through the right ventricle.

7. Tie a nylon or silk suture thread loosely around the trachea of the mouse.

8. Make a small incision in the trachea using scissors.

9. Feed a cannula attached to a 20 gauge needle on a 1 cc syringe through the lumen of the trachea, and direct it caudally. Deliver the fixative (1 mL of 10% neutral-buffered formalin).

10. Tie the knot in the suture thread tightly while removing the cannula to keep fixative from escaping from the lungs.

11. Remove the inflated lungs en bloc from the body cavity of the mouse, and place them into a conical tube containing 10% neutral-buffered formalin overnight.

12. The lungs are then dehydrated by placing the inflated lung in 70% ethanol. At this point the lungs can be delivered to a histology processing laboratory for paraffin embedding. Once

embedded, 3 μm lung slices are prepared and stained using hematoxylin and eosin to assess inflammatory infiltrates (e.g., Fig 2) and Masson's trichrome to assess collagen accumulation (e.g., Fig. 3). *See* Chapter 20 for lung histological methods.

3.6   Assessment of Hydroxyproline Content in the Lungs

1. Following euthanasia, record the weight of the mouse.

2. Perfuse the lungs with 2–3 mL of PBS (*see* **Note 4**).

3. Place lung lobes in 1 mL of 1× PBS and homogenize with a tissue homogenizer. Rinse homogenizer thoroughly between each sample.

4. Remove 500 μL of homogenized lung and place in a glass screw cap tube. Add 500 μL of 12 N HCl. Screw the lids on very tightly.

5. Place the tubes in an oven or water bath set at 120 °C for 8 h to overnight (*see* **Note 5**).

6. Next day: prepare "chloramine-T" and "Ehrlich's reagent" fresh each time. Ensure you wear protective gloves.

7. In a 96-well plate (flat bottom, ELISA plate), add 5 μL of lung sample or standard to each well (run samples in duplicate or triplicate). It is also a good idea to dilute the lung samples 1:2 and test this dilution as well.

8. Add 5 μL citrate/acetate buffer to all wells and tap the edges of the plate to mix well.

9. Add 100 μL of chloramine-T solution to all wells (samples + standards) and tap plate to mix.

10. Incubate for 20 min at room temperature and turn on an oven to 65 °C.

11. Add 100 μL of Ehrlich's solution to each well (samples + standards), and incubate for 15–20 min at 65 °C.

12. Read plate in a plate reader set to monitor absorbance at 550 nm.

13. Calculate the amount of hydroxyproline per mL in your lung sample by comparison to standard curve.

4   Notes

1. The three most common strains utilized in fibrosis studies are C57Bl/6, CBA/J, and BALB/c (all are commercially available) for studies on mice 6–12 weeks of age. These are specific pathogen-free (SPF) mice. Housing units should be maintained in positive-pressure, air-conditioned units at 25 °C with 50% relative humidity, and a 12 h day-night cycle is recommended. The use of matched strain, age, and gender animals is

important, as responses may be variable. BALB/c mice will often require higher doses of bleomycin to induce comparable fibrosis compared to other strains. Aged mice should be at least 15 months old and are available in limited quantities through the National Institute on Aging (NIA) (https://www.nia.nih.gov/research/dab/aged-rodent-colonies-handbook/ordering-instructions).

2. Bleomycin has been used in multiple animal models and has been administered by varied routes including intraperitoneal injection, intravenous injection, subcutaneous injection, and intratracheal injection [43, 51, 52]. Of these methods, the intratracheal or trans-oral application of bleomycin has emerged as the most common. Bleomycin is available from several vendors and can also be ordered as clinical-grade Blenoxane. The dose needed to induce robust fibrosis can vary greatly when using bleomycin from different vendors, and thus, users should test multiple concentrations to determine an effective dose. In our laboratory, we typically purchase bleomycin from Sigma-Aldrich and dilute the stock in sterile saline before delivering 0.025–0.05 U per mouse. Bleomycin should be kept on ice until it is administered. Bleomycin stock should be stored in aliquots for single use, as repeated freeze-thaw cycles will reduce its efficacy.

3. FITC is administered by intratracheal or trans-oral injection which results in alveolar damage and increased vascular permeability. FITC (fluorescein isothiocyanate isomer I, Sigma) must be made fresh just prior to inoculation. The doses for mice range from 0.0007 mg per gram body weight dissolved in PBS to an intratracheal delivery of 50 μL of a 1.4 mg/mL solution [41, 42, 53]. The preparation of FITC requires its addition to PBS in a conical tube and vortexing to ensure the formation of an equal solution. The mixture is then sonicated at 50% power for 30 s. The mixture is transferred to multiuse vials; each vial should be vortexed prior to instillation to ensure uniformity of delivery and injury. The sonication time is important, as prolonged sonication results in smaller particle size and increased toxicity which may lead to early death [2], whereas too little sonication results in particle sizes that do not produce extensive lung injury.

4. This step can be skipped if you want to weigh the lungs prior to perfusion. You can also weigh the lungs post-perfusion, but this is less accurate. We generally do not normalize to lung weight at all.

5. It is easy for your sample to boil away; be very careful about securing your caps.

References

1. Raghu G, Collard HR, Egan JJ, Martinez FJ, Behr J, Brown KK, Colby TV, Cordier JF, Flaherty KR, Lasky JA, Lynch DA, Ryu JH, Swigris JJ, Wells AU, Ancochea J, Bouros D, Carvalho C, Costabel U, Ebina M, Hansell DM, Johkoh T, Kim DS, King TE Jr, Kondoh Y, Myers J, Muller NL, Nicholson AG, Richeldi L, Selman M, Dudden RF, Griss BS, Protzko SL, Schunemann HJ, Fibrosis AEJACoIP (2011) An official ATS/ERS/JRS/ALAT statement: idiopathic pulmonary fibrosis: evidence-based guidelines for diagnosis and management. Am J Respir Crit Care Med 183(6):788–824. https://doi.org/10.1164/rccm.2009-040GL

2. Moore BB, Hogaboam CM (2008) Murine models of pulmonary fibrosis. Am J Physiol Lung Cell Mol Physiol 294(2):L152–L160. https://doi.org/10.1152/ajplung.00313.2007

3. Moore BB, Lawson WE, Oury TD, Sisson TH, Raghavendran K, Hogaboam CM (2013) Animal models of fibrotic lung disease. Am J Respir Cell Mol Biol 49(2):167–179. https://doi.org/10.1165/rcmb.2013-0094TR

4. Peng R, Sridhar S, Tyagi G, Phillips JE, Garrido R, Harris P, Burns L, Renteria L, Woods J, Chen L, Allard J, Ravindran P, Bitter H, Liang Z, Hogaboam CM, Kitson C, Budd DC, Fine JS, Bauer CM, Stevenson CS (2013) Bleomycin induces molecular changes directly relevant to idiopathic pulmonary fibrosis: a model for "active" disease. PLoS One 8(4):e59348. https://doi.org/10.1371/journal.pone.0059348

5. Sime PJ, Xing Z, Graham FL, Csaky KG, Gauldie J (1997) Adenovector-mediated gene transfer of active transforming growth factor-beta1 induces prolonged severe fibrosis in rat lung. J Clin Invest 100(4):768–776. https://doi.org/10.1172/JCI119590

6. Lee CG, Cho SJ, Kang MJ, Chapoval SP, Lee PJ, Noble PW, Yehualaeshet T, Lu B, Flavell RA, Milbrandt J, Homer RJ, Elias JA (2004) Early growth response gene 1-mediated apoptosis is essential for transforming growth factor beta1-induced pulmonary fibrosis. J Exp Med 200(3):377–389. https://doi.org/10.1084/jem.20040104

7. Korfhagen TR, Swantz RJ, Wert SE, McCarty JM, Kerlakian CB, Glasser SW, Whitsett JA (1994) Respiratory epithelial cell expression of human transforming growth factor-alpha induces lung fibrosis in transgenic mice. J Clin Invest 93(4):1691–1699. https://doi.org/10.1172/JCI117152

8. Hardie WD, Korfhagen TR, Sartor MA, Prestridge A, Medvedovic M, Le Cras TD, Ikegami M, Wesselkamper SC, Davidson C, Dietsch M, Nichols W, Whitsett JA, Leikauf GD (2007) Genomic profile of matrix and vasculature remodeling in TGF-alpha induced pulmonary fibrosis. Am J Respir Cell Mol Biol 37(3):309–321. https://doi.org/10.1165/rcmb.2006-0455OC

9. Lee CG, Homer RJ, Zhu Z, Lanone S, Wang X, Koteliansky V, Shipley JM, Gotwals P, Noble P, Chen Q, Senior RM, Elias JA (2001) Interleukin-13 induces tissue fibrosis by selectively stimulating and activating transforming growth factor beta(1). J Exp Med 194(6):809–821

10. Kolb M, Margetts PJ, Anthony DC, Pitossi F, Gauldie J (2001) Transient expression of IL-1beta induces acute lung injury and chronic repair leading to pulmonary fibrosis. J Clin Invest 107(12):1529–1536. https://doi.org/10.1172/JCI12568

11. Sime PJ, Marr RA, Gauldie D, Xing Z, Hewlett BR, Graham FL, Gauldie J (1998) Transfer of tumor necrosis factor-alpha to rat lung induces severe pulmonary inflammation and patchy interstitial fibrogenesis with induction of transforming growth factor-beta1 and myofibroblasts. Am J Pathol 153(3):825–832

12. Sisson TH, Mendez M, Choi K, Subbotina N, Courey A, Cunningham A, Dave A, Engelhardt JF, Liu X, White ES, Thannickal VJ, Moore BB, Christensen PJ, Simon RH (2010) Targeted injury of type II alveolar epithelial cells induces pulmonary fibrosis. Am J Respir Crit Care Med 181(3):254–263. https://doi.org/10.1164/rccm.200810-1615OC

13. Pierce EM, Carpenter K, Jakubzick C, Kunkel SL, Flaherty KR, Martinez FJ, Hogaboam CM (2007) Therapeutic targeting of CC ligand 21 or CC chemokine receptor 7 abrogates pulmonary fibrosis induced by the adoptive transfer of human pulmonary fibroblasts to immunodeficient mice. Am J Pathol 170(4):1152–1164. https://doi.org/10.2353/ajpath.2007.060649

14. Trujillo G, Meneghin A, Flaherty KR, Sholl LM, Myers JL, Kazerooni EA, Gross BH, Oak SR, Coelho AL, Evanoff H, Day E, Toews GB, Joshi AD, Schaller MA, Waters B, Jarai G, Westwick J, Kunkel SL, Martinez FJ, Hogaboam CM (2010) TLR9 differentiates rapidly from slowly progressing forms of idiopathic pulmonary fibrosis. Sci Transl Med

2(57):57ra82. https://doi.org/10.1126/scitranslmed.3001510

15. Blackwell TS, Tager AM, Borok Z, Moore BB, Schwartz DA, Anstrom KJ, Bar-Joseph Z, Bitterman P, Blackburn MR, Bradford W, Brown KK, Chapman HA, Collard HR, Cosgrove GP, Deterding R, Doyle R, Flaherty KR, Garcia CK, Hagood JS, Henke CA, Herzog E, Hogaboam CM, Horowitz JC, King TE Jr, Loyd JE, Lawson WE, Marsh CB, Noble PW, Noth I, Sheppard D, Olsson J, Ortiz LA, O'Riordan TG, Oury TD, Raghu G, Roman J, Sime PJ, Sisson TH, Tschumperlin D, Violette SM, Weaver TE, Wells RG, White ES, Kaminski N, Martinez FJ, Wynn TA, Thannickal VJ, Eu JP (2014) Future directions in idiopathic pulmonary fibrosis research. An NHLBI workshop report. Am J Respir Crit Care Med 189(2):214–222. https://doi.org/10.1164/rccm.201306-1141WS

16. Collard HR, Moore BB, Flaherty KR, Brown KK, Kaner RJ, King TE Jr, Lasky JA, Loyd JE, Noth I, Olman MA, Raghu G, Roman J, Ryu JH, Zisman DA, Hunninghake GW, Colby TV, Egan JJ, Hansell DM, Johkoh T, Kaminski N, Kim DS, Kondoh Y, Lynch DA, Muller-Quernheim J, Myers JL, Nicholson AG, Selman M, Toews GB, Wells AU, Martinez FJ, Idiopathic Pulmonary Fibrosis Clinical Research Network Investigators (2007) Acute exacerbations of idiopathic pulmonary fibrosis. Am J Respir Crit Care Med 176(7):636–643. https://doi.org/10.1164/rccm.200703-463PP

17. Fell CD, Martinez FJ, Liu LX, Murray S, Han MK, Kazerooni EA, Gross BH, Myers J, Travis WD, Colby TV, Toews GB, Flaherty KR (2010) Clinical predictors of a diagnosis of idiopathic pulmonary fibrosis. Am J Respir Crit Care Med 181(8):832–837. https://doi.org/10.1164/rccm.200906-0959OC

18. Geifman N, Rubin E (2013) The mouse age phenome knowledgebase and disease-specific inter-species age mapping. PLoS One 8(12):e81114. https://doi.org/10.1371/journal.pone.0081114

19. Fox JG, Barthold S, Davisson M, Newcomer C, Quimby F, Smith A (2007) The mouse in aging research, 2nd edn. American College Laboratory Animal Medicine, Burlington, MA, pp 637–672

20. Redente EF, Jacobsen KM, Solomon JJ, Lara AR, Faubel S, Keith RC, Henson PM, Downey GP, Riches DW (2011) Age and sex dimorphisms contribute to the severity of bleomycin-induced lung injury and fibrosis. Am J Physiol Lung Cell Mol Physiol 301(4):L510–L518. https://doi.org/10.1152/ajplung.00122.2011

21. Stout-Delgado HW, Cho SJ, Chu SG, Mitzel DN, Villalba J, El-Chemaly S, Ryter SW, Choi AM, Rosas IO (2016) Age-dependent susceptibility to pulmonary fibrosis is associated with NLRP3 inflammasome activation. Am J Respir Cell Mol Biol 55(2):252–263. https://doi.org/10.1165/rcmb.2015-0222OC

22. Sueblinvong V, Neujahr DC, Mills ST, Roser-Page S, Ritzenthaler JD, Guidot D, Rojas M, Roman J (2012) Predisposition for disrepair in the aged lung. Am J Med Sci 344(1):41–51. https://doi.org/10.1097/MAJ.0b013e318234c132

23. Sueblinvong V, Neveu WA, Neujahr DC, Mills ST, Rojas M, Roman J, Guidot DM (2014) Aging promotes pro-fibrotic matrix production and increases fibrocyte recruitment during acute lung injury. Adv Biosci Biotechnol 5(1):19–30. https://doi.org/10.4236/abb.2014.51004

24. Naik PN, Horowitz JC, Moore TA, Wilke CA, Toews GB, Moore BB (2012) Pulmonary fibrosis induced by gamma-herpesvirus in aged mice is associated with increased fibroblast responsiveness to transforming growth factor-beta. J Gerontol A Biol Sci Med Sci 67(7):714–725. https://doi.org/10.1093/gerona/glr211

25. Torres-Gonzalez E, Bueno M, Tanaka A, Krug LT, Cheng DS, Polosukhin VV, Sorescu D, Lawson WE, Blackwell TS, Rojas M, Mora AL (2012) Role of endoplasmic reticulum stress in age-related susceptibility to lung fibrosis. Am J Respir Cell Mol Biol 46(6):748–756. https://doi.org/10.1165/rcmb.2011-0224OC

26. McMillan TR, Moore BB, Weinberg JB, Vannella KM, Fields WB, Christensen PJ, van Dyk LF, Toews GB (2008) Exacerbation of established pulmonary fibrosis in a murine model by gammaherpesvirus. Am J Respir Crit Care Med 177(7):771–780. https://doi.org/10.1164/rccm.200708-1184OC

27. Umezawa H, Maeda K, Takeuchi T, Okami Y (1966) New antibiotics, bleomycin A and B. J Antibiot 19(5):200–209

28. Schein PS, VT DV Jr, Hubbard S, Chabner BA, Canellos GP, Berard C, Young RC (1976) Bleomycin, adriamycin, cyclophosphamide, vincristine, and prednisone (BACOP) combination chemotherapy in the treatment of advanced diffuse histiocytic lymphoma. Ann Intern Med 85(4):417–422

29. Suzuki Y, Miyake H, Sakai M, Inuyama Y, Matsukawa J (1969) Bleomycin in malignant tumors of head and neck. Keio J Med 18(3):153–162

30. Oka S, Sato K, Nakai Y, Kurita K, Hashimoto K, Oshibe M (1969) Treatment of lung cancer

with bleomycin. The science reports of the research institutes, Tohoku University Ser C, Medicine. Tohoku Daigaku 16 (1):30–36

31. Clinical Screening Co-operative Group of the European Organization for Research on the Treatment of Cancer (1970) Study of the clinical efficiency of bleomycin in human cancer. Br Med J 2(5710):643–645

32. O'Sullivan JM, Huddart RA, Norman AR, Nicholls J, Dearnaley DP, Horwich A (2003) Predicting the risk of bleomycin lung toxicity in patients with germ-cell tumours. Ann Oncol 14(1):91–96

33. Sleijfer S (2001) Bleomycin-induced pneumonitis. Chest 120(2):617–624

34. Huang CH, Mirabelli CK, Jan Y, Crooke ST (1981) Single-strand and double-strand deoxyribonucleic acid breaks produced by several bleomycin analogues. Biochemistry 20(2):233–238

35. Tounekti O, Kenani A, Foray N, Orlowski S, Mir LM (2001) The ratio of single- to double-strand DNA breaks and their absolute values determine cell death pathway. Br J Cancer 84(9):1272–1279. https://doi.org/10.1054/bjoc.2001.1786

36. Sugiura Y, Kikuchi T (1978) Formation of superoxide and hydroxy radicals in iron(II)-bleomycin-oxygen system: electron spin resonance detection by spin trapping. J Antibiot 31(12):1310–1312

37. Liu LV, Bell CB 3rd, Wong SD, Wilson SA, Kwak Y, Chow MS, Zhao J, Hodgson KO, Hedman B, Solomon EI (2010) Definition of the intermediates and mechanism of the anti-cancer drug bleomycin using nuclear resonance vibrational spectroscopy and related methods. Proc Natl Acad Sci U S A 107(52):22419–22424. https://doi.org/10.1073/pnas.1016323107

38. Decker A, Chow MS, Kemsley JN, Lehnert N, Solomon EI (2006) Direct hydrogen-atom abstraction by activated bleomycin: an experimental and computational study. J Am Chem Soc 128(14):4719–4733. https://doi.org/10.1021/ja057378n

39. Izbicki G, Segel MJ, Christensen TG, Conner MW, Breuer R (2002) Time course of bleomycin-induced lung fibrosis. Int J Exp Pathol 83(3):111–119

40. Kumar RK, Watkins SG, Lykke AW (1985) Pulmonary responses to bleomycin-induced injury: an immunomorphologic and electron microscopic study. Exp Pathol 28(1):33–43

41. Roberts SN, Howie SE, Wallace WA, Brown DM, Lamb D, Ramage EA, Donaldson K (1995) A novel model for human interstitial lung disease: hapten-driven lung fibrosis in rodents. J Pathol 176(3):309–318. https://doi.org/10.1002/path.1711760313

42. Christensen PJ, Goodman RE, Pastoriza L, Moore B, Toews GB (1999) Induction of lung fibrosis in the mouse by intratracheal instillation of fluorescein isothiocyanate is not T-cell-dependent. Am J Pathol 155(5):1773–1779. https://doi.org/10.1016/S0002-9440(10)65493-4

43. Fleischman RW, Baker JR, Thompson GR, Schaeppi UH, Illievski VR, Cooney DA, Davis RD (1971) Bleomycin-induced interstitial pneumonia in dogs. Thorax 26(6):675–682

44. Organ L, Bacci B, Koumoundouros E, Barcham G, Milne M, Kimpton W, Samuel C, Snibson K (2015) Structural and functional correlations in a large animal model of bleomycin-induced pulmonary fibrosis. BMC Pulm Med 15:81. https://doi.org/10.1186/s12890-015-0071-6

45. Snider GL, Hayes JA, Korthy AL (1978) Chronic interstitial pulmonary fibrosis produced in hamsters by endotracheal bleomycin: pathology and stereology. Am Rev Respir Dis 117(6):1099–1108. https://doi.org/10.1164/arrd.1978.117.6.1099

46. Thrall RS, McCormick JR, Jack RM, McReynolds RA, Ward PA (1979) Bleomycin-induced pulmonary fibrosis in the rat: inhibition by indomethacin. Am J Pathol 95(1):117–130

47. Williams KJ, Robinson NE, Lim A, Brandenberger C, Maes R, Behan A, Bolin SR (2013) Experimental induction of pulmonary fibrosis in horses with the gammaherpesvirus equine herpesvirus 5. PLoS One 8(10):e77754. https://doi.org/10.1371/journal.pone.0077754

48. Miele A, Dhaliwal K, Du Toit N, Murchison JT, Dhaliwal C, Brooks H, Smith SH, Hirani N, Schwarz T, Haslett C, Wallace WA, McGorum BC (2014) Chronic pleuropulmonary fibrosis and elastosis of aged donkeys: similarities to human pleuroparenchymal fibroelastosis. Chest 145(6):1325–1332. https://doi.org/10.1378/chest.13-1306

49. Williams K, Malarkey D, Cohn L, Patrick D, Dye J, Toews G (2004) Identification of spontaneous feline idiopathic pulmonary fibrosis: morphology and ultrastructural evidence for a type II pneumocyte defect. Chest 125(6):2278–2288

50. Corcoran BM, Cobb M, Martin MW, Dukes-McEwan J, French A, Fuentes VL, Boswood A, Rhind S (1999) Chronic pulmonary disease in West Highland white terriers. Vet Rec 144(22):611–616

378   David N. O'Dwyer and Bethany B. Moore

51. Adamson IY, Bowden DH (1974) The pathogenesis of bleomycin-induced pulmonary fibrosis in mice. Am J Pathol 77(2):185–197

52. McCullough B, Collins JF, Johanson WG Jr, Grover FL (1978) Bleomycin-induced diffuse interstitial pulmonary fibrosis in baboons. J Clin Invest 61(1):79–88. https://doi.org/10.1172/JCI108928

53. Moore BB, Kolodsick JE, Thannickal VJ, Cooke K, Moore TA, Hogaboam C, Wilke CA, Toews GB (2005) CCR2-mediated recruitment of fibrocytes to the alveolar space after fibrotic injury. Am J Pathol 166(3):675–684. https://doi.org/10.1016/S0002-9440(10)62289-4

Chapter 25

Mouse Models of COPD

Karina A. Serban and Irina Petrache

Abstract

Elastase and chronic cigarette smoke exposure animal models are commonly used to study lung morphologic and functional changes associated with emphysema-like airspace enlargement in various animal species. This chapter describes the rationale for using these two models to study mechanisms of COPD pathogenesis and provides protocols for their implementation. E-cigarettes are an emerging health concern and may also contribute to lung disease. Accordingly, approaches to study e-cigarette vapors are provided. This chapter also includes methods and tools necessary to assess lung morphologic and functional changes in animals with emphysema-like airspace enlargement.

Key words COPD, Mouse, Emphysema, Cigarette smoke, Elastase, Mean linear intercept, E-cigarette

1 Introduction

Animal models of COPD are instrumental for interrogating molecular mechanisms of disease suggested by findings in patients or cell culture models of cigarette smoke (CS) exposure. From the classical discovery that instillation of porcine pancreatic elastase or human neutrophil elastase into the lungs of mice and hamsters culminated in the development of emphysema-like alveolar destruction [1, 2], to the airspace enlargement induced by chronic CS exposure in mice, guinea pigs, and rats [3, 4], animal models have enabled a better understanding of the mechanisms that underlie COPD pathogenesis.

In humans, COPD comprises several anatomic lesions within the lung: emphysema, small airway remodeling, vascular remodeling, and large airway epithelial injury (bronchitis). Moreover, there are significant systemic effects associated with human COPD, including acute infectious exacerbations, weight loss that may culminate in cachexia, myopathy, bone marrow suppression, and cardiovascular disease. Whether animal models can fully recapitulate the anatomical and functional changes of the human COPD is still an open question. Comparative anatomy studies have demonstrated differences in lung

Scott Alper and William J. Janssen (eds.), *Lung Innate Immunity and Inflammation: Methods and Protocols*,
Methods in Molecular Biology, vol. 1809, https://doi.org/10.1007/978-1-4939-8570-8_25,
© Springer Science+Business Media, LLC, part of Springer Nature 2018

development, maturation, and responses to injury between animal species and strains. These differences explain and account for variability between different animal models of COPD pathobiology and can be leveraged to select the best model for a particular research question. In this chapter we will focus on two well-established and commonly used animal models of COPD, the elastase and CS exposure models.

Elastase-induced emphysema is a relatively fast, low-cost rodent model, since a single administration may induce histological changes similar to panacinar emphysema over a period of 2–3 weeks. Moreover, the severity of lung and systemic changes (e.g., inflammation, oxidative stress, apoptosis, body weight loss, reduced endurance) can be easily modulated by titrating the enzyme dose. Due to perceived highest relevance to human COPD (which is most frequently caused by exposure to CS), murine models of emphysema caused by chronic CS exposure are considered the gold standard. Other models of COPD include 1. the tight skin and 2. the pallid mouse in which emphysema develops spontaneously and 3. various pharmacological or transgenic manipulations (e.g., VEGF receptor blockade, IL-13 overexpression, or TERT deletion, among many others) that induce airspace enlargement. While these models have provided unique and valuable discoveries in the field, they also have notable limitations. The elastase-induced model is limited by the artificial nature of enzyme exposure, a sudden onset of severe inflammatory injury which does not model human COPD, and the relatively limited initiation of pathways of lung injury and systemic involvement. In comparison, the CS model is a more accurate mimic of human exposure, but is cumbersome, requires extensive duration of experimentation (e.g. 6 months of exposure to CS), and produces only a mild emphysema phenotype with minimal large airway involvement, modest remodeling of the pulmonary microvasculature, few systemic manifestations, and lack of spontaneous disease exacerbations. Moreover shorter (minutes to days) CS or e-cigarette vape exposure allows us to investigate the molecular mechanisms responsible for the initiation and propagation of lung and systemic injury. Despite these concerns, the elastase and CS exposure models have informed our knowledge about the major mechanistic paradigms leading to COPD: inflammation, oxidative stress, protease/antiprotease balance, alveolar cell apoptosis, early senescence, and autophagy. Herein we describe the materials and methods used to induce emphysema-like changes in commonly used inbred mouse strains following exposure to elastase or CS. In addition, methods to assess lung function and histopathology are also provided.

2 Materials

2.1 Elastase and CS Exposures

1. Porcine pancreatic elastase.

2. Research-grade cigarettes. Use filtered research-grade cigarettes (3R4F, 0.73 mg nicotine/cigarette) or nicotine-free cigarettes (1R6F, 0.16 mg nicotine/cigarette) from the Kentucky Tobacco Research and Development Center (University of Kentucky).

3. Nose-only smoke exposure device. The inExpose nose-only tower (SCIREQ, Scientific Respiratory Equipment) ensures uniform CS inhalation to individual animals and prevents rebreathing of air among animals.

4. "Homemade" vacuum trap (*see* Subheading 3.4, **step 2** below).

5. Nebulizer unit (Aeroneb micropump, 2.5–4.0 VMD, Aerogen).

6. Inhalation restraint devices for nose-only inhalation exposure.

7. *SoftRestraints* (Scientific Respiratory Equipment) made from nylon-coated stainless steel wires, suitable for rodents from 15 to 30 g.

8. Small acrylic platform at a 60° angle.

9. Whole-body smoke exposure device. The two most commonly used devices are the TE-10 (Fig. 1) and SIU24 supplied by Teague (Woodland, California) and the Promech Lab Holding AB (Vintrie, Sweden). Inside the TE-10 smoking machine, the animal is exposed to variable mixtures of 89% sidestream and 11% mainstream smoke.

10. Enzyme-linked immunosorbent assay kit for measurement of serum cotinine levels.

11. For CO level monitoring, use the in-line gas analyzer for carbon monoxide and particulate matter provided by the supplier to measure real-time estimates of cigarette smoke intensity and ensure accurate and safe exposure of animals. Maintain CO concentrations of 190 ppm.

12. Pallflex Air Monitoring Filters for TSP concentration monitoring (Emfab), 25 mm placed in the in-line gas analyzer.

13. Animals. The most common inbred mouse strains used are C57Bl/6, DBA2/J, Balb/C, A/J, and AKR/J; however, strain susceptibility to CS injury should be considered (*see* **Note 1**).

2.2 Sedatives and Anesthetics

1. Ketamine (100 mg/kg) administered intraperitoneally.

2. Xylazine (10 mg/kg) administered intraperitoneally.

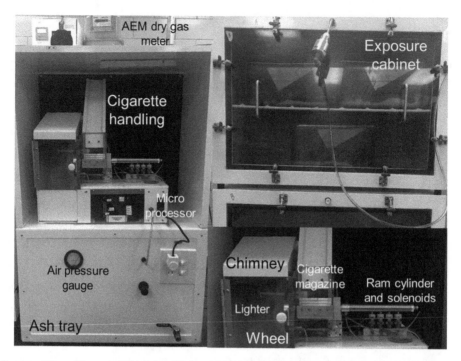

Fig. 1 Teague-10 smoking machine setup. The machine has four sections: cabinet, cigarette handling, chimney, and ash collection tray. This microprocessor-controlled machine generates sidestream and mainstream cigarette smoke from research-grade cigarettes automatically loaded into a wheel. The cigarettes are lit, puffed, and produce smoke at a constant rate. The smoke is mixed, diluted, and metered to the exposure cabinet at a constant rate. To maintain a target concentration of total suspended particulates (TSP), the level is monitored by an in-line gas analyzer and AEM dry gas meter

2.3 Tracheostomy

1. Intramedic polyethylene tubing (I.D. 0.045″, O.D. 0.062″; Becton Dickinson, New Jersey).

2. 18 gauge needle.

2.4 Tissue Processing and Analysis

1. 0.25% agarose in 10% neutral-buffered formalin for fixation.

2. Paraffin for embedding.

3. Multi-fit glass syringe with Luer-Lock tip, 30 mL.

4. Air balance (Hydrogen Balance air 103L, Air Liquide).

5. Gas chromatograph.

6. Glass syringe with extension tube and Luer-lock.

7. Suture silk.

8. Blunt-tipped scissors.

9. Cold-buffered saline.

10. Hematoxylin and eosin to stain tissue sections.

Fig. 2 Functional analysis of emphysema in mice. An anesthetized mouse is intubated and ventilated using the Flexivent. The computer screen enables real-time display of respiratory rate and airflow resistance, calibration parameters, and physiologic outputs such as lung compliance and elastance

2.5 Devices for Functional and Morphological Assessment

1. Plethysmograph (SCIREQ, Scientific Respiratory Equipment, Montreal) to detect cough.

2. Flexivent (Fig. 2 SCIREQ, Scientific Respiratory Equipment, Montreal).

3. Stereology for calculation of mean linear intercept.

4. Micro-computed tomography (microCT).

3 Methods

3.1 Porcine Pancreatic Elastase (PPE) Administration

1. Sedate the mouse using xylazine (10 mg/kg) and ketamine (100 mg/kg).

2. After the animal has lost footpad-pinch response, place it face up on a small acrylic platform (60° angle) while the head is held stationary and in alignment with the body by a rubber band underneath the upper teeth.

3. Pull the tongue aside to view the trachea, and use a pipette tip to cannulate the trachea as distally as possible.

4. Administer elastase solution (30 μg in a total volume of less than 75 μL) or vehicle control intratracheally. Monitor the mouse as it breathes in the substance.

5. Administer 100–200 μL of air. When no more liquid is visible, the animal is removed from the restraint board and monitored until it recovers.

6. Measure physiologic parameters (*see* Subheadings 3.5 and 3.6), and isolate tissue for lung histology 21 days after PPE administration (*see* **Note 2**).

3.2 Chronic Nose-Only CS Exposure

1. Inside the inExpose nose-only tower, the animals are exposed to firsthand CS that is similar to firsthand CS exposure that occurs in humans that smoke.

2. Place animals in *SoftRestraints* or inhalation restraint device.

3. Load 24 3R4F cigarettes into the carousel. The smoke is automatically generated and delivered as mainstream smoke into an individual inhalation restraint device.

4. Expose the mice to four cigarettes, for 60 min, twice a day, 5 days/week for 6 months (*see* **Note 3**).

3.3 Chronic Whole-Body CS Exposure

1. Transfer mice in standard cages to TE-10c smoking chamber.

2. Load five 3R4F cigarettes into the two available hood ports, and push them into the "O" ring. The sidestream smoke is generated inside the TE-10c hood and pumped through the mixing and dilution chamber into the exposure chamber.

3. Monitor carbon monoxide levels using the in-line gas analyzer provided by the supplier to obtain real-time estimates of cigarette smoke intensity, and ensure accurate and safe exposure of animals. Maintain CO concentration of 190 ppm or less (*see* **Note 4**). An adequately calibrated flow meter for the TE-10c chamber will ensure CO concentration within the recommended range.

4. Monitor total suspended particulate (TSP) concentrations using Pallflex Air Monitoring Filters placed in the in-line gas analyzer (*see* **Note 5**). First, record dry filter weight. Then, run the sample and record the time, flow, readings on the AEM dry gas meter attached to the TE-10c setup, and the final filter weight. Finally, calculate TSP by dividing the difference in filter weight by the volume collected. Maintain TSP within concentrations of 90–110 mg/m^3.

5. Expose mice to CS for up to 5 h per day.

6. Acclimate mice to normal air for an additional hour, and then return them to their normal cages in the animal facility.

7. Serum and urine cotinine, a nicotine metabolite, or blood carboxyhemoglobin (COHgb) levels may be monitored during the experiment to ensure "physiologic" nicotine levels (<50 ng/mL). Use ELISA to measure serum or urine cotinine (expected ~5 ng/mL) and blood-gas analyzer to measure COHgb concentration (expected ~5%).

8. Continuous CS exposure for 5 h/day, 5 days/week for 6 months will induce emphysematous changes in susceptible

strains of mice (*see* **Note 6**). Alternatively, for acute smoke exposure studies, mice can be exposed to cigarette smoke from 5 min up to 10 days (5 days/week).

3.4 Acute e-Cigarette Exposure

1. Acute cigarette exposure models are insufficient to cause immediate functional or morphological changes suggestive of emphysema/COPD, even in susceptible mouse strains. However, acute (minutes to days) cigarette smoke or e-cigarette vape exposure results in oxidative stress (lower reduced: oxidized glutathione ratio, increased level of inducible nitric oxide synthase or endothelial nitric oxide synthase), loss of epithelial and endothelial barriers, endothelial exosome shedding, and leukocyte adhesion to the lung microvasculature.

2. Electronic cigarette (e-cigarette) vape is obtained from vaporization of commercially available electronic cigarette solution in a "homemade" vacuum trap (Fig. 3) and collection of the post-vaporized condensate. Alternatively, one can use the e-cigarette extension designed for the inExpose (SCIREQ, Scientific Respiratory Equipment) CS exposure unit that offers automated activation and custom puff profiles of the e-cigarette cartridge.

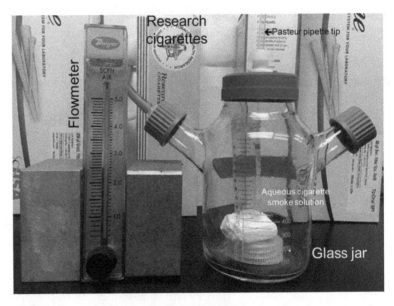

Fig. 3 Custom-designed smoke trap to collect cigarette smoke or vape condensate. The in-line flowmeter is set at 1.5 cc/min. It will burn research cigarettes at a rate of one cigarette/min to 0.5 cm above the filter. The smoke is trapped inside the glass jar and bubbled into 20 mL sterile PBS. We burn two cigarettes (the filtered end inside the glass Pasteur pipette tip) per final 20 mL aqueous cigarette smoke solution. Alternatively, the mouthpiece of a commercially available vaporizer is connected in-line, and the e-cigarette vape is trapped and condensed directly into the glass jar

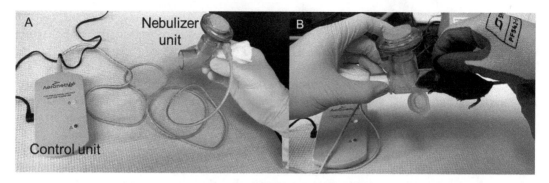

Fig. 4 Nebulization setup. Electronic cigarette vape loaded in the nebulizer unit (**a**) is nebulized to the animal as demonstrated in the right panel (**b**)

3. Pour the e-cigarette vape (~2 µg nicotine/200 µL) into the nebulizer unit (Aeroneb micropump, 2.5–4.0 VMD, Aerogen).

4. Connect the nebulizer unit to the control module (Fig. 4), and follow the manufacturer's instructions to turn on the system.

5. Restrain the mouse by grabbing the scruff of the neck and tail.

6. Place the animal's head inside the tube extension that is connected to the nebulizer unit. Turn on the nebulizer unit. The nebulization procedure takes 1–2 min.

7. Return the animal to the cage.

3.5 Physiologic Assessment of Emphysema

1. Perform functional assessment of emphysema on mice 21 days after elastase administration or chronic exposure to CS. Although emphysematous changes can be detected at earlier time points (e.g., D10), measurements at D21 after elastase administration provide a more consistent increase in lung compliance.

2. Sedate the mouse using xylazine (10 mg/kg) and ketamine (100 mg/kg).

3. Verify depth of sedation using footpad-pinch response. If needed, animals can be given a supplemental dose of ketamine (10 mg/kg) and xylazine (1 mg/kg).

4. Place the mouse face up on a small acrylic platform. Position a rubber band underneath the upper teeth to hold the head stationary and in alignment with the body.

5. Spray the anterior neck with 100% ethanol, to wet and clean the fur for easier access to the skin overlaying the trachea.

6. Use sterile scissors to make an anterior cervical incision. Cut vertically to expose the underlying soft tissue. Spread the soft tissue laterally to expose the trachea.

7. Carefully cut a small puncture in the anterior trachea.

8. Insert a cannula made from a truncated 18 gauge needle into the hole and connect. Be careful not to transect or perforate the posterior tracheal wall.

9. Intubate the trachea with a cannula fabricated from a 2.5–3 cm long piece of intramedic polyethylene tubing attached to the end of the truncated 18 gauge needle.

10. Connect the cannula to the Flexivent (Fig. 2). Parameters for the Flexivent ventilation are a positive end-expiratory pressure (PEEP) of 2.5–5 cm H_2O, 200 breaths per minute, and tidal volume of 0.3 mL.

11. The respiratory machine also contains an in-line (or side-stream) nebulizer connected to the Y-tubing by which drugs or inhaled anesthetics can be administered directly to the unconscious animal's lungs without release to the surrounding environment.

12. One can use the Flexivent to measure resistance, impedance, elastance, or pressure-volume curves. Changes in resistance and respiratory impedance are measurements of airway obstruction, whereas a decrease in elastance or a shift in the pressure-volume curves indicates increased compliance consistent with emphysematous changes. Please *see* Chap. 19 for additional details about using Flexivent to monitor lung function.

13. When this is a terminal procedure, the mice are then euthanized by bilateral pneumothorax and exsanguination, and the lungs are processed for morphologic and morphometric measurements.

3.6 Measurement of Diffusion Factor for Carbon Monoxide (DF$_{CO}$)

1. The DF_{CO} measurement involves inhaling a gas mixture containing a small amount of carbon monoxide and an inert gas (e.g., neon in mice) [5].

2. Sedate and intubate the mouse using the steps described in Subheading 3.5.

3. Use a 3 mL syringe to draw 0.8 mL gas mixture from the tank containing the gas mixture, 0.3% Ne, 0.3% CO, in an air balance (hydrogen balance air103L, Air Liquide).

4. Connect the tracheostomy cannula to the syringe containing the gas mixture.

5. Quickly inflate the lungs with 0.8 mL of gas mixture.

6. Hold inflation for 9 s, and then withdraw 0.8 mL from the lungs. Dilute the returning gas mixture with 2 mL room air inside the 3 mL syringe.

7. Immediately use a gas chromatograph to measure the change in gas concentrations. Alternatively, one may use polypropylene gas sample bags to store the gas sample for delayed measurements.

8. Compare uptake of CO to dilution of Ne. Use the following formula: $1 - (CO9/COc)/(Ne9/Nec)$ to define DF_{CO} (c, the calibration gas, and 9, the gas from the 9 s measurement time). A value of 1 for DF_{CO} reflects 100% CO uptake and normal diffusion factor for CO.

9. This is a terminal procedure; the mice are euthanized post DL_{CO} measurement (*see* **Note 7**).

3.7 Preparation of Lung Tissue for Histology

1. Heat the 0.25% agarose in 10% formalin solution for 10 min and until it is melted and has clear appearance.

2. Prepare a glass syringe by wrapping a heating cord evenly around it. This will keep the syringe warm and prevent the agarose from cooling and thickening.

3. Attach the syringe to a ring holder. Attach an extension tube to the syringe using a Luer-lock.

4. Set the heat box at 37 °C (Fig. 5), and center a heat lamp on the syringe to prevent agarose from hardening.

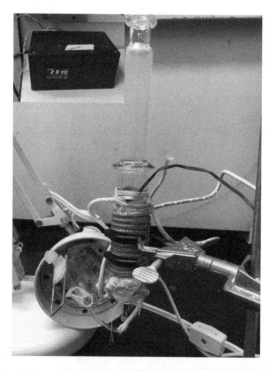

Fig. 5 Custom-designed setup for lung inflation and fixation. The melted agarose-formalin solution is poured in the glass syringe that is warmed at 37 °C by the heating cord and heat lamp. The syringe is filled with fixation solution. An extension tube with Luer-lock connectors is attached to the distal end of the syringe

5. Fill syringe with agarose-formalin solution to the top. Adjust the height of the fluid column so that it is 20 cm above the chest of the animal. This will ensure that the lungs inflate at a constant pressure of 20 cm H_2O (Fig. 6).

6. Open the stopcock to allow air to purge from the tubing, and then connect the extension tube to the tracheostomy cannula. Make sure the solution is still liquid before you let the lung fill. Keep the tubing system straight to promote flow of the agarose solution, and work quickly through all steps to prevent the agarose from cooling.

7. Put suture silk around the lung. Tighten knot around the lung.

8. Using blunt-tipped scissors carefully, cut the lung away from the chest. Avoid puncturing the lung.

9. Dip the lung into cold-buffered saline until the agarose-formalin solution hardens.

10. Measure lung volume (V) using the water displacement method. Perform three volume measurements; make sure you dry the lung between the three measurements.

11. Place the dry lung in a histology cassette, and submerge it in a 10% formalin with 0.1% glutaraldehyde solution.

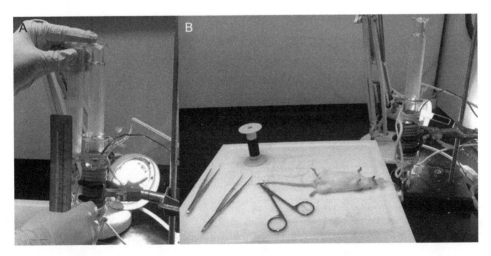

Fig. 6 Lung inflation method. (**a**) The syringe plunger is set at 10 cm from the counter to ensure a constant pressure of 20 cm water during lung inflation. In the next step (**b**) the mouse tracheostomy tube is connected to the extension tube. The mouse lungs require less than 3 mL warm fixation solution (~1.5 mL for one lung) to be fully inflated at constant pressure. Next we use the suture silk to tighten a knot around the trachea or left main bronchus (if only one lung is inflated and fixed). The dissected lung is dipped into cold-buffered saline until the fixation solution hardens

3.8 Morphologic and Morphometric Analysis of Emphysema

Stereology can be used to obtain unbiased estimates of alveolar number, alveolar surface, and lung volume [6, 7]. For systematic uniform random sampling, use a tissue slicer to cut the lung into slabs of equal thickness.

3.8.1 Stereology

3.8.2 Cavalieri Method

1. Use the Cavalieri method to calculate the total lung volume.

2. Begin "bread loafing" the tissue by sectioning. Start the sectioning outside the tissue margins (i.e., cut through paraffin to get close to the lung surface). Once the lung surface has been reached, take every second section (or third or fourth, as decided a priori by your laboratory) for further processing.

3. Place a point grid with a known area randomly over the slabs (cut face up). Count all grid points that are within the cut surface of the tissue.

4. Calculate total lung volume by applying the formula $V_{lung} = a(p) \times T \times \Sigma P$ where $a(p)$ is the area per point, T is the tissue slab thickness, and ΣP is the number of points hitting the tissue cut face of each particular slab.

3.8.3 Calculation of Total Alveolar Surface Area Per Lung

1. To calculate the total alveolar surface area per lung, use surface estimation by intersection counting.

2. Section the tissue (5 μm sections) using a microtome.

3. Stain tissue sections with hematoxylin and eosin.

4. Image 20 fields at 10× to 20×, and take random pictures from random lung sections.

5. Superimpose randomly over each picture a test system of two test lines of known length (50 μm), each associated with two points.

6. Count the number of points that hit the alveolar surface. This can be done manually or with automated software packages such as STEPanizer or Metamorph.

7. Calculate the mean linear intercept (Lm) using the following formula. $Lm = d \times P_{air}/I_{alv}$ where d is the length of single test line, P_{air} is the number of points hitting airspaces inside alveoli and ducts, and I_{alv} is the number intercepts with alveolar septal walls.

8. The total alveolar surface area, S, can be quantified using the following formula: $S = 4V \times p/Lm$. One can use the lung volume (V) measured via water displacement (Subheading 3.7, **step 10**) or Cavalieri (Subheading 3.8.2) methods.

3.9 Micro-computed Tomography (microCT)

1. MicroCT allows recurrent volumetric measurements in a sedated and ventilated mouse.

2. Images at end-inspiration and end-expiration are used to compute total lung volumes [8].

3. Emphysema assessment using this method should be performed in collaboration with the radiology research core at your institution as the image acquisition, processing, and analysis protocols require qualified expertise.

4 Notes

1. The AKR/J mouse strain is the most susceptible to the development of emphysema; the C57BL/6/J, A/J, DBA2/J, Balb/C, and SJ/L strains are mildly susceptible when exposed to CS [9]. The earliest time point when AKR/J mice develop airspace enlargement after CS exposure is 6–7 weeks [10]. In common inbred strains of mice, such as C57Bl/6 or DBA2/J, chronic (6 months) CS exposure causes airspace enlargement associated with various levels of inflammation, oxidative stress, increased proteolysis, parenchyma cell apoptosis, markers of cellular stress, elevated lung static compliance, weight loss, and decreased exercise endurance [11]. However, the degree of large airway involvement is relatively mild; therefore, if CS-induced alteration in bronchial glands and epithelial cell metaplasia are the main end points of your study, other rodent models should be considered (e.g., rats, guinea pig, hamsters, rabbits, or ferrets) [12, 13]. One might consider guinea pig or rat models of CS exposure to study vascular remodeling and secondary pulmonary hypertension, because the common mouse strains do not develop marked increase in mean pulmonary arterial pressure despite significant alterations and endothelial cell apoptosis within the lung microvasculature [14].

2. Compared to the CS exposure mouse model, the elastase model is of shorter duration and less labor intensive. Following elastase instillation, the airspace destruction resembles panacinar emphysema associated with the genetic form of emphysema due to alpha-1 antitrypsin deficiency [15]. The more commonly used elastase in this model is the porcine pancreatic elastase (PPE), which is cleared from the lungs within 24 h of administration. However, the airspace enlargement continues for 10 up to 21 days. Lesion severity is dependent on the PPE dose, and the progression of emphysema is related to the degree of inflammatory response generated by PPE administration [2, 16]. Repeated PPE instillations (5 PPE administrations 1 week apart) induce more severe airspace enlargement, body weight loss, increased right ventricular mass, and dia-

phragmatic dysfunction, all systemic manifestation of severe human COPD. These changes are perpetuated for up to 6 months in the elastase mouse model of emphysema [17].

3. One needs to monitor smoking chamber atmosphere for nicotine, carbon monoxide (CO), and total suspended particulates (TSP).

4. Carbon monoxide (CO) levels in the whole-body exposure chambers (150–400 ppm) are 40–400% higher than the levels experienced by individuals exposed to secondhand CS. Active smokers are exposed to CO levels of 10–26 ppm, while secondhand CS-exposed individuals experience CO levels of 5–20 ppm [18] that lead to COHgb levels of 4.43% [19]. Furthermore, animals may reach COHgb saturations of ~10.5% [20] that are typically higher than even those of actively smoking individuals (5% on average), while secondhand CS-exposed individuals may reach 0.5–2% COHgb saturations. Monitor animals carefully to avoid CO poisoning; watch for symptoms like uncoordinated movements, lethargy, seizures, and coma.

5. Plasma nicotine concentrations in the CS exposure mouse models range from 38.5 to 188 ng/mL. By comparison, actively smoking individuals maintain nicotine blood levels between 20 and 50 ng/mL [21], and those with moderate secondhand smoke exposure have plasma nicotine concentrations of ~6 ng/mL [22, 23].

6. There is a synergistic effect of co-exposure to CS and viral pathogen-associated molecular patterns (PAMPs). C57BL/6J or BALB/c mice exposed to CS for 2 weeks and four doses of 15–50 μg of polyinosine-polycytidylic acid [poly(I:C)] develop airspace enlargement, enhanced parenchymal inflammation and apoptosis, and airway fibrosis [24].

7. DF_{CO} analysis can be done without sacrificing the mouse if the animal is maintained under continuous anesthesia, and intubation is performed under direct fiber-optic visualization using a 0.5 mm fiber-optic cable (Edmund Optics) to intubate the trachea with a 20 gauge IV cannula (BD Insyte) [25]. After DL_{CO} measurements, the animal is extubated and monitored until recovery from anesthesia.

Acknowledgments

Funding sources are RO1HL077328 (IP) and 2014 Alpha-1 Foundation Gordon L. Snider Scholar Award (KAS).

References

1. Lucey EC, Goldstein RH, Stone PJ, Snider GL (1998) Remodeling of alveolar walls after elastase treatment of hamsters. Results of elastin and collagen mRNA in situ hybridization. Am J Respir Crit Care Med 158(2):555–564

2. Snider GL, Lucey EC, Stone PJ (1986) Animal models of emphysema. Am Rev Respir Dis 133(1):149–169

3. Wright JL, Churg A (1990) Cigarette smoke causes physiologic and morphologic changes of emphysema in the Guinea pig. Am Rev Respir Dis 142(6 Pt 1):1422–1428

4. Takubo Y, Guerassimov A, Ghezzo H, Triantafillopoulos A, Bates JH, Hoidal JR et al (2002) Alpha1-antitrypsin determines the pattern of emphysema and function in tobacco smoke-exposed mice: parallels with human disease. Am J Respir Crit Care Med 166(12 Pt 1):1596–1603

5. Fallica J, Das S, Horton M, Mitzner W (2011) Application of carbon monoxide diffusing capacity in the mouse lung. J Appl Physiol (1985) 110(5):1455–1459

6. Weibel ER, Hsia CC, Ochs M (2007) How much is there really? Why stereology is essential in lung morphometry. J Appl Physiol (1985) 102(1):459–467

7. Ochs M (2006) A brief update on lung stereology. J Microsc 222(Pt 3):188–200

8. Vasilescu DM, Klinge C, Knudsen L, Yin L, Wang G, Weibel ER et al (2013) Stereological assessment of mouse lung parenchyma via nondestructive, multiscale micro-CT imaging validated by light microscopic histology. J Appl Physiol (1985) 114(6):716–724

9. Guerassimov A, Hoshino Y, Takubo Y, Turcotte A, Yamamoto M, Ghezzo H et al (2004) The development of emphysema in cigarette smoke-exposed mice is strain dependent. Am J Respir Crit Care Med 170(9):974–980

10. Podowski M, Calvi C, Metzger S, Misono K, Poonyagariyagorn H, Lopez-Mercado A et al (2012) Angiotensin receptor blockade attenuates cigarette smoke-induced lung injury and rescues lung architecture in mice. J Clin Invest 122(1):229–240

11. Gosker HR, Langen RC, Bracke KR, Joos GF, Brusselle GG, Steele C et al (2009) Extrapulmonary manifestations of chronic obstructive pulmonary disease in a mouse model of chronic cigarette smoke exposure. Am J Respir Cell Mol Biol 40(6):710–716

12. Churg A, Wright JL (2009) Testing drugs in animal models of cigarette smoke-induced chronic obstructive pulmonary disease. Proc Am Thorac Soc 6(6):550–552

13. Ni K, Serban KA, Batra C, Petrache I (2016) Alpha-1 antitrypsin investigations using animal models of emphysema. Ann Am Thorac Soc 13(Suppl 4):S311–S316

14. Yamato H, Sun JP, Churg A, Wright JL (1997) Guinea pig pulmonary hypertension caused by cigarette smoke cannot be explained by capillary bed destruction. J Appl Physiol (1985) 82(5):1644–1653

15. Antunes MA, Rocco PR (2011) Elastase-induced pulmonary emphysema: insights from experimental models. An Acad Bras Cienc 83(4):1385–1396

16. Sawada M, Ohno Y, La BL, Funaguchi N, Asai T, Yuhgetsu H et al (2007) The Fas/Fas-ligand pathway does not mediate the apoptosis in elastase-induced emphysema in mice. Exp Lung Res 33(6):277–288

17. Luthje L, Raupach T, Michels H, Unsold B, Hasenfuss G, Kogler H et al (2009) Exercise intolerance and systemic manifestations of pulmonary emphysema in a mouse model. Respir Res 10:7

18. Cohen A, George O (2013) Animal models of nicotine exposure: relevance to second-hand smoking, electronic cigarette use, and compulsive smoking. Front Psych 4:41

19. Yee BE, Ahmed MI, Brugge D, Farrell M, Lozada G, Idupaganthi R et al (2010) Second-hand smoking and carboxyhemoglobin levels in children: a prospective observational study. Paediatr Anaesth 20(1):82–89

20. Harris AC, Mattson C, Lesage MG, Keyler DE, Pentel PR (2010) Comparison of the behavioral effects of cigarette smoke and pure nicotine in rats. Pharmacol Biochem Behav 96(2):217–227

21. Benowitz NL, Jacob P 3rd (1984) Daily intake of nicotine during cigarette smoking. Clin Pharmacol Ther 35(4):499–504

22. Brody AL, Mandelkern MA, London ED, Khan A, Kozman D, Costello MR et al (2011) Effect of secondhand smoke on occupancy of nicotinic acetylcholine receptors in brain. Arch Gen Psychiatry 68(9):953–960

23. Brody AL, Mandelkern MA, London ED, Olmstead RE, Farahi J, Scheibal D et al (2006) Cigarette smoking saturates brain alpha 4 beta

2 nicotinic acetylcholine receptors. Arch Gen Psychiatry 63(8):907–915

24. Kang MJ, Lee CG, Lee JY, Dela Cruz CS, Chen ZJ, Enelow R et al (2008) Cigarette smoke selectively enhances viral PAMP- and virus-induced pulmonary innate immune and remodeling responses in mice. J Clin Invest 118(8):2771–2784

25. Das S, MacDonald K, Chang HY, Mitzner W (2013) A simple method of mouse lung intubation. J Vis Exp 21(73):e50318. PMC 3639692

Chapter 26

Mouse Models of Viral Infection

Kerry M. Empey, R. Stokes Peebles Jr, and William J. Janssen

Abstract

Viral respiratory tract infections are common in both children and adults. Mouse models of viral infection enable the characterization of host immune factors that protect against or promote virus infection; thus, mouse models are essential for interrogation of potential therapeutic targets. Moreover, they serve as critical models for the development of novel vaccine strategies. In this chapter, we describe methods for establishing mouse models of respiratory syncytial virus (RSV) and H1N1 influenza A virus infection. Protocols are provided for viral culture and expansion, plaque-forming assays for viral quantification, and infection of mice. Alternate modifications to the models are also described, and their potential impact is discussed.

Key words Respiratory syncytial virus (RSV), Influenza A virus, H1N1, Mouse model, Infection, Lung injury

1 Introduction

Lower respiratory tract infections, including bronchitis and pneumonia, affect over 100 million individuals in the United States each year and are a leading cause of health-care encounters, particularly in the late fall and winter [1]. Bronchitis results from infection of the bronchial mucosa. It has been estimated that over 60% of cases are caused by viruses, of which the most common are influenza A and B, respiratory syncytial virus (RSV), parainfluenza, coronaviruses, rhinoviruses, and human metapneumovirus [2, 3]. In comparison to bronchitis, pneumonia affects the more distal portions of the lung including the alveoli. Although it is widely accepted that the majority of pneumonia cases are caused by bacteria, recent evidence suggests that respiratory viruses are also common, either as the single cause or as a coinfecting pathogen [4–6]. Common viral pathogens in adults include influenza A and B, parainfluenza virus, rhinovirus, and coronavirus [4–7]. These pathogens are also common causes of pediatric pneumonia, as is RSV [7, 8].

Murine models of pulmonary viral infection have significantly advanced the characterization of immune targets, our understanding

Scott Alper and William J. Janssen (eds.), *Lung Innate Immunity and Inflammation: Methods and Protocols*,
Methods in Molecular Biology, vol. 1809, https://doi.org/10.1007/978-1-4939-8570-8_26,
© Springer Science+Business Media, LLC, part of Springer Nature 2018

of viral evasion mechanisms, and the development of vaccines and therapeutics [9–11]. Despite the significant contributions of these models, inconsistencies among published reports call into question the impact of variable model approaches and how they affect outcome measures. Moreover, the lack of published methodologic details makes it impossible to address the impact of variability among murine models of viral infection [9, 12–14]. Here we provide detailed methodologic procedures for generation of RSV and influenza A virus working stock, and efficient viral plaque purification. Moreover, we provide methods for inducing RSV and H1N1 influenza A infection in adult mice and quantification of viral lung titers. Alternative approaches to the described methods are included where available.

2 Materials

RSV and influenza A are BSL2 level organisms. Proper precautions should be followed and proper personal protective equipment should be worn when working with RSV and influenza A viruses and virus-infected animals.

2.1 RSV Master Stock Preparation (see Note 1)

1. Hep-2 media: To a 500 mL bottle of 1× MEM + L-glutamine + Earle's salts (with phenol red) (10% EMEM), add 5 mL of penicillin-streptomycin (p-10,000 U/mL, s-10,000 µg/mL), 0.5 mL of Fungizone (amphotericin B, 250 µg/mL), and 50 mL of heat-inactivated fetal bovine serum (HI-FBS). To heat inactivate the FBS, thaw at room temperature, and then place in a 56 °C water bath for 30 min (see **Note 2**).

2. 1× phosphate-buffered saline (PBS).

3. 0.25% trypsin (Gibco).

4. 1% agarose: To a 400 mL of deionized water, add 4 g of agarose. Autoclave the mixture for 30 minutes at 120 °C.

5. 2× Hep-2 media = 1× Hep-2 media with 2× additives: To a 500 mL bottle of 1× MEM + L-glutamine + Earle's salts (with phenol red), add 100 mL of HI-FBS (do not remove 50 mL of media), 1 mL gentamicin (50 mg/mL concentration), 1 mL Fungizone, 10 mL penicillin-streptomycin, and 5 mL of L-glutamine (see **Note 3**).

6. Dry ice/alcohol bath: Dry ice should be broken into small pieces. To this, add 2-propanol; replenish with dry ice and 2-propanol as needed to maintain bubbles (will look as though it is boiling). Note that it is important to wear appropriate protective gloves and eyewear to prevent thermal injury from the ethanol-dry ice solution.

7. Plastic/vials: T-162 tissue culture flasks, 6- and 12-well tissue culture plates, autoclaved 4 mL 1-dram glass vials with printed labels and assembled cap and septum, electrical tape.

2.2 RSV Working Stock Preparation

1. Plastics/glass: T-162 tissue culture flasks, 50 autoclaved 4 mL 1-dram glass vials with printed labels and assembled cap and septum, 50 mL conical centrifuge tubes, 15 mL conical centrifuge tubes, sterile glass flasks, 10 mL pipettes.

2. 10% EMEM: To a 500 mL bottle of EMEM, add 10 mL of HI-FBS.

3. Dry ice/alcohol bath: Dry ice is broken into small pieces and placed in a Styrofoam container. Add 100% isopropyl alcohol to cover the dry ice; the alcohol will bubble as though boiling. A rack to hold the glass vials must be placed in the dry ice bath such that the bottoms of the vials are immersed but the caps are not (*see* **Note 4**).

4. Electrical tape.

5. 70% ethanol.

6. Sonicator/cell disruptor with probe and pulse operation.

2.3 RSV Hematoxylin and Eosin Plaque Assay

1. Hep-2 media (10% EMEM) as described in Subheading 2.1.

2. 10% EMEM+0.75% methyl cellulose: Put 3.75 g of methyl cellulose and a stir bar in a 500 mL bottle and sterilize in the autoclave. When cool, put on a stir plate and add 500 mL of room temperature Hep-2 in a sterile hood. Stir mixture for 1–2 days at room temperature. Make sure to monitor its progress. Store at 4 °C, dissolves better at cooler temperature.

3. Gill modified hematoxylin solution 1 (Fisher).

4. Eosin yellowish solution 1% w/v (Fisher).

5. 10% buffered formalin solution.

6. 1× phosphate-buffered saline (PBS).

7. Trypsin/EDTA (0.25%), phenol red.

8. Plastic/glass wear: 12-well tissue culture plates, T175 tissue culture flasks, 5 mL polystyrene tubes.

2.4 Preparation of H1N1 Influenza A Stock

1. Influenza A, strain A/PR/8/34 (American Type Culture Collection, ATCC, VR-95).

2. Freshly fertilized specific pathogen-free (SPF) chicken eggs (Charles River Laboratories).

3. Humidified egg incubator, preferably with automatic egg turner.

4. BSL-2 biosafety hood.

5. Egg candler.

6. 70% ethanol.

7. Glue gun or paraffin wax.

8. Sterile 18-guage needle.

9. Sterile 22-gauge needle, 1 in. long.

10. Sterile spatula or spoon.

11. Sharp sterile scissors.

12. 50 mL conical tube, sterile.

13. Sterile 0.5 mL microfuge tubes.

14. Liquid nitrogen for flash-freezing microfuge tubes.

2.5 Influenza A Plaque Assay

1. MDCK cells (American Type Culture Collection, ATCC).

2. PBS without Ca^{2+} and Mg^{2+}.

3. Complete Eagle's minimum essential medium (cMEM): Mix 432.5 mL MEM, 50 mL heat-inactivated fetal bovine serum, 12.5 mL 1 M HEPES, and 5 mL penicillin/streptomycin (5000 U/mL each). Mix thoroughly and then filter using a 0.2 μm filtration system. Store at 4 °C.

4. 2× MEM: Mix 325 mL deionized water, 100 mL 10× MEM, 40 mL 7.5% BSA, 25 mL 1 M HEPES, and 5 mL penicillin/streptomycin (5000 U/mL each). Mix thoroughly and then filter using a 0.2 μm filtration system. Store at 4 °C.

5. 2% agarose solution: Add 2 g low melting point agarose powder to 100 mL deionized water. Mix in a 250 mL Erlenmeyer flask and autoclave at 120 °C, 20 psi for 30 min. Store at 4 °C.

6. TPCK-trypsin (Thermo Fisher Scientific).

7. 10% formalin.

8. 12-well tissue culture plates.

9. Sterile 1 mL microfuge tubes.

2.6 Viral Lung Titers

1. Screw thread, 15 × 45 mm 1 dram vials with rubber-lined lids.

2. Ceramic mortar and pestle that has been autoclaved and stored at 4 °C.

3. Glass beads.

4. Hep-2 media.

5. Electrical tape.

6. Histologic grade 2-propanol.

7. Dry ice.

8. 12-well tissue culture plates.

9. 10% EMEM + 0.75% methyl cellulose.

2.7 Murine Infection

1. For RSV infection, use female or male BALB/c mice 8–10 weeks old (Jackson Laboratory, Bar Harbor, ME) (*see*

Note 5). For H1N1 influenza A infection, mice on any genetic background can be used.

2. Balance: mice should be weighed at baseline and then daily following infection to assess/document weight loss from baseline values (*see* **Note 6**).

3. BSL-2 hood.

4. Anesthesia vaporizer.

5. Isoflurane *or* ketamine/xylazine cocktail: Mix 30 mL bacteriostatic sodium chloride injection (30 mL vials), 3 mL ketamine (100 mg/mL, store at room temperature), and 0.54 mL xylazine (100 mg/mL—AnaSed Injection—store at room temperature). Inject 100 μL intramuscular (I.M.) per 20 g mouse using 1 mL syringes and 26-gauge 3/8 needles. Allow 2–3 min for drug to take effect. Monitor animals until they are mobile again.

6. Oxygen tank.

7. Induction chamber.

8. Charcoal filter.

9. Pipette.

3   Methods

3.1   Generation of RSV Master Stock

Day (− 1): Plating Cells

1. Grow Hep-2 cells to ~80% confluence in T-162 flasks in 50 mL of Hep-2 media (*see* **Note 7**).

2. Remove old media and wash cells aggressively twice with 10 mL of 1 × PBS pre-warmed to 37 °C.

3. Add 2 mL of 0.25% trypsin and aggressively rock back and forth to allow removal of adhered cells from the flask (*see* **Note 8**).

4. Neutralize the 0.25% trypsin by adding 8 mL of Hep-2 medium (total volume of 10 mL) and pipette up and down several times.

5. Take an aliquot for cell counts, and then dilute the cell suspension in Hep-2 media to achieve a concentration of 5×10^5 cells/mL (*see* **Note 9**).

6. Aliquot 2 mL of diluted Hep-2 cells into each well of a 6-well tissue culture plate to achieve 1×10^6 cells per well, and immediately shake the plate front to back and side to side to disperse the cells evenly (*see* **Note 10**).

7. Place the plate in the 37 °C 5% CO_2 incubator to be ready the following day.

Day (0): Infecting Cells

8. Cells should be 80% confluent before infection with RSV.

9. Make RSV dilutions in 1× Hep-2 media warmed to 37 °C. Prepare tenfold serial dilutions (1:1–1:1,000,000) using 1× Hep-2 media; be sure to vortex between each dilution. Vortex again before adding 100 μL to each well of a 6-well plate in duplicate.

10. Rock the plates on a plate rocker, at 37 °C at a fast speed for 1 h; rock the plates by hand in the opposite direction of the plate rocker every 20 min during the 1 h incubation to ensure even distribution of virus.

11. After the 1 h incubation, mix 1% agarose and 2× Hep-2 media (warmed to 37 °C) immediately before adding 2 mL to each well; add to the sides of the wells overlaying the inoculum that is already in the wells (*see* **Note 11**).

12. Place the 6-well plates in a 37 °C, 5% CO_2 incubator for 2–6 days until syncytia are visible; monitor closely (*see* **Note 12**).

Day 4–6: Picking Plaques

13. When virus is ready, pick at least five *well-isolated plaques*. Circle the plaques under an inverted microscope (4× objective works best) (*see* **Note 13**).

14. Remove plaque and agarose plug using a 1 mL or 2 mL pipette without sliding the pipette tip across plate. After removing the plaque, check the plate under the microscope to ensure plaque was successfully removed.

15. Place plaques in individual cryotubes filled with 0.5 mL of 1× Hep-2 media; pipette up and down several times to break up the agarose. Reserve a well-isolated plaque for the next round of plaque purification and freeze the rest at −80 °C.

16. Initiate the next round of plaque purification with the well-isolated plaque from **step 15**, making tenfold serial dilutions as described in **step 9**. Add 100 μL of each dilution to each well of 6-well plate in duplicate (cells should be 80% confluent). Overlay with agarose as described above.

17. Perform four rounds of plaque purification by repeating the steps above.

18. To generate master stock when the fourth round of plaques are picked, place the plaques in 0.5 mL of 1× Hep-2 media. Be sure to break up the agarose by pipetting up and down, and use the entire solution to infect an 80% confluent T-162 flask of Hep-2 cells. Pick several fourth round plaques that can be frozen as well.

19. Grow and harvest stock as per working stock protocol.

3.2   Generation of RSV Working Stock

Day 0: Splitting Cells

1. Split Hep-2 Cells into two T-162 flasks so that they will be 80% confluent on Day 0.

2. Infect two 80% confluent, T-162 flasks of cells. Remove the media from flasks, leaving 3 mL/flask; add the complete contents of one vial of RSV master stock (about 0.5 mL) to each of the two flasks (one vial of RSV master stock should be used for each flask).

3. Incubate both flasks for 1 h at 37 °C on a rocker table at a fast setting, rocking the flasks in the opposite direction every 20 min during the 1 h incubation to distribute virus (*see* **Note 14**).

4. Add 46 mL of 10% EMEM to each flask, and then incubate at 37 °C + 5% CO_2. Different viral subtypes induce cytopathology at different rates; check for cytopathic effects (CPE) 2–3 days after infection (Fig. 1).

Day 3: Viral Harvest

5. First thing in the morning: Set the centrifuge temperature to 4 °C; put dry ice in the dry ice/alcohol bath; put all refrigerator-cooled supplies on ice (except 10 mL pipettes).

6. Scrape flasks with cell scraper (do not remove media). The cell scraper does not need to be cold. Rinse the cells by pipetting media up and down and detach any remaining cells.

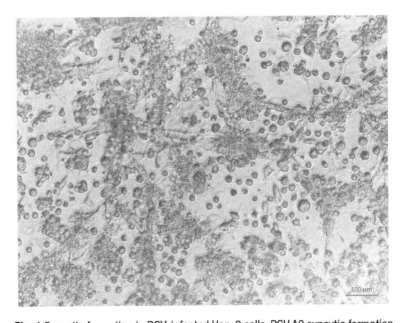

Fig. 1 Syncytia formation in RSV-infected Hep-2 cells. RSV A2 syncytia formation Day 2 after Hep-2 infection using a 20× objective. Syncytia extensions and fusion of cells can be observed

7. Harvest the entire contents of each flask by dividing into two 50 mL conical centrifuge tubes/flasks (on ice). There will be approximately 25 mL per tube.

8. Sterilize the sonicator overnight (or on the day of use) under a UV lamp, and clean the sonicator probe with 70% ethanol. Rinse probe three times with 10% Hep-2 media and then submerge probe tip in ice-cold media in a 15 mL conical tube until ready for use.

9. Sonicate each tube (on ice) at an amplitude of 50% with one pulse per second for each ml of volume (25 times). Be sure to keep the virus suspension cold at all times.

10. Centrifuge at $800 \times g$ for 10 min at 4 °C.

11. Pool the contents of the four tubes into a sterile glass flask on ice; mix by swirling.

12. Aliquot 1.0 mL into each precooled (on ice) glass dram vial. To one of the glass dram vials, add only 500 µL to be used later for viral titers.

13. Tighten each vial carefully and seal with electrical tape before placing in the dry ice/alcohol bath for approximately 5 min. Store vials at −80 °C and quick thaw when ready to use.

14. Follow the RSV hematoxylin/eosin plaque assay protocol (Subheading 3.3) to titer viral stocks. Prepare eight dilution tubes (450 µL each) and two 12-well plates with 1×10^5 Hep-2 cells per ml per well. Quick thaw the 500 µL vial of viral stock and inoculate 50 µL/well using dilutions ranging from 10^{-2} to 10^{-8}, leaving the last row of wells for control. After 5 days, fix, stain, and count plaques. Calculate the virus titer using the dilution having at least five plaques per well (Subheading 3.3).

15. As an infection control, live virus is placed in a petri dish approximately 4 in. from a 40-W UV light on ice for 4 h. To ensure inactivation, UV-inactivated virus should be titrated on Hep-2 cells as described in Subheading 3.3 (*see* **Note 15**).

3.3 RSV Hematoxylin and Eosin Plaque Assay

Day (−1): Plating Cells

1. Allow Hep-2 cells to become about 80% confluent in a T-162 flask, and check under microscope using a 10× objective to verify. When cells are ready, remove media using a Pasteur pipet; wash the flask twice with 10 mL 37 °C 1× PBS.

2. Add 2 mL of trypsin/EDTA, incubate until cells start to detach, and give the flask a tap occasionally. Once cells are detached, neutralize the trypsin with 8 mL of Hep-2 media, and pipet up and down to assure all cells are detached and in solution (*see* **Note 8**).

3. Count cells and dilute the cell suspension to a concentration of 1×10^5 cells/mL (*see* **Note 9**). Add 1 mL of cell suspension to every well of a 12-well plate. Gently mix the cells in the plate in an (+) shaped pattern (15–20 times) to prevent the cells from settling in the middle of the plate. Label plate with date and contents (*see* **Note 16**).

4. Cells should be ready (70–80% confluent) the next day. Be sure to monitor the confluence several times a day so that they do not become over-confluent.

Day (0): Infecting Plates

5. Thaw virus sample quickly in a 37 °C water bath and place on ice. Keep samples on ice during assay.

6. Make three dilution tubes/sample (Falcon #2054 polystyrene); add 450 μL of 37 °C 10% EMEM/tube.

7. To titer viral stocks: Prepare eight dilution tubes (450 μL) each and two 12-well plates of 80% confluent Hep-2 cells. Quick thaw a 500 μL vial of viral stock and inoculate 50 μL/well using dilutions from 10^{-2} to 10^{-8}, leaving the last row of wells for control.

8. To titer infected lungs: Prepare three dilution tubes of lung homogenate (*see* Subheading 3.7) with 450 μL each and one 12-well plate of 80% confluent Hep-2 cells. Inoculate with 50 μL/well using dilutions from neat to 10^{-3}.

9. Decant media from 12-well plate into biohazard bag and add dilutions beginning with the most dilute samples.

10. Incubate for 1 h at 37 °C on a rocker table at a fast speed; hand rock every 20 min in the opposite direction.

11. After 1 h incubation, overlay cells with 1 mL of 10% EMEM + 0.75% methyl cellulose pre-warmed to 37 °C; do not remove inoculum.

12. Incubate in 5% CO_2 incubator at 37 °C for 5 days.

Day (5): Staining Plates

13. Do not remove the media. Add 1 mL of 10% formalin, and then wait 1 h.

14. Invert plates and dump formalin out. Wash plate three times in room temperature water: run water in a bucket and let it continually overflow so that the water is refreshed. Dunk the plate vertically in the water bucket and scoop water to fill all wells and dump. Repeat this two more times and dump out the water.

15. Add 1 mL hematoxylin to every well with a 10 mL pipette and let sit for 20 min. Then invert plate and dump liquid.

16. Wash the plates three times as in **step 14**.

17. Add 1 mL of eosin Y to each well, wait 3 min, invert plate and dump, and then wash wells as in **step 14**. Shake excess water from plates and let dry upside down overnight.

18. Count plaques at the dilution with no less than five plaques/well. Total the number of plaques counted at that dilution in each replicate well and add them together, divide by the number of replicates (3), multiply by total volume in the stock vial (1000 μL) divided by the amount added to each well (50 μL), and then multiply by the dilution. This will give you the number of plaque-forming units per ml. Calculate the $\log_{10}$ pfu/mL = [(Avg count) × (total volume)]/[(volume added/well) × (dilution)].

3.4 Preparation of H1N1 Influenza A Stock

H1N1 influenza A virus may be purchased from a commercial vendor or obtained from a trusted collaborator. If the H1N1 is obtained from a collaborator, it most likely has already been propagated and ideally has been sent to you in aliquots that can be used for each set of experiments. If this is the case, proceed to Subheading 3.5 for methods to quantify the virus. It is critical that aliquots remain frozen at −80 °C until use. Each freeze-thaw cycle results in loss of virus.

3.4.1 Egg Inoculation

1. Obtain freshly fertilized specific pathogen-free (SPF) chicken eggs from a commercial vendor.

2. Immediately place the eggs in a humidified egg incubator at 37 °C with 55–60% humidity. Most incubators have an automatic egg turner. If yours does not have one, gently rotate the eggs several times each day.

3. Incubate the eggs for 7–8 days.

4. Candle the eggs to ensure that they are fertilized. Egg candlers can be purchased from poultry supply vendors or can be homemade (*see* **Note 17**) and consist of a bright light that is held against the egg for transillumination. Clean the candler with 70% ethanol to avoid contaminating the eggs. Remove the eggs from the incubator and place them in an empty egg carton. Turn off the lights; the room must be very dark. Hold the egg by the larger end and place it against the candle to transilluminate the contents. Fertilized eggs will have a thin network of blood vessels that lead to a pea-sized embryo. Unfertilized eggs will not have blood vessels, and a yolk will be readily apparent. Return the eggs to the incubator. Do not leave the eggs outside the incubator for more than 20 min.

5. Continue to incubate the eggs until Day 10 or 11.

6. Prepare virus for inoculation. Remove the virus from the freezer and thaw on ice. If the virus warms or sits for an extended period of time, it will degrade. Working in a BSL-2

biosafety hood, dilute virus stock to 10^3–10^4 PFU/mL in sterile DPBS. The ideal volume for injection is 1 mL. Keep the diluted virus on ice at all times.

7. Inoculate the allantoic cavity of each egg as outlined in **steps 8–16**.

8. Candle each egg to ensure that the embryo is viable by checking for movement. Discard eggs with dead embryos.

9. Select three eggs and mark the margins of the air sac with a permanent marker. The air sac will be on the fat end of the egg and is readily identifiable as a translucent area when you are candling the egg.

10. Put the eggs in a container with the air sac up and place the container in a BSL-2 biosafety hood.

11. Clean the egg shell above the airspace with 70% ethanol.

12. Punch a small hole in the shell of each egg over the air sac. This can be achieved using a sterile 18-gauge needle. Some investigators prefer to use a high-speed rotary drill (such as a Dremel). Take care not to insert the needle too deeply to avoid damage to the yolk or embryo.

13. Draw up the diluted influenza virus in a sterile 1 mL syringe.

14. Attach a sterile 22-gauge 1 in. long needle to the syringe and advance the needle through the hole at a 45° angle into the allantoic cavity, the large fluid filled space just below the air sac.

15. Inject 0.2 mL of the virus into each egg.

16. Seal the holes in the eggs. This can be achieved using a small drop of glue from a glue gun, melted paraffin wax, or sterile Parafilm.

17. Place the eggs back in the incubator with the air sac pointing up.

18. Incubate the infected eggs for an additional 48 h.

3.4.2 Influenza Virus Harvest

1. Chill the eggs at 4 °C overnight, or for a minimum of 2 h. This kills the embryo, constricts the blood vessels, and prevents erythrocytes from absorbing the virus.

2. Transfer the eggs to a biosafety hood and wipe their shells with 70% ethanol.

3. Carefully open the shell around the air sac using sharp sterile scissors.

4. Holding the egg upright, gently open the allantoic membrane with sterile forceps.

5. Use a sterile spatula or small spoon to gently move the yolk to the side without rupturing it.

6. Gently aspirate the clear allantoic fluid using a sterile transfer pipette or sterile pipette tip. You may need to tip the egg slightly to the side to maximize the yield of fluid.

7. Place the fluid into a 50 mL conical tube. Place the conical tube in ice.

8. Pool the fluid from all three eggs.

9. Centrifuge the fluid at $1500 \times g$ at 4 °C for 10–12 min.

10. Working in the hood, transfer the supernatant to a fresh 50 mL conical tube and then aliquot the fluid into sterile microfuge tubes.

11. Immediately snap freeze the fluid in liquid nitrogen.

12. Store aliquots of the virus at −80 °C for future use.

3.5   H1N1 Influenza A Plaque Assay

1. Make a solution of MDCK cells in cDMEM.

2. Seed 12-well plates with MDCK cells in cDMEM at a concentration of 2.5×10^5 cells/well. 16 wells will be needed in total.

3. Incubate at 37 °C in 5% CO_2 for 24 h.

4. Look at cells under a microscope. Cells are ready to be infected when they are 90–100% confluent.

5. Prepare serial virus dilutions (10^{-1} to 10^{-8}): Fill eight sterile microfuge tubes with 1080 μL of DMEM *without* FBS. Thaw virus stock on ice. Remove 120 μL of virus stock and add to the first tube. This has a dilution of 10^{-1}. Mix the contents of the tube thoroughly by vortexing. Remove 120 μL and add to the next tube. This one will have a dilution of 10^{-2}. Repeat steps until serial dilutions (10^{-1} to 10^{-8}) have been prepared for all tube.

6. Infect monolayers: Aspirate the media from each well. Wash twice with PBS. It is important to get all of the FBS-containing media out of the wells. Add 500 μL of each virus dilution to each well. Do this in duplicate for each dilution. Gently shake the plates back and forth to make sure the virus is evenly distributed in each well. Incubate monolayers at 37 °C in 5% CO_2 for 1 h.

7. Prepare agarose solution: Warm 2× DMEM to 37 °C. Melt 2% agarose in a microwave or using a hot plate. Do not overheat, just melt. Once liquefied maintain at 45 °C. Mix 2× DMEM and 2% agarose in 1:1 ratio in a sterile container. You will need a minimum of 8 mL. Cool DMEM-agarose mixture at room temperature for 5–10 min. Add TPCK-trypsin (final concentration: 1 μg/mL) to the DMEM-agarose mixture. Maintain at 37 °C until ready to use, but do not allow the solution to sit too long, since the activity of the TPCK-trypsin will wane over time.

8. Aspirate media from cells.

9. Rinse cells with DPBS.

10. Add 2 mL of the agarose-DMEM-trypsin mixture to each well.

11. Let plates sit for 15 min with the lids off to allow the agarose to solidify.

12. Turn plates upside down and return them to the incubator at 37 °C, 5% CO_2.

13. Check for plaques daily. They will be present after 48–72 h.

14. Once plaques appear, fix cells by adding 10% formalin on top of the agarose for 1–2 h at room temperature.

15. Aspirate formalin and agarose. Discard in accordance with institutional safety measures.

16. Wash plates under water.

17. Stain 5 min with crystal violet (0.5%).

18. Wash plates thoroughly to remove dye.

19. Allow plates to air dry.

20. Count plaques at the dilution with no less than five plaques/ well. PFU/mL (of original stock) = # plaques/(dilution factor x ml of inoculum/well).

3.6 Infection with RSV or Influenza A

Virus should always be handled in a BSL2 biosafety cabinet, including when creating murine infections. Virus should be rapidly thawed immediately prior to use and stored on ice throughout the infection period.

1. Use the appropriate number of mice (8–10 weeks of age) per experimental and control groups (*see* **Notes 18** and **19**) for each experiment. BALB/c mice should be used for RSV infections. Any strain may be used for influenza A infection.

2. Prior to viral inoculation, weigh all mice to accurately determine inoculum required.

3. Lightly anesthetize the mice with 2% isoflurane in an induction chamber equipped with a charcoal filter. To avoid over- or under-sedation, it is recommended that no more than four mice be placed in the induction chamber at one time. Initially the mice will demonstrate rapid breathing. They are ready to inoculate when they establish a slowed, rhythmic breathing pattern (~6 min) (*see* **Note 20**).

4. Remove one mouse from the induction chamber at a time by grabbing the nape of the neck. Hold the mouse vertically, then slightly tip it back to a 20-degree angle, and deliver the calculated dose intranasally using a standard 100–200 μL pipette. Keep the mouse at the 20-degree angle until the entire inoculum is inhaled, and then return it to the cage. Monitor mice for 30 min to assure breathing and activity returns to normal.

5. For RSV infection, inoculate BALB/c mice with 5×10^5 PFU of RSV per gram of body weight (inoculum for an average 20 g mouse is 1×10^7 PFU) in 100 μL of MEM (*see* **Notes 15** and **21**).

6. For influenza A infection, use 50 PFU per mouse for BALB/c and C57BL/6 mice. The dose will need to be titrated to achieve the desired effect (*see* **Note 22**).

BALB/c RSV Lung Titers

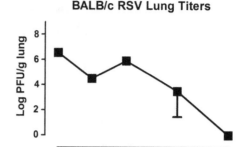

Fig. 2 Viral kinetics in adult BALB/c mice. RSV Line 19 was delivered intranasally in 100 μL of vehicle. Viral lung titers were quantified by H&E plaque assay at indicated times postinfection

7. To confirm successful infection, monitor mice for weight loss and harvest left lungs to quantify viral lung titers as described in Subheading 3.7 on various days postinfection (Fig. 2).

8. Perform viral quantification by hematoxylin and eosin plaque assays as described in Subheadings 3.3 or 3.5. Expected findings at the following times postinfection for RSV are listed below. Findings for influenza A are generally similar but vary by mouse strain.

   (a) 1 h: Estimates viral titer delivered to the airway.

   (b) 2–4 days: Captures ongoing viral replication.

   (c) 4 days: Peak viral titers.

   (d) 5–7 days: Captures viral clearance.

   (e) 10 days: Below limit of detection.

3.7 Harvesting Lungs for Viral Plaque Assay

1. Label two 1-dram glass 4 mL sample vials with rubber-lined lids per mouse and add 2 mL of 1× MEM to each vial. Weigh each vial before and after adding left lung lobes.

2. Animals are euthanized with pentobarbital at 250 mg/kg or 3× the anesthetic dose followed by cervical dislocation and exsanguination.

3. Collect and weigh harvested lungs, and then seal the cap with electrical tape. Snap freeze in the 2-propanol/dry ice bath for 5 min, and then place in −80 °C freezer.

4. Quick thaw lungs in 37 °C water bath when ready to process.

5. Use small, ceramic mortar and pestle that has been autoclaved and stored at 4 °C.

6. Add a small amount of sterile glass beads to the mortar.

7. When grinding lung with mortar and pestle, add both the lung and media from the glass vials.

8. After grinding, add contents to a 15 mL conical tube and spin at $8000 \times g$ for 15 min at 4 °C.

9. After centrifuging, remove supernatant and return to original glass vials.

10. Proceed to Subheading 3.3 for RSV plaque assay or Subheading 3.5 for influenza A plaque assay. Fewer dilutions (3–4) will be required for viral lung titers than when titering stock virus. Thus, one 12-well plate should be prepared for each lung sample

4 Notes

1. Plaque purification is important for purifying a clonal population of virus or "master stock."

2. FBS is heat inactivated at 56 °C for 30 min to reduce the concentration of heat-labile complement proteins which contribute to the serum's hemolytic activity [15]. Though some studies have shown that heat inactivation does not impact in vitro studies with cultured lymphocytes [16]. Data published by Smith and colleagues [17] suggests that complement activation promotes the lysis of RSV-infected Hep-2 cells. Thus, heat inactivation of serum is recommended for growing RSV in Hep-2 cells.

3. Only add L-glutamine if it is not included in the base MEM media.

4. Vials must remain in the dry ice bath for no less than 5 min. A peak formed in the center and yellow color will indicate successful snap freezing of each vial. When vials are thawed, the color of the medium should be pink. If the medium is yellow after thawing, this usually indicates that ethanol contaminated the contents of the vial and that the vial should not be used for infection as the virus will not infect cells or mice.

5. With the surge of genetically modified mice on a C57BL/6 background, many investigators now are establishing RSV infection models using C57BL/6 mice. The key difference between the two mouse strains relevant to RSV host immunity is that BALB/c mice express a predominantly Th2 phenotype whereas C57BL/6 mice have a more balanced Th1/Th2 response.

6. In the adult mouse model of RSV infection, weight loss is used as a measure of illness. A bimodal weight loss pattern is typically observed at 2 and 6 days postinfection [18].

7. The use of early passage Hep-2 cells improves viral growth and produces higher viral titers.

8. If aggressive rocking of the trypsin over the cells does not elicit cell detachment within 1–2 min, placing at 37 °C may promote trypsin-mediated removal of adherent cells.

9. When the T-162 flask is at 80% confluence, cell counts are typically performed at a 1:10 dilution.

10. Hep-2 cells adhere to the plate quickly; thus, the shaking step must be done immediately after plating.

11. When preparing agarose, start by microwaving 1% agarose until completely melted (may boil), let cool on bench slightly, and then maintain at 55–60 °C in a water bath or dry oven. Make sure agarose has cooled enough so as not to kill the virus during the overlay as RSV is very temperature sensitive. If the agarose cools too much before addition to the plate, the agarose will solidify too early.

12. Agarose will solidify while in the incubator. When grown in agarose, syncytia look more rounded and clumpy, with little to no fusions compared to growing in a flask with medium alone.

13. To ensure optimal plaque isolation, circle and pick plaques from plates with less than 50 plaques.

14. To prepare for the viral harvest, print labels with ink that won't smudge in alcohol with date and place on glass dram vials. Precool supplies in refrigerator over the weekend (e.g., vials, sterile glass flasks, 10 mL pipettes, six 50 mL centrifuge tubes).

15. Various RSV infection controls have been reported in the literature, including the use of cell lysate [13], vehicle [19], PBS [20], or UV-inactivated virus [21]. In our hands, the use of cell lysate elicits early, subtle increases in macrophage MHC class II expression, whereas MEM (vehicle) does not. Depending on the question being asked in your model, UV-inactivated virus may serve as a more suitable control to determine disease pathology associated with live, replicating virus.

16. This step is critical for even distribution of the cells. If the cells are not distributed evenly throughout the well, areas of cell accumulation and overgrowth will occur. Hep-2 cells tend to accumulate in the center of the wells creating a characteristic "donut-hole" appearance and making viral purification and titers impossible (Fig. 3).

17. A candler can be constructed from a small LED flashlight (lamp size approximately 1.5 in. diameter), a small medicine cup, and electrical tape. Using a sharp blade, cut the bottom off the base of the medicine cup. Place the base of the cup over the lamp of the flashlight and cover it completely with electrical tape. The open end of the medicine cup should fit snugly over the end of the egg. Light should pass freely from the flashlight through the medicine cup and transilluminate the

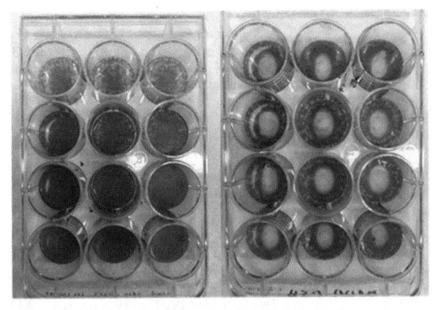

Fig. 3 The effects of poorly dispersed Hep-2 cells on RSV Line 19 plaque formation. When Hep-2 cells are plated evenly, H&E staining will be uniform, and plaques will be countable; indicated here by plaque dots (left). If Hep-2 cells are overgrown in the middle of the well, they will stain with a "donut hole" or "crescent moon" making plaques impossible to count (right)

egg. If a medicine cup is not available, cardboard may be used but this is less sturdy and more difficult to keep clean.

18. BALB/c and C57BL/6 mice are both commonly used in mouse models of RSV infection. BALB/c mice display a more Th2-type cytokine response compared to C57BL/6 mice warranting their use in RSV-based studies. However, C57BL/6 mice have a wider variety of genetically modified mice, enabling more mechanistic-based questions.

19. Pilot experiments using at least five mice per group should be performed to determine the effect size of a particular intervention, allowing for a power analysis to be performed to determine the appropriate number of mice per group for future experiments.

20. Anesthesia: In order to deliver 2% isoflurane/oxygen, a vaporizer and induction chamber are required. Isoflurane induces a light level of sedation to overcome the animals' resistance to inhaling the inoculum and allows for a rapid recovery time. Alternatively, ketamine (50 mg/kg)/xylazine (9 mg/kg) delivered intramuscularly provides a deeper level of sedation than isoflurane and is good option if you do not have an induction chamber and vaporizer.

21. In adult mice (6–12 weeks of age), a variety of RSV viral strains, inoculum size, volumes, and routes of delivery are reported.

(a) Viral strain: The RSV A2, available through American Type Culture Collection (ATCC), is a commonly used viral strain for established murine models of RSV infection. The A2 strain has been used extensively in murine studies to model the immune response to infection, disease pathogenesis, and to establish critical vaccine parameters. Alternatively, RSV Line 19, originally isolated from an infant at the University of Michigan Hospital [22], has been used to establish murine models of severe RSV disease with significant mucus production and development of airway hyperresponsiveness [23]. Unlike RSV A2 or Long strains, RSV Line 19 elicits extensive IL-13 and mucin production, which are critical in assessing the role of RSV in allergic asthma [24].

(b) Viral inoculum: Reported viral inoculums range from 1×10^5 PFU to 1.5×10^7 PFU. The size of the viral inoculum will depend on the desired model. Despite reduced viral replication of RSV Line 19 compared to A2 in vivo, a dose-dependent increase in Gob5 and Muc5A expression along with increased IL-13 has been reported, suggesting the use of lower Line 19 inoculums in models designed to understand the role of RSV in allergic asthma [23]. Alternatively, higher doses may be warranted in models designed to assess disease pathology during acute primary infection.

(c) Route: Inhaled and intratracheal viral inoculums are the most common forms of delivery reported. Inhaled viral delivery is easy and quick and achieves viral inoculation in the nose and airway [9]. Intratracheal delivery may be used to deliver virus to the lower airway [25].

(d) Volume: The viral inoculum is most frequently delivered in 100 μL based on work published by Graham et al. [9] demonstrating more efficient delivery of 99m technetium sulfur colloid to the lungs upon inhalation of higher volumes of the inoculum.

22. The pathogenicity and lethality of H1N1 influenza A vary between mouse strains and may also vary with the sex of the animals. A dose of 50 PFU per mouse is a good starting dose; however, before proceeding with a full experiment, a pilot experiment should be performed to determine the optimal dose. Several doses, varied by two- to tenfold, may be tried. For studies of acute inflammation and lung injury, we chose a dose that induces a 10–15% drop in body weight. Weight loss is generally greatest 7 days postinfection. Weight loss greater than 20% is commonly premorbid.

References

1. Harris AM, Hicks LA, Qaseem A, High Value Care Task Force of the American College of Physicians, for the Centers for Disease Control and Prevention (2016) Appropriate antibiotic use for acute respiratory tract infection in adults: advice for high-value care from the American College of Physicians and the Centers for Disease Control and Prevention. Ann Intern Med 164(6):425–434. https://doi.org/10.7326/M15-1840

2. Clark TW, Medina MJ, Batham S, Curran MD, Parmar S, Nicholson KG (2014) Adults hospitalised with acute respiratory illness rarely have detectable bacteria in the absence of COPD or pneumonia; viral infection predominates in a large prospective UK sample. J Infect 69(5):507–515. https://doi.org/10.1016/j.jinf.2014.07.023

3. Creer DD, Dilworth JP, Gillespie SH, Johnston AR, Johnston SL, Ling C, Patel S, Sanderson G, Wallace PG, McHugh TD (2006) Aetiological role of viral and bacterial infections in acute adult lower respiratory tract infection (LRTI) in primary care. Thorax 61(1):75–79. https://doi.org/10.1136/thx.2004.027441

4. Johansson N, Kalin M, Tiveljung-Lindell A, Giske CG, Hedlund J (2010) Etiology of community-acquired pneumonia: increased microbiological yield with new diagnostic methods. Clin Infect Dis 50(2):202–209. https://doi.org/10.1086/648678

5. Jennings LC, Anderson TP, Beynon KA, Chua A, Laing RT, Werno AM, Young SA, Chambers ST, Murdoch DR (2008) Incidence and characteristics of viral community-acquired pneumonia in adults. Thorax 63(1):42–48. https://doi.org/10.1136/thx.2006.075077

6. Gadsby NJ, Russell CD, McHugh MP, Mark H, Conway Morris A, Laurenson IF, Hill AT, Templeton KE (2016) Comprehensive molecular testing for respiratory pathogens in community-acquired pneumonia. Clin Infect Dis 62(7):817–823. https://doi.org/10.1093/cid/civ1214

7. Jain S, Self WH, Wunderink RG, Fakhran S, Balk R, Bramley AM, Reed C, Grijalva CG, Anderson EJ, Courtney DM, Chappell JD, Qi C, Hart EM, Carroll F, Trabue C, Donnelly HK, Williams DJ, Zhu Y, Arnold SR, Ampofo K, Waterer GW, Levine M, Lindstrom S, Winchell JM, Katz JM, Erdman D, Schneider E, Hicks LA, McCullers JA, Pavia AT, Edwards KM, Finelli L, Team CES (2015) Community-acquired pneumonia requiring hospitalization among U.S. adults. N Engl J Med 373(5):415–427. https://doi.org/10.1056/NEJMoa1500245

8. Ruuskanen O, Lahti E, Jennings LC, Murdoch DR (2011) Viral pneumonia. Lancet 377(9773):1264–1275. https://doi.org/10.1016/S0140-6736(10)61459-6

9. Graham BS, Perkins MD, Wright PF, Karzon DT (1988) Primary respiratory syncytial virus infection in mice. J Med Virol 26(2):153–162

10. Le Nouen C, Brock LG, Luongo C, McCarty T, Yang L, Mehedi M, Wimmer E, Mueller S, Collins PL, Buchholz UJ, DiNapoli JM (2014) Attenuation of human respiratory syncytial virus by genome-scale codon-pair deoptimization. Proc Natl Acad Sci U S A 111(36):13169–13174. https://doi.org/10.1073/pnas.1411290111

11. Hotard AL, Lee S, Currier MG, Crowe JE Jr, Sakamoto K, Newcomb DC, Peebles RS Jr, Plemper RK, Moore ML (2015) Identification of residues in the human respiratory syncytial virus fusion protein that modulate fusion activity and pathogenesis. J Virol 89(1):512–522. https://doi.org/10.1128/JVI.02472-14

12. Stier MT, Bloodworth MH, Toki S, Newcomb DC, Goleniewska K, Boyd KL, Quitalig M, Hotard AL, Moore ML, Hartert TV, Zhou B, McKenzie AN, Peebles RS Jr (2016) Respiratory syncytial virus infection activates IL-13-producing group 2 innate lymphoid cells through thymic stromal lymphopoietin. J Allergy Clin Immunol 138(3):814–824. e811. https://doi.org/10.1016/j.jaci.2016.01.050

13. Long X, Li S, Xie J, Li W, Zang N, Ren L, Deng Y, Xie X, Wang L, Fu Z, Liu E (2015) MMP-12-mediated by SARM-TRIF signaling pathway contributes to IFN-gamma-independent airway inflammation and AHR post RSV infection in nude mice. Respir Res 16:11. https://doi.org/10.1186/s12931-015-0176-8

14. Saravia J, You D, Shrestha B, Jaligama S, Siefker D, Lee GI, Harding JN, Jones TL, Rovnaghi C, Bagga B, DeVincenzo JP, Cormier SA (2015) Respiratory syncytial virus disease is mediated by age-variable IL-33. PLoS Pathog 11(10):e1005217. https://doi.org/10.1371/journal.ppat.1005217

15. Soltis RD, Hasz D, Morris MJ, Wilson ID (1979) The effect of heat inactivation of serum on aggregation of immunoglobulins. Immunology 36(1):37–45

16. Leshem B, Yogev D, Fiorentini D (1999) Heat inactivation of fetal calf serum is not required for in vitro measurement of lymphocyte functions. J Immunol Methods 223(2):249–254

17. Smith TF, McIntosh K, Fishaut M, Henson PM (1981) Activation of complement by cells infected with respiratory syncytial virus. Infect Immun 33(1):43–48

18. Empey KM, Orend JG, Peebles RS Jr, Egana L, Norris KA, Oury TD, Kolls JK (2012) Stimulation of immature lung macrophages with intranasal interferon gamma in a novel neonatal mouse model of respiratory syncytial virus infection. PLoS One 7(7):e40499. https://doi.org/10.1371/journal.pone.0040499

19. Ohol YM, Wang Z, Kemble G, Duke G (2015) Direct inhibition of cellular fatty acid synthase impairs replication of respiratory syncytial virus and other respiratory viruses. PLoS One 10(12):e0144648. https://doi.org/10.1371/journal.pone.0144648

20. Goritzka M, Makris S, Kausar F, Durant LR, Pereira C, Kumagai Y, Culley FJ, Mack M, Akira S, Johansson C (2015) Alveolar macrophage-derived type I interferons orchestrate innate immunity to RSV through recruitment of antiviral monocytes. J Exp Med 212(5):699–714. https://doi.org/10.1084/jem.20140825

21. Stokes KL, Chi MH, Sakamoto K, Newcomb DC, Currier MG, Huckabee MM, Lee S, Goleniewska K, Pretto C, Williams JV, Hotard A, Sherrill TP, Peebles RS Jr, Moore ML (2011) Differential pathogenesis of respiratory syncytial virus clinical isolates in BALB/c mice.

J Virol 85(12):5782–5793. https://doi.org/10.1128/JVI.01693-10

22. Herlocher ML, Ewasyshyn M, Sambhara S, Gharaee-Kermani M, Cho D, Lai J, Klein M, Maassab HF (1999) Immunological properties of plaque purified strains of live attenuated respiratory syncytial virus (RSV) for human vaccine. Vaccine 17(2):172–181

23. Lukacs NW, Moore ML, Rudd BD, Berlin AA, Collins RD, Olson SJ, Ho SB, Peebles RS Jr (2006) Differential immune responses and pulmonary pathophysiology are induced by two different strains of respiratory syncytial virus. Am J Pathol 169(3):977–986. https://doi.org/10.2353/ajpath.2006.051055

24. Moore ML, Chi MH, Luongo C, Lukacs NW, Polosukhin VV, Huckabee MM, Newcomb DC, Buchholz UJ, Crowe JE Jr, Goleniewska K, Williams JV, Collins PL, Peebles RS Jr (2009) A chimeric A2 strain of respiratory syncytial virus (RSV) with the fusion protein of RSV strain line 19 exhibits enhanced viral load, mucus, and airway dysfunction. J Virol 83(9):4185–4194. https://doi.org/10.1128/JVI.01853-08

25. Lukacs NW, Smit JJ, Mukherjee S, Morris SB, Nunez G, Lindell DM (2010) Respiratory virus-induced TLR7 activation controls IL-17-associated increased mucus via IL-23 regulation. J Immunol 185(4):2231–2239. https://doi.org/10.4049/jimmunol.1000733

Chapter 27

Delivery of Therapeutics to the Lung

Dominique N. Price and Pavan Muttil

Abstract

Pulmonary delivery in animal models can be performed using either direct administration methods or by passive inhalation. Direct pulmonary delivery requires the animal to be endotracheally intubated, whereas passive delivery uses a nose-only or a whole-body chamber. Endotracheal delivery of therapeutics and vaccines allows investigators to deliver the payload directly into the lung without the limitations associated with passive pulmonary administration methods. Additionally, endotracheal delivery can achieve deep lung delivery without the involvement of other exposure routes and is more reproducible and quantitative than passive pulmonary delivery in terms of accurate dosing. Here we describe the endotracheal delivery of both liquids and dry powders for preclinical models of treatment and exposure.

Key words Pulmonary delivery, Endotracheal delivery, Intratracheal delivery, Intubation, Insufflation, MicroSprayer, Dry powder

1 Introduction

Delivery of therapeutics to the lung has become increasingly important over the last several decades [1–3]. Pulmonary administered therapeutics and vaccines are currently being used to treat cystic fibrosis [2, 4], neonatal respiratory distress syndrome [5], influenza [6], asthma [7, 8], chronic obstructive pulmonary disease [9, 10], and most recently, diabetes [11, 12]. Additionally, clinical and preclinical research continues to explore the pulmonary route of administration for the treatment of cancer and infectious diseases [13–23].

Pulmonary delivery is ideal for local therapeutic delivery and has also shown to be an adequate route for systemic delivery [24]. The lungs are well-suited for therapeutic delivery into the systemic circulation due to their large surface area [1]. Therapeutics delivered by the pulmonary route can achieve rapid systemic absorption, but without the risks associated with needlestick injuries when parenteral routes of administration are employed [25]. Additionally, the respiratory system is one of the main mucosal surfaces within

Scott Alper and William J. Janssen (eds.), *Lung Innate Immunity and Inflammation: Methods and Protocols*,
Methods in Molecular Biology, vol. 1809, https://doi.org/10.1007/978-1-4939-8570-8_27,
© Springer Science+Business Media, LLC, part of Springer Nature 2018

the body, and mucosal delivery has become an important topic in the pharmaceutical and immunological research fields [26–28].

1.1 Pulmonary Delivery Devices: Hurdles from Preclinical to Human Studies

In humans, devices for pulmonary delivery vary with drug formulation and disease state. Liquid therapeutics are delivered using pressurized metered-dose inhalers, nebulizers, or soft mist sprays. Alternatively, dry powder therapeutics are delivered using dry powder inhalers. However, pulmonary delivery to small animals during preclinical therapeutic testing can be more challenging, which affects the translational potential of many novel inhaled therapies. Currently, animal models are used to study the safety, efficacy, and the pharmacokinetic and pharmacodynamic behaviors of inhaled therapeutics. Despite significant research being done in the preclinical pulmonary delivery field, very few inhaled therapeutics have made it to the market. This may be related to a lack of standardized pulmonary administration methodologies in small and large animal models of research.

Similarly, formulation plays a significant role in the successful delivery of therapeutics into the respiratory tract [3]. Deep lung delivery requires aerosol droplets or particles to be in the 1–3 μm range [29]. Larger particles will impact on the walls of the trachea and upper bronchus, whereas smaller particles less than 0.5 μm are known to be exhaled. For this reason, viscous liquid and cohesive or "sticky" particles should be avoided for pulmonary delivery as they exhibit poor flowability and aerosolization performance.

1.2 Pulmonary Delivery in Small Animals

The current noninvasive methodologies for preclinical pulmonary delivery in animals can be divided into two distinct groups, passive and direct inhalation. Passive pulmonary administration includes the use of whole-body and nose-only chambers. These devices do not require the animal to be sedated and therefore mimic physiological breathing conditions in humans [25, 30, 31]. However, both have limitations. Whole-body chambers (Fig. 1a) expose the entire animal to the aerosolized therapeutic, and although they are ideal for environmental aerosol exposures, surfaces other than the lungs are also exposed, such as the oropharynx and skin. Another drawback of whole-body exposure is that it requires large amounts of the therapeutic to deliver the correct dose into the lung. In comparison, nose- or head-only chambers (Fig. 1b) require less therapeutic to achieve the correct dose in the lung because only the nose or head is exposed to the aerosolized dose. While this delivery method limits drug exposure to the whole body, it can also cause significant stress to the animal, which must be constrained within the holding chamber for the duration of exposure (Fig. 1b). The animals do not have access to food or water and may suffocate if the system generates too much heat or if the animal turns itself around within the holding tube. Another significant limitation of the nose- and head-only chambers is that the aerosols are mainly deposited into the naso-

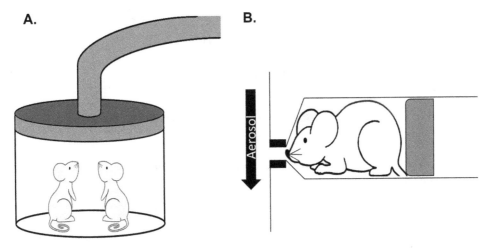

Fig. 1 Passive inhalation devices. A pictorial representation of the passive inhalation devices used for small animal models. (**a**) The whole-body chamber. (**b**) The nose-only chamber

pharynx where mucociliary clearance is active, thus delivering the therapeutic into the GI tract rather than into the lung [25].

Endotracheal (also referred to as intratracheal) administration devices deliver the therapeutic directly into the lung. With these devices, the animal is intubated by the insertion of a tube into the trachea. Liquid and dry powders can be administered directly into the lung, making precise dosing easier to achieve and preventing drug exposure by other routes. The limitations of endotracheal delivery are mostly related to the intubation procedure; the animal must be anesthetized in order to intubate the trachea, which is not physiologically similar to humans and makes the administration process difficult for multiple dosing studies. Furthermore, while intubation of larger rodents (rats, guinea pigs) is not difficult, intubation of mice can be challenging due to their small size. Therefore, endotracheal intubation in smaller rodents requires training of personnel.

In this chapter, we will limit our discussion to four different methodologies for endotracheal delivery of therapeutics into the lungs of small animals. Importantly, we will address both liquid and dry powder delivery. The reader is directed to other literature for discussion on the whole-body and nose-only devices for pulmonary delivery [25, 32, 33].

2   Materials

2.1 Intubation Preparation

1. Intubation platform.

2. Cotton-tipped applicators.

3. Lidocaine (optional).

4. Catheter tubing for lidocaine (optional).

5. Ophthalmic ointment.

6. Dental mirror (optional).

2.2 Liquid
Therapeutic Delivery:
MicroSprayer Method

1. MicroSprayer (Fig. 2b).

2. Hamilton syringe.

3. Laryngoscope: including tongue depressor (animal size appropriate) and otoscope attachment.

2.3 Liquid
Therapeutic Delivery:
Fiber-Optic Method

1. Fiber-optic cable (Fig. 2d) including light source.

2. Silicone tubing (optional) including cable tie.

3. Plastic hollow IV cannula (Fig. 2e).

4. Dropper pipette (Fig. 2c).

5. Diabetic or tuberculin syringe.

2.4 Dry Powder
Therapeutic Delivery:
Dry Powder
Insufflation Method

1. Insufflator (Fig. 2a): including intubation tube and 3 mL syringe.

2.5 Dry Powder
Therapeutic Delivery:
Fiber-Optic Method

1. Fiber-optic cable (Fig. 2d) including light source.

2. Silicone tubing (optional) including cable tie.

3. Plastic hollow IV cannula (Fig. 2e).

4. Dropper pipette (Fig. 2c).

3 Methods

3.1 Liquid
Therapeutic Delivery

3.1.1 Liquid Therapeutic
Delivery: Endotracheal
Delivery Using
a MicroSprayer

Two types of devices are used for the pulmonary delivery of a *liquid therapeutic*. Both devices require intubation of the mouse under anesthesia. These two methods will be referred to by the intubation device used, the *MicroSprayer* (*see* **Note 1**) [34](Fig. 2b) and the *Fiber-Optic and Cannula* [35, 36] (Fig. 2c–e).

MicroSprayer Therapeutic
Loading Protocol

1. Insert the plunger into the proximal end of the syringe while that end of the syringe is kept under running water to remove air that is trapped in the syringe.

2. Once the air has been eliminated from the syringe, fill the syringe with water by inserting the distal end into a water reservoir and pulling the plunger back (the MicroSprayer delivery tube is not attached yet).

3. Attach the MicroSprayer® delivery tube to the water-filled syringe (*see* **Note 2**).

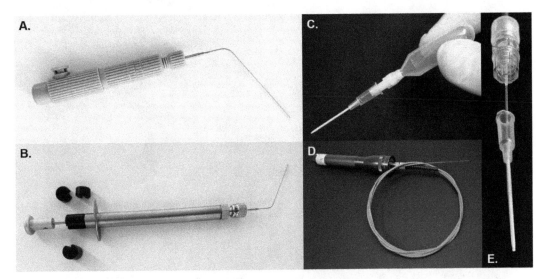

Fig. 2 Endotracheal delivery devices. The components of the three endotracheal delivery devices are shown here. (**a**) The Penn-Century Insufflator—consisting of the delivery tube (right side) attached to the powder-holding chamber (left side). (**b**) The MicroSprayer, consisting of the delivery tube (right side) attached to the liquid holding Hamilton syringe (left side). Two sizes of stoppers/dose volume spacers are used to measure the doses to be administered. The smaller stoppers (above the syringe) are for a 25 μL dose and the larger stoppers are for a 50 μL dose (shown on the syringe plunger and below the syringe). (**c–e**) Braintree Scientific Biolite fiber-optic intubation system with the dropper pipette (**c**) and fiber-optic intubation guide (**d**) and hollow I.V. cannula (**e**)

4. Push the plunger quickly and with force to aerosolize the water from the MicroSprayer completely. The plume from the MicroSprayer should disperse without a water stream. As the plunger is depressed, there should be a back pressure felt by the operator as the water is aerosolized from the device (*see* **Note 2**).

5. Remove the delivery tube from the MicroSprayer and fill the syringe with the liquid therapeutic as described in **step 2**.

6. Reattach the MicroSprayer delivery tube. Apply stoppers/dose volume spacers to the plunger (each spacer measures a dose of 25 or 50 μL) (Fig. 2b).

Intubation/Pulmonary Administration via MicroSprayer Protocol

1. Administer the anesthesia to the animal (*see* **Note 3**).

2. Add a generous amount of eye ointment to the lower eyelids of the mouse using a cotton-tipped applicator.

3. Check pedal reflexes and breathing rate to ensure the mouse is properly sedated.

4. Place the mouse on the workstand and suspend it by the incisors. Raise the stand to 60° (Fig. 3a) (*see* **Note 4**).

5. Use a cotton-tipped applicator to open the mouth and roll the tongue to either the right or left side (Fig. 3b).

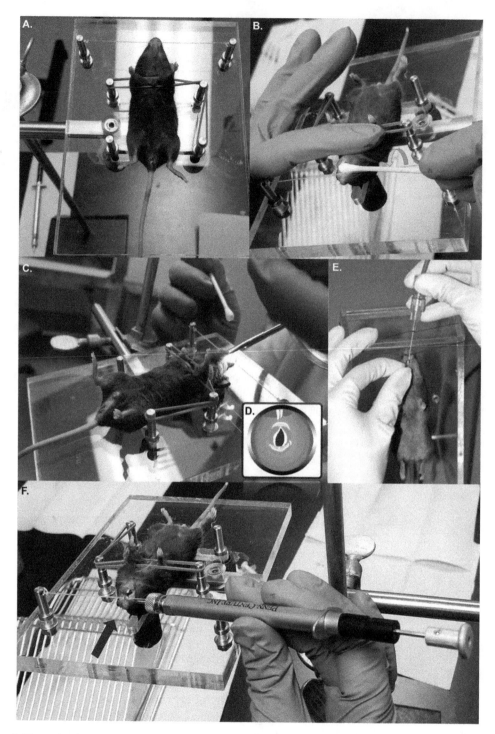

Fig. 3 Mouse intubation. (**a**) Mouse positioned on the intubation work stand, hanging by the incisors and additionally held in place with a rubber band positioned across the body. (**b**) The tongue is rolled out of the mouth using a cotton swab. (**c**) A laryngoscope with a tongue depressor is positioned in the rodent's mouth and used to visualize the tracheal opening. (**d**) Cartoon visualization of the tracheal opening of a mouse (inset of **c**). (f) Mouse intubated using the MicroSprayer (Insufflator is similar), with the delivery tube bend positioned (shown with arrow) at the Fig. 3 (continued) animal incisors to ensure the appropriate intubation depth. (e) Intubation using the Braintree Scientific Biolite fiber-optic intubation system, which uses a fiber-optic guide instead of the classical laryngoscope

6. Use the laryngoscope to obtain a clear view of the trachea. The tracheal opening consists of a white arytenoid cartilage that is continuously opening and closing at the back of the throat (*see* **Note 5**).

7. Apply two drops of lidocaine (if using) directly to the arytenoid cartilage region. Wait for a few seconds for the lidocaine to take effect (*see* **Note 6**).

8. Carefully remove one stopper/dose volume spacer (the one that is closest to the end of the plunger) before intubation.

9. The laryngoscope should be introduced into the oral cavity with the nondominant hand. With a clear view of the tracheal opening, insert the MicroSprayer delivery tube (with the syringe loaded with the liquid therapeutic and attached to the delivery tube) using the dominant hand into the tracheal opening until the curved/bent region of the delivery tube reaches the incisors. This ensures that the tip of the MicroSprayer delivery tube is within a few millimeters from the carina of the lungs (first bronchial bifurcation) (Fig. 3c) (*see* **Note 7**).

10. Carefully remove the laryngoscope.

11. To administer the dose, push the plunger on the syringe firmly without jarring the MicroSprayer tip. Each spacer will deliver 25/50 μL volume of the therapeutic directly into the respiratory tract (Fig. 2b) (*see* **Note 8**).

12. Remove the MicroSprayer (along with the attached delivery tube) slowly from the mouse trachea.

13. Remove the mouse from the work stand, transfer onto a heat blanket, and monitor breathing until the mouse regains consciousness.

3.2 Liquid Therapeutic Delivery: Endotracheal Delivery Using the Fiber-Optic Cable and a Catheter for Intubation

3.2.1 Fiber-Optic and Cannula Preparation

1. If you are making your own fiber-optic intubation device, instead of using the commercially available version (*see* **Note 9**), then complete the following: Attach a short piece of silicone rubber tubing (0.8 mm inner diameter and 4 mm outer diameter, about 5 mm in length) to the fiber-optic cable. This will prevent the cannula from moving up the length of the fiber-optic cable during intubation (optional) (*see* **Note 10**).

2. Place a hollow plastic IV cannula (20–22 gauge) over the fiber-optic cable, such that the Luer lock faces the rubber tubing (Fig. 2e). The fiber-optic cable will guide the cannula during endotracheal intubation.

3. Position the cannula such that its tip is 4 mm from the end of the fiber-optic cable tip. If you are constructing your own, tie a piece of silicone tubing (with a cable tie or a strong string) behind the cannula to prevent it from moving (*see* **Note 10**).

4. Connect a 0.5 mm fiber-optic cable to a light source.

5. Have a syringe filled with the liquid therapeutic ready.

*3.2.2 Intubation/
Pulmonary Administration
via Fiber-Optic Cable
and Cannula Protocol*

1. Administer the anesthesia to the animal (*see* **Note 3**).

2. Add a generous amount of eye ointment to the lower eyelids of the mouse using the cotton-tipped applicator.

3. Check pedal reflexes and breathing rate to ensure that the mouse is properly sedated.

4. Place the mouse on the workstand and suspend it by the incisors. Raise the stand to 60° (Fig. 3a) (*see* **Note 4**).

5. Use a cotton-tipped applicator to open the mouth and roll the tongue either to the left or right side (Fig. 3b).

6. Obtain a clear view of the trachea using the light from the fiber-optic cable. The tracheal opening consists of the white arytenoid cartilage that is continuously opening and closing at the back of the throat (*see* **Note 5**).

7. Apply two drops of lidocaine (if using) directly to the arytenoid cartilage. Wait for a few seconds for the lidocaine to take effect (*see* **Note 6**).

8. Insert and push the fiber-optic cable into the tracheal opening just proximal to the main carina of the lungs (first bronchial bifurcation). The fiber-optic cable should reach the carina when the Luer lock of the cannula is at the level of the incisors (Fig. 3c) (*see* **Notes 5, 7, and 10**).

9. Carefully remove the fiber-optic cable from the cannula. Applying tape to the Luer lock of the catheter tube where it enters the mouth will hold the cannula in place.

10. Ensure proper intubation by looking for condensation on a dental mirror (hold the mirror next to the Luer lock end of the cannula) (optional).

11. Gently attach an insulin (mouse) or tuberculin (rat) syringe to the cannula while being careful not to jar the animal. These syringes allow you to administer a 50 μL or 1 mL dose.

12. Compress the plunger on the syringe to deliver the therapeutics without pushing the cannula further into the trachea (this process may require two people) (*see* **Note 7**).

13. Slowly remove the cannula with the attached syringe.

14. Remove the mouse from the work stand, transfer it onto a heat blanket, and monitor breathing until the mouse regains consciousness.

3.3 Dry Powder Therapeutic Delivery

Two types of devices are used for the pulmonary delivery of a *dry powder therapeutic*. Both devices require intubation of the mouse under anesthesia. These two methods will be referred to as the *Insufflator* (*see* **Note 1**) [37] and the *Fiber-Optic Cable and Cannula* [35, 36] based on the intubation device used.

3.3.1 Dry Powder Therapeutic Delivery: Endotracheal Delivery Using an Insufflator

Insufflator Powder Loading Protocol

1. Weigh the empty Insufflator.

2. Load the Insufflator with a dry powder therapeutic by placing 1–2 mg into the powder chamber of the device (Fig. 2a) (*see* **Note 11**).

3. Weigh the powder-loaded Insufflator and subtract from the weight of the empty device in **step 1** to obtain the weight of the loaded powder.

Intubation/Pulmonary Administration via Insufflator Protocol

1. Administer the anesthesia to the animal (*see* **Note 3**).

2. Add a generous amount of eye ointment to the lower eyelids of the mouse using a cotton-tipped applicator.

3. Check for pedal reflexes and breathing rate to ensure the mouse is completely sedated.

4. Place the mouse on the workstand and suspend it by the incisors. Raise the stand to 60° (Fig. 3a) (*see* **Note 4**).

5. Use a cotton-tipped applicator to open the mouth and roll the tongue to either the left or right side (Fig. 3b).

6. Apply about two drops of lidocaine (if using) directly to the arytenoid cartilage region. Wait for a few seconds for lidocaine to take effect (*see* **Note 6**).

7. The laryngoscope will be introduced into the oral cavity with the nondominant hand. With a clear view of the tracheal opening, insert the Insufflator delivery tube using the dominant hand (with the Insufflator loaded with the therapeutic dry powder and attached to the delivery tube) into the tracheal opening until the curved/bent piece of the delivery tube reaches the incisors (Fig. 3c). This ensures that the tip of the delivery tube is within a few millimeters from the carina of the lungs (*see* **Note 7**).

8. Carefully remove the laryngoscope.

9. Attach a 3 mL syringe to the Insufflator (with 1 mL air drawn into it), making sure not to jar the animal. Push the plunger firmly to aerosolize the powder loaded in the Insufflator directly into the lung (*see* **Note 8**).

10. Remove the syringe and repeat **step 9** seven to nine more times (with 1 mL of air drawn into the syringe each time) to help aerosolize any residual powder from the device (*see* **Note 12**).

11. Remove the Insufflator slowly.

12. Weigh the Insufflator after dosing the animal to determine the amount of powder remaining in the device, and that delivered into the lungs.

13. Remove the mouse from the work stand, transfer onto a heat blanket, and monitor breathing until the mouse regains consciousness.

3.4 Dry Powder Therapeutic Delivery: Cannula and Drop Pipettor Intubation Device

3.4.1 Fiber-Optic and Cannula Preparation

1. If you are making your own fiber-optic intubation device, instead of using the commercially available version (*see* **Note 9**), complete the following: Attach a short piece of silicone rubber tubing (0.8 mm inner diameter and 4 mm outer diameter, about 5 mm in length) to the fiber-optic cable. This will prevent the cannula from moving up the length of the fiber-optic cable during intubation (optional).

2. Place a hollow plastic IV cannula (20–22 gauge) over the fiber-optic cable (Fig. 2e), such that the Luer lock faces the rubber tubing. The fiber-optic cable will guide the cannula during endotracheal intubation.

3. Position the cannula such that its tip is 4 mm from the end of the fiber-optic cable tip. If you are constructing your own, tie a piece of silicone tubing (with a cable tie or a strong string) behind the cannula to prevent it from moving (*see* **Note 10**).

4. Connect a 0.5 mm fiber-optic cable to a light source.

3.4.2 Dropper Pipette Loading Protocol

1. Weigh an empty dropper pipette (Fig. 2c).

2. Tap the distal end of the pipette into the powder vial and weigh again to determine the amount of powder loaded in the pipette. Adjust the dose as needed by continuing to tap the distal end of the pipette into the powder until the target dose is achieved.

3. Set aside the pipette horizontally on the table until ready for use.

3.4.3 Intubation/ Pulmonary Administration via Fiber-Optic and Cannula Protocol

1. Administer the anesthesia to the animal (*see* **Note 3**).

2. Add a generous amount of eye ointment to the lower eyelids of the mouse using the cotton-tipped applicator.

3. Check pedal reflexes and breathing rate to ensure the mouse is completely sedated.

4. Place the mouse on the workstand, suspending it by the incisors. Raise the stand to 60° (Fig. 3a) (*see* **Note 4**).

5. Use a cotton-tipped applicator to open the mouth and roll the tongue to either the left or right side (Fig. 3b).

6. Obtain a clear view of the trachea by using the light from the fiber-optic cable. The tracheal opening consists of the white arytenoid cartilage that is continuously opening and closing at the back of the throat (*see* **Note 5**).

7. Apply two drops of lidocaine (if using) directly to the arytenoid cartilage. Wait a few seconds for lidocaine to take effect (*see* **Note 6**).

8. Insert and push the fiber-optic cable into the tracheal opening just proximal to the carina of the lungs (first bronchial bifurcation). The fiber-optic cable reaches the carina when the Luer

lock of the cannula is positioned at the incisors (Fig. 3d) (*see* **Notes 5**, **7**, and **10**).

9. Carefully remove the fiber-optic cable from the cannula. Applying tape to the Luer lock of the catheter tube where it enters the mouth will hold the cannula in place.

10. Ensure proper intubation by looking for condensation on a dental mirror (hold the mirror next to the Luer lock end of the cannula) (optional).

11. Attach a dropper pipette (previously loaded with the therapeutic powder) to the cannula while being careful not to jar the animal.

12. Compress firmly and quickly the proximal end of the pipettor to deliver the dry powder into the mouse lung. Make sure not to push the cannula further into the trachea (this process may require two people) (*see* **Note 7**).

13. Remove the cannula carefully from the animal trachea.

14. Weigh the dropper pipette postdelivery to determine the amount of powder delivered into the respiratory tract.

15. Remove the mouse from the work stand, transfer it onto a heat blanket, and monitor breathing until the mouse gains consciousness.

4 Notes

1. The MicroSprayer and the Insufflator devices are no longer available for purchase from Penn-Century, Inc., but may be available from other sources (as Penn-Century devices have been widely disturbed throughout the world) or specially fabricated. If you cannot access the MicroSprayer or Insufflator devices or fabricate your own, you should use the fiber-optic and cannula intubation method (liquid therapeutic) (*see* Subheading 3.2) or catheter and drop pipettor method (powder therapeutic) (*see* Subheading 3.4).

2. If a liquid stream is visible at the end of the aerosol spray from the MicroSprayer, then there is air trapped in the syringe. Repeat **steps 1** and **2** again, making sure that the running water used to displace the air from the syringe in **step 1** does not have air bubbles within the water stream.

3. All anesthesia methods should be approved by the Institutional Animal Care and Use Committee. The anesthesia should deeply sedate the animal for several minutes. We use a combination of ketamine and xylazine given by the intraperitoneal (I.P.) injection route at the following concentrations: mouse, ketamine, 90–100 mg/kg, and xylazine, 9–10 mg/kg, and rat, ketamine, 45–90 mg/kg, and xylazine, 4.5–9 mg/kg. After

endotracheal delivery of the drug or treatment, follow with atipamezole (0.1–1 mg/kg) by the I.P. route to shorten recovery time and 300 μL of subcutaneous 0.9% saline to maintain hydration.

4. Commercially available mouse and rat stands are available from several vendors. These provide easily maneuverable platforms to intubate rodents. Alternatively, a platform can easily be fabricated from materials available at most hardware stores.

5. Morello and colleagues [37] published a similar dry powder insufflation protocol which shows the tracheal opening pictorially in a mouse and should be used as an additional reference.

6. Lidocaine prevents the overstimulation of the trachea during the intubation process, which can sometimes lead to animal death. This step may not be required if the researcher is experienced in intubating rodents. However, it should be noted that lidocaine has been associated with an anti-inflammatory immune response in many tissues and may dampen pro-inflammatory migration and signaling [38].

7. The depth of intubation is critical for deep lung delivery. The carina is the region where the trachea bifurcates into the right and left lung. The intubation tube should deliver the drug payload at this depth into the trachea; this would lead to minimal impaction on the tracheal walls and maximum drug delivery to the lungs. There should not be any resistance during the animal intubation process. If resistance is felt, withdraw the tube or the fiber-optic cable and try the procedure again.

8. This step may require an extra pair of hands, or a platform to stabilize the user's elbow. Too much hand movement during the intubation process may traumatize the animal. During pulmonary delivery, it is recommended that one person stabilize the syringe attached to the intubation tube and the other push the syringe plunger to deliver the dose.

9. The fiber-optic intubation kit can be purchased from Braintree Scientific, Inc., or it can be assembled as originally done by Mitzner's group [35, 36].

10. Accurate positioning of the tie (**step 1a**—Fiber-Optic and Cannula Preparation) will ensure that the fiber-optic cable and the intubation tube reach the carina of the lungs. Training and optimization by the user will be required since the insertion length of the fiber-optic cable will vary for different-sized animals.

11. Penn-Century designed the Insufflator to deliver up to a maximum of 2 mg of powder into the lungs of mice. However, Morello et al. [37] reported that mice that were administered more than 0.35 mg died immediately. For this reason, it may be necessary to limit doses to less than 0.5 mg of powder per mouse.

12. It takes several puffs of air (1 mL) from the syringe to aerosolize all the powder from the Insufflator. We found that eight to ten actuations are necessary to achieve adequate dosing using the dry powder Insufflator [39]. Additionally, the Insufflator delivery tube can sometimes clog due to powder aggregation. One of the reasons for Insufflator clogging is the humid environment in the animal lungs, which affects the ability of the powder to flow freely through the Insufflator tube. The clogging of the Insufflator can be minimized by cleaning and drying the device after each insufflation.

References

1. Patton JS, Byron PR (2007) Inhaling medicines: delivering drugs to the body through the lungs. Nat Rev Drug Discov 6:67–74. https://doi.org/10.1038/nrd2153

2. Hickey AJ (2013) Back to the future: inhaled drug products. J Pharm Sci 102:1165–1172. https://doi.org/10.1002/jps.23465

3. Price DN, Kunda NK, McBride AA, Muttil P (2016) Vaccine preparation: past, present, and future. In: Drug delivery systems for tuberculosis prevention and treatment. John Wiley & Sons, Ltd, Chichester, pp 67–90. https://doi.org/10.1002/9781118943182.ch4

4. Garcia-Contreras L, Hickey AJ (2003) Aerosol treatment of cystic fibrosis. Crit Rev Ther Drug Carrier Syst 20:317–356. https://doi.org/10.1615/CRITREVTHERDRUGCARRIERSYST.V20.I5.10

5. Polin RA, Carlo WA, Committee on Fetus and Newborn, American Academy of Pediatrics (2014) Surfactant replacement therapy for preterm and term neonates with respiratory distress. Pediatrics 133:156–163. https://doi.org/10.1542/peds.2013-3443

6. Monto AS, Paul Robinson D, Louise Herlocher M, Hinson JM Jr, Elliott MJ, Crisp A (1999) Zanamivir in the prevention of influenza among healthy adults. JAMA 282:31. https://doi.org/10.1001/jama.282.1.31

7. Chan HK, Chew NY (2003) Novel alternative methods for the delivery of drugs for the treatment of asthma. Adv Drug Deliv Rev 55:793–805

8. Johnson MA, Newman SP, Bloom R, Talaee N, Clarke SW (1989) Delivery of albuterol and ipratropium bromide from two nebulizer systems in chronic stable asthma. Chest 96:6–10. https://doi.org/10.1378/chest.96.1.6

9. Mahler DA, Wire P, Horstman D, Chang C-N, Yates J, Fischer T, Shah T (2002) Effectiveness of fluticasone propionate and salmeterol combination delivered via the diskus device in the treatment of chronic obstructive pulmonary disease. Am J Respir Crit Care Med 166:1084–1091. https://doi.org/10.1164/rccm.2112055

10. Acerbi D, Brambilla G, Kottakis I (2007) Advances in asthma and COPD management: delivering CFC-free inhaled therapy using Modulite® technology. Pulm Pharmacol Ther 20:290–303. https://doi.org/10.1016/j.pupt.2006.05.005

11. Santos Cavaiola T, Edelman S (2014) Inhaled insulin: a breath of fresh air? A review of inhaled insulin. Clin Ther 36:1275–1289. https://doi.org/10.1016/j.clinthera.2014.06.025

12. Traynor K (2014) Inhaled insulin product approved. Am J Health-Syst Pharm 71:1238. https://doi.org/10.2146/news140052

13. Price DN, Kusewitt DF, Lino CA, McBride AA, Muttil P (2016) Oral tolerance to environmental mycobacteria interferes with intradermal, but not pulmonary, immunization against tuberculosis. PLOS Pathog 12:e1005614. https://doi.org/10.1371/journal.ppat.1005614

14. McBride AA, Price DN, Lamoureux LR, Elmaoued AA, Vargas JM, Adolphi NL, Muttil P (2013) Preparation and characterization of novel magnetic nano-in-microparticles for site-specific pulmonary drug delivery. Mol Pharm 10:3574–3581. https://doi.org/10.1021/mp3007264

15. Price DN, Muttil P (2016) Directed intervention and immunomodulation against pulmonary tuberculosis. In: Drug delivery systems for tuberculosis prevention and treatment. John Wiley & Sons, Ltd, Chichester, pp 346–377. https://doi.org/10.1002/9781118943182.ch18

16. Yi D, Wiedmann TS (2010) Inhalation adjuvant therapy for lung cancer. J Aerosol Med Pulm Drug Deliv 23:181–187. https://doi.org/10.1089/jamp.2009.0787

17. Zarogoulidis P, Darwiche K, Krauss L, Huang H, Zachariadis GA, Katsavou A, Hohenforst-Schmidt W et al (2013) Inhaled cisplatin deposition and distribution in lymph nodes in stage II lung cancer patients. Future Oncol 9:1307–1313. https://doi.org/10.2217/fon.13.111

18. Tatsumura T, Koyama S, Tsujimoto M, Kitagawa M, Kagamimori S (1993) Further study of nebulisation chemotherapy, a new chemotherapeutic method in the treatment of lung carcinomas: fundamental and clinical. Br J Cancer 68:1146–1149

19. Gagnadoux F, Hureaux J, Vecellio L, Urban T, Le Pape A, Valo I, Montharu J et al (2008) Aerosolized chemotherapy. J Aerosol Med Pulm Drug Deliv 21:61–70. https://doi.org/10.1089/jamp.2007.0656

20. Lemarie E, Vecellio L, Hureaux J, Prunier C, Valat C, Grimbert D, Boidron-Celle M et al (2011) Aerosolized gemcitabine in patients with carcinoma of the lung: feasibility and safety study. J Aerosol Med Pulm Drug Deliv 24:261–270. https://doi.org/10.1089/jamp.2010.0872

21. Garcia-Contreras L, Wong Y-L, Muttil P, Padilla D, Sadoff J, Derousse J, Germishuizen WA et al (2008) Immunization by a bacterial aerosol. Proc Natl Acad Sci U S A 105:4656–4660. https://doi.org/10.1073/pnas.0800043105

22. Otterson GA, Villalona-Calero MA, Sharma S, Kris MG, Imondi A, Gerber M, White DA et al (2007) Phase I study of inhaled doxorubicin for patients with metastatic tumors to the lungs. Clin Cancer Res 13:1246–1252. https://doi.org/10.1158/1078-0432.CCR-06-1096

23. Otterson GA, Villalona-Calero MA, Hicks W, Pan X, Ellerton JA, Gettinger SN, Murren JR (2010) Phase I/II study of inhaled doxorubicin combined with platinum-based therapy for advanced non-small cell lung cancer. Clin Cancer Res 16:2466–2473. https://doi.org/10.1158/1078-0432.CCR-09-3015

24. Gonda I (2006) Systemic delivery of drugs to humans via inhalation. J Aerosol Med 19:47–53. https://doi.org/10.1089/jam.2006.19.47

25. Nahar K, Gupta N, Gauvin R, Absar S, Patel B, Gupta V, Khademhosseini A, Ahsan F (2013) In vitro, in vivo and ex vivo models for studying particle deposition and drug absorption of inhaled pharmaceuticals. Eur J Pharm Sci 49:805–818. https://doi.org/10.1016/j.ejps.2013.06.004

26. Boegh M, Foged C, Müllertz A, Mørck Nielsen H (2013) Mucosal drug delivery: barriers, in vitro models and formulation strategies. J Drug Deliv Sci Technol 23:383–391. https://doi.org/10.1016/S1773-2247(13)50055-4

27. Mathias NR, Hussain MA (2010) Non-invasive systemic drug delivery: developability considerations for alternate routes of administration. J Pharm Sci 99:1–20. https://doi.org/10.1002/jps.21793

28. Courrier HM, Butz N, Vandamme TF (2002) Pulmonary drug delivery systems: recent developments and prospects. Crit Rev Ther Drug Carrier Syst 19:425–498. https://doi.org/10.1615/CritRevTherDrugCarrierSyst.v19.i45.40

29. Labiris NR, Dolovich MB (2003) Pulmonary drug delivery. Part I: physiological factors affecting therapeutic effectiveness of aerosolized medications. Br J Clin Pharmacol 56:588–599. https://doi.org/10.1046/j.1365-2125.2003.01892.x

30. Dorato MA (1990) Overview of inhalation toxicology. Environ Health Perspect 85:163–170

31. Cryan S-A, Sivadas N, Garcia-Contreras L (2007) In vivo animal models for drug delivery across the lung mucosal barrier. Adv Drug Deliv Rev 59:1133–1151. https://doi.org/10.1016/j.addr.2007.08.023

32. Wong BA (2007) Inhalation exposure systems: design, methods and operation. Toxicol Pathol 35:3–14. https://doi.org/10.1080/01926230601060017

33. Forbes B, Asgharian B, Dailey LA, Ferguson D, Gerde P, Gumbleton M, Gustavsson L et al (2011) Challenges in inhaled product development and opportunities for open innovation. Adv Drug Deliv Rev 63:69–87. https://doi.org/10.1016/j.addr.2010.11.004

34. Bivas-Benita M, Zwier R, Junginger H, Borchard G (2005) Non-invasive pulmonary aerosol delivery in mice by the endotracheal route. Eur J Pharm Biopharm 61:214–218. https://doi.org/10.1016/j.ejpb.2005.04.009

35. Das S, MacDonald K, Chang H-YS, Mitzner W (2013) A simple method of mouse lung intubation, J Vis Exp:e50318. https://doi.org/10.3791/50318

36. MacDonald KD, Chang H-YS, Mitzner W (2009) An improved simple method of mouse lung intubation. J Appl Physiol (1985) 106:984–987. https://doi.org/10.1152/japplphysiol.91376.2008

37. Morello M, Krone CL, Dickerson S, Howerth E, Germishuizen WA, Wong Y-LL, Edwards D, Bloom BR, Hondalus MK (2009) Dry-powder pulmonary insufflation in the mouse for

application to vaccine or drug studies. Tuberculosis (Edinb) 89:371–377. https://doi.org/10.1016/j.tube.2009.07.001

38. Caracas HC, Maciel JV, Martins PM, de Souza MM, Maia LC (2009) The use of lidocaine as an anti-inflammatory substance: a systematic review. J Dent 37:93–97. https://doi.org/10.1016/j.jdent.2008.10.005

39. Price DN, Stromberg LR, Kunda NK, Muttil P (2017) In vivo pulmonary delivery and magnetic-targeting of dry powder nano-in-microparticles. Mol Pharm. https://doi.org/10.1021/acs.molpharmaceut.7b00532. [Epub ahead of print]

INDEX

Scott Alper and William J. Janssen (eds.), *Lung Innate Immunity and Inflammation: Methods and Protocols*,
Methods in Molecular Biology, vol. 1809, https://doi.org/10.1007/978-1-4939-8570-8,
© Springer Science+Business Media, LLC, part of Springer Nature 2018

CPSIA information can be obtained
at www.ICGtesting.com
Printed in the USA
LVHW06*1942290718
585291LV00002B/3/P

9 781493 985692